Springer Theses

Recognizing Outstanding Ph.D. Research

For further volumes:
http://www.springer.com/series/8790

Aims and Scope

The series "Springer Theses" brings together a selection of the very best Ph.D. theses from around the world and across the physical sciences. Nominated and endorsed by two recognized specialists, each published volume has been selected for its scientific excellence and the high impact of its contents for the pertinent field of research. For greater accessibility to non-specialists, the published versions include an extended introduction, as well as a foreword by the student's supervisor explaining the special relevance of the work for the field. As a whole, the series will provide a valuable resource both for newcomers to the research fields described, and for other scientists seeking detailed background information on special questions. Finally, it provides an accredited documentation of the valuable contributions made by today's younger generation of scientists.

Theses are accepted into the series by invited nomination only and must fulfill all of the following criteria

- They must be written in good English.
- The topic should fall within the confines of Chemistry, Physics and related interdisciplinary fields such as Materials, Nanoscience, Chemical Engineering, Complex Systems and Biophysics.
- The work reported in the thesis must represent a significant scientific advance.
- If the thesis includes previously published material, permission to reproduce this must be gained from the respective copyright holder.
- They must have been examined and passed during the 12 months prior to nomination.
- Each thesis should include a foreword by the supervisor outlining the significance of its content.
- The theses should have a clearly defined structure including an introduction accessible to scientists not expert in that particular field.

Sebastian Klein

Charm Production in Deep Inelastic Scattering

Mellin Moments of Heavy Flavor Contributions to $F_2(x, Q^2)$ at NNLO

Doctoral Thesis accepted by
Technical University Dortmund, Germany

Springer

Author
Dr. Sebastian Klein
Institute for Theoretical Particle Physics
and Cosmology, RWTH Aachen
52056 Aachen
Germany
e-mail: sklein@physik.rwth-aachen.de

Supervisor
Prof. Johannes Blümlein
Department of Physics
Technical University Dortmund
Otto-Hahn-Str. 4
44227 Dortmund
Germany
e-mail: johannes.bluemlein@desy.de

ISSN 2190-5053
e-ISSN 2190-5061
ISBN 978-3-642-27030-7
ISBN 978-3-642-23286-2 (eBook)
DOI 10.1007/978-3-642-23286-2
Springer Heidelberg Dordrecht London New York

Cover design: eStudio Calamar, Berlin/Figueres

Printed on acid-free paper

Springer is part of Springer Science+Business Media (www.springer.com)

This dissertation contains contributions to the following publications:

(i) *Journal articles*

1. **$O(\alpha^2_s)$ and $O(\alpha^3_s)$ Heavy Flavor Contributions to Transversity at $Q^2 \gg m^2$** J. Blümlein, S. Klein and B. Tödtli Phys. Rev. **D80** (2009) 094010 [arXiv:hep-ph/0909.1547].

2. **Mellin Moments of the $O(\alpha^3_s)$ Heavy Flavor Contributions to unpolarized Deep-Inelastic Scattering at $Q^2 \gg m^2$ and Anomalous Dimensions** I. Bierenbaum, J. Blümlein and S. Klein Nucl. Phys. **B820** (2009) 417 [arXiv:hep-ph/0904.3563].

3. **Determining the closed forms of the $O(a^3_s)$ anomalous dimensions and Wilson coefficients from Mellin moments by means of computer algebra** J. Blümlein, M. Kauers, S. Klein and C. Schneider Comput. Phys. Commun. **180** (2009) 2143 [arXiv:hep-ph/0902.4091].

4. **The Gluonic Operator Matrix Elements at $O(\alpha^2_s)$ for DIS Heavy Flavor Production** I. Bierenbaum, J. Blümlein and S. Klein Phys. Lett. **B672**, 401 (2009) [arXiv:hep-ph/0901.0669].

5. **2-Loop Massive Operator Matrix Elements for Unpolarized Heavy Flavor Production to $O(\epsilon)$** I. Bierenbaum, J. Blümlein, S. Klein and C. Schneider Nucl. Phys. **B803**, 1 (2008) [arXiv:hep-ph/0803.0273].

(ii) *Contributions to conference proceedings*

6. **2-and 3-loop heavy flavor contributions to $F_2(x, Q^2)$, $FL(x, Q^2)$ and $g_{1,2}(x, Q^2)$** I. Bierenbaum, J. Blümlein and S. Klein in: Proceedings of the workshop: HERA and the LHC workshop series on the implications of HERA for LHC physics, H. Jung and A. De Roeck, eds., (2006–2008, Hamburg, Geneve), DESY-PROC-2009-02, March 2009, pp. 363 [arXiv:hep-ph/0903.3861].

7. **From Moments to Functions in Quantum Chromodynamics** J. Blümlein, M. Kauers, S. Klein and C. Schneider PoS ACAT 2008, 106 (2008) [arXiv:hep-ph/0902.4095].

8. **Heavy flavor operator matrix elements at $O(a^3_s)$** I. Bierenbaum, J. Blümlein and S. Klein PoS Confinement8, 185 (2008) [arXiv:hep-ph/0812.2427].

9. **First $O(\alpha^3_s)$ heavy flavor contributions to deeply inelastic scattering** I. Bierenbaum, J. Blümlein and S. Klein Nucl. Phys. Proc. Suppl. **183**, 162 (2008) [arXiv:hep-ph/0806.4613].

10. **Higher order corrections to heavy flavour production in deep inelastic scattering** I. Bierenbaum, J. Blümlein and S. Klein Acta Phys. Polon. **B39**, 1531 (2008) [arXiv:hep-ph/0806.0451].

11. **Heavy flavour production in deep-inelastic scattering—2-loop massive operator matrix elements and beyond** I. Bierenbaum, J. Blümlein and S. Klein Acta Phys. Polon. **B38**, 3543 (2007); PoS RADCOR 2007, 034 (2007) [arXiv:hepph/0710.3348].

12. **Difference Equations in Massive Higher Order Calculations** I. Bierenbaum, J. Blümlein, S. Klein and C. Schneider PoS ACAT 2007, 082 (2007) [arXiv:math-ph/0707.4659].

13. **2-loop massive operator matrix elements for polarized and unpolarized deep-inelastic scattering** I. Bierenbaum, J. Blümlein and S. Klein in: Proc. of 15th International Workshop On Deep-Inelastic Scattering And Related Subjects (DIS2007), G. Grindhammer, K. Sachs, eds., (16–20 April 2007, Munich), Vol. **2**, pp. 821 [arXiv:hep-ph/0706.2738].

14. **Structural Relations between Harmonic Sums up to w=6** J. Blümlein and S. Klein PoS **ACAT**, 084 (2007) [arXiv/hep-ph:0706.2426].

15. **2-loop massive operator matrix elements for polarized and unpolarized deep-inelastic scattering** I. Bierenbaum, J. Blümlein and S. Klein PoS **ACAT**, 070 (2007).

Supervisor's Foreword

Quantum corrections in relativistic renormalizable quantum field theories are of importance to explore the fundamental properties of elementary particles at high precision. A central issue concerns the accurate measurement of the fundamental parameters of the Standard Model, like the couplings of the different forces, the matter and gauge-particle masses, and mixing parameters. The corresponding corrections to cope with the current experimental precision are those of the three-loop level for the strong interactions, as described by Quantum Chromodynamics for high energy scattering experiments. Here an important process is deep-inelastic lepton-nucleon scattering through which the very short-distance structure of nucleons is probed. Some years ago, the dynamical contributions of massless quarks have been calculated to this level, while corrections due to heavy quarks, such as charm and bottom, were known to 1-loop level less only. In particular the contribution due to charm quarks is rather large, requiring the treatment at the same loop order.

In the present Thesis essential steps are undertaken to determine the massive Wilson coefficients for deep-inelastic scattering at three-loop order. Currently the computation of these quantities in the full kinematic region, i.e. also including low values of the virtuality Q^2 of the exchanged gauge boson, cannot be performed. However, already at scales $Q^2 \gtrsim m_c^2$, with m_c the mass of the charm quark, technologies are available to calculate these quantities even in analytic form. This kinematic condition forms no essential restriction, but rather serves the purpose to apply the theory to a range of better validity. At these scales the Wilson coefficients for the structure functions $F_{2,3}(x, Q^2)$ and $g1(x, Q^2)$ factorize into the known massless Wilson coefficients and the process-independent massive operator matrix elements, which are computed to three-loop for the first time.

The renormalization of these quantities is worked out. In case of charge renormalization a new element had to be incorporated. It is advised to refer first to a MOM-scheme before the desired expressions in the $\overline{\mathsf{MS}}$-scheme can be obtained. Lower loop order operator matrix elements contribute. They were all calculated for general values of the Mellin variable N up to the corresponding order in the dimensional parameter ε. Here modern summation technologies as developed by the Linz group have been applied using the package Sigma. For the computation of the three-loop massive operator matrix elements emphasis was laid on the corresponding Mellin moments. An extension of the Feynman diagram generator QGRAF allowing for local operator insertions was given. The corresponding moments became calculable using the FORM-code MATAD after applying corresponding projectors of the spin-structures which emerge for the respective moments. With these technologies all operator matrix elements, including those of the gluonic channels, needed to establish the variable flavor scheme, were computed for the moments $N = 10,\ldots,14$ depending on the respective complexity. These moments provide a unique and necessary basis for all upcoming calculations aiming on the result for general values of the Mellin variable N. An important check consists in comparing the moments of the three-loop anomalous dimensions, obtained as a by-product of the present calculation, to the results in the literature, which were confirmed. The computations required about one CPU year using machines with a memory of up to 96 Gbyte and 1.5 Tbyte fast disc. Further applications concerned the heavy flavor contributions to the transversity distribution to three-loop order and the polarized heavy flavor Wilson coefficients to two-loops. Finally, first steps were undertaken to compute a first non-trivial massive three-loop topology for general values of N analytically.

The Thesis of Sebastian Klein represents an essential step in the field of analytic three-loop calculations in Quantum Chromodynamics for single differential quantities in the presence of a single mass and is an impressive piece of work, given the short period of time through which it has been performed. Currently it forms the basis of calculations determining the massive Wilson coefficients for deep-inelastic scattering at high virtualities for general values of the Mellin variable N.

Zeuthen, August 2011 Dr. Johannes Blümlein

Acknowledgments

Foremost, I would like to thank Johannes Blümlein for his constant support during the last years, putting very much time and effort into supervising and teaching me.

Further I would like to thank Prof. Reya for giving me the opportunity of getting my Ph.D. at the University of Dortmund.

I am particularly grateful to I. Bierenbaum for her advice and friendship during the last years and for reading the manuscript.

I would like to thank D. Broadhurst, K. Chetyrkin, J. Kallarackal, M. Kauers, D. Renner, C. Schneider, J. Smith, F. Stan, M. Steinhauser and J. Vermaseren for useful discussions.

Additionally, I would like to thank M. Steinhauser for help with the use of MATAD, J. Vermaseren for help with the use of FORM, and C. Schneider for help with the use of Sigma.

Further thanks go to B. Tödtli and K. Litten for reading parts of the manuscript.

Finally, I also would like to thank my family, Frank, Katharina, Max and Ute for moral support.

This work was supported in part by DFG Sonderforschungsbereich Transregio 9, Computergestützte Theoretische Teilchenphysik, Studienstiftung des Deutschen Volkes, the European Commission MRTN HEPTOOLS under Contract No. MRTN-CT-2006-035505, and DESY. I thank both IT groups of DESY providing me access to special facilities to perform the calculations involved in this thesis.

Contents

Chapter 1
Introduction

Quantum Chromodynamics (QCD) is the well-established theory of the strong interaction. It explains the properties and the structure of hadrons, such as the proton or the neutron, in particular at short distances. Together with the electroweak interaction it forms the Standard Model[1] of elementary particle physics, which up to date explains the microscopic world to a very good accuracy. The low energy manifestations of the electroweak interaction are the electromagnetic and the weak interaction. The former has been known the longest since it is also observable in the everyday world, whereas the weak interaction concerns only microscopic phenomena such as the radioactive β-decay. These theories have three features in common: (1) the matter content is described in terms of elementary particles with definite properties such as mass, charge or spin, (2) the interaction between these particles is mediated by gauge bosons, (3) the strength of the interaction is given by a coupling parameter. In the electroweak sector, there are six particles (and their respective anti-particles), the electron, the muon, the tau and the neutrinos associated to them, the ν_e, ν_μ and ν_τ. They are referred to as leptons and the electroweak force is described in terms of four gauge bosons, the photon and the $W^\pm$ and Z-boson, which are exchanged between these particles. In case of QCD, the matter content is given by six different quarks (also called flavors)—the up, down, strange, charm, bottom and top quark—which are bound together by the mediator of the strong force, the gluons, to form hadrons. Contrary to leptons, only these hadrons are observable in nature, because the size of the strong interaction at small energies prevents the existence of free quark states. This is one of the most important properties of QCD and referred to as confinement.

The path to the discovery of QCD started in the 1960s. By that time, a large amount of hadrons had been observed in cosmic ray and accelerator experiments, which occur as mesons (spin $= 0, 1$) or baryons (spin $= 1/2, 3/2$). In the early 1960s investigations were undertaken to classify all hadrons, based on their properties such as flavor—and spin quantum numbers and masses. In 1964, Gell-Mann [1] and Zweig [2], proposed the quark model as a mathematical description for these hadrons. Three fractionally charged quark flavors, up (u), down (d) and strange (s), known as valence

[1] Gravity, the 4th fundamental force known today, is not yet incorporated in the Standard Model.

S. Klein, *Charm Production in Deep Inelastic Scattering*, Springer Theses,
DOI: 10.1007/978-3-642-23286-2_1,

quarks, were sufficient to describe the quantum numbers of the hadron spectrum which had been discovered by then. Baryons are thus considered as bound states of three quarks and mesons of a quark-antiquark pair. Assuming an approximate $SU(3)$ flavor symmetry, "the eightfold way" [3–5], mass formulas for hadrons built on the basis of quark states could be derived. A great success for the quark model was marked by the prediction of the mass of the Ω^--baryon before it was finally observed [6]. In the same year, Gürsey and Radicati [7], introduced spin into the model and proposed a larger $SU(6)_{spin-flavor} = SU(2)_{spin} \otimes SU(3)_{flavor}$ symmetry. This allowed the unification of the mass formulas for the spin-1/2 and spin-3/2 baryons and provided the tool to calculate the ratio of the magnetic moments of the proton and the neutron to be $\approx -3/2$, which is in agreement with experiment within 3% [8, 9]. However, this theory required the quarks that gave the correct low-lying baryons to be in a symmetric state under permutations, which contradicts the spin-statistics theorem [10], since quarks have to be fermions. Greenberg [11], resolved this contradiction by introducing a "symmetric quark model". It allows quarks to have a new hidden three-valued charge, called color, which is expressed in terms of parafermi statistics. Finally, in 1965, Nambu [12], and Han and Nambu [13], proposed a new symmetry, $SU(3)_{color}$, which makes the hidden three-valued charge degree of freedom explicit and is equivalent to Greenberg's description. Since there was no explicit experimental evidence of this new degree of freedom, the assumption was made that all physical bound states must be color-neutral [12–14].

The possibility to study the substructure of nucleons arose at the end of the 1960s with the advent of the Stanford Linear Accelerator (SLAC)[15, 16]. This facility allowed to perform deeply inelastic lepton-nucleon scattering (DIS) experiments at much higher resolutions than previously possible. The cross section can be parameterized quite generally in terms of several structure functions F_i [17, 18], which can be thought of as generalizations of the elastic nucleon form factors. The former were measured for the proton by the SLAC-MIT experiments and depend both on the energy transfer ν and the four-momentum transfer $q^2 = -Q^2$ from the lepton to the nucleon in the nucleon's rest frame. In the Bjorken limit, $\{Q^2, \nu \to \infty, Q^2/\nu = \text{fixed}\}$ [19], it was found that the structure functions depend on the ratio of Q^2 and ν only, $F_i(\nu, Q^2) = F_i(Q^2/\nu)$. This phenomenon was called scaling [20–23] cf. also [24], and had been predicted by Bjorken in his field theoretic analysis based on current algebra [19]. As the relevant parameter in the deep-inelastic limit he introduced the Bjorken-scaling variable $x = Q^2/2M\nu$, where M is the mass of the nucleon. After scaling was discovered, Feynman gave a phenomenological explanation for this behavior of the structure functions within the parton model [25–29]. According to this model, the proton consists of several point-like constituents, the partons. His assumption was that during the interaction time—which is very short since high energies are involved—these partons behave as free particles off which the electrons scatter elastically. Therefore, the total cross section is just the incoherent sum of the individual electron-parton cross-sections, weighted by the probability to find the particular parton inside the proton. The latter is described by the parton density $f_i(z)$. It denotes the probability to find parton i in the proton, carrying the fraction z of the total proton momentum P. In the limit considered by Feynman,

z becomes equal to x, giving an explanation for scaling. This is a direct consequence of the rigid correlation $M\nu = q.P$, as observed in experiment. Even more important for the acceptance of the quark parton model was the observation that the Callan–Gross relation [30], holds, namely that the longitudinal structure function F_L vanishes in the situation of strict scaling. This experimental result favored the idea of the proton containing spin-1/2, point-like constituents and ruled out different approaches, such as the algebra of fields [31], or explanations assuming vector–meson dominance [32–35]. Finally, Bjorken and Paschos [36], linked the parton model to the group theoretic approach by identifying quarks and partons.

Today QCD forms one part of the Standard Model of elementary particle physics, supplementing the electroweak $SU_L(2) \times U_Y(1)$ sector, which had been proposed by Weinberg in 1967 [37], extending earlier work by Glashow [38], cf. also [39, 40], for the leptonic sector. This theory was proved to be renormalizable by t'Hooft and Veltman in 1972 [41], see also [42–45], if anomalies are canceled [46–48], requiring an appropriate representation for all fermions. t' Hooft also proved renormalization for massless Yang–Mills theories [49]. These gauge theories had first been studied by Yang and Mills in 1954 [50], and have the distinctive property that their gauge group is non-abelian, leading to interactions between the gauge and bosons [14], contrary to the case of Quantum Electrodynamics. In 1972/73, Gell-Mann, Fritzsch and Leutwyler [51], cf. also [12], proposed to gauge color which led to an extension of the Standard Model to $SU_L(2) \times U_Y(1) \times SU_c(3)$, including the strongly interacting sector. The dynamical theory of quarks and gluons, Quantum Chromodynamics, is thus a massless Yang–Mills theory which describes the interaction of different quark flavors via massless gluons. Among the semi-simple compact Lie-groups, $SU(3)_c$ turns out to be the only possible gauge group for this theory, cf. [52, 53]. In 1973, Gross and Wilczek [54], and Politzer [55], showed in a 1-loop calculation that Quantum Chromodynamics is an asymptotically free gauge theory, cf. also G. t'Hooft 1972 (unpublished), which allows to perform perturbative calculations for processes at large enough scales. There, the strong coupling constant α_s becomes a sufficiently small perturbative parameter, meaning that for a given physical observable $\mathsf{A}(\alpha_s)$ the expansion around $\alpha_s = 0$

$$\mathsf{A}(\alpha_s) = \mathsf{A}^{(0)} + \alpha_s \mathsf{A}^{(1)} + \alpha_s^2 \mathsf{A}^{(2)} + \ldots \tag{1.1}$$

gives a good approximation for $\mathsf{A}(\alpha_s)$. In contrast if $\alpha_s \geq 1$—as is the case for QCD for small energies—the expansion Eq. 1.1 diverges and one has to use different methods to calculate $\mathsf{A}(\alpha_s)$.

In the beginning, QCD was not an experimentally well-established theory, which was mainly due to its non-perturbative nature. In the course of performing precision tests of QCD, the operator product expansion near the light-cone, the light-cone expansion (LCE) [56–59], was important. By applying it to deep-inelastic processes in the Bjorken-limit, one facilitates a separation of hadronic bound state effects and the short distance effects. This is possible, since the cross sections of deeply inelastic processes receive contributions from two different resolution scales μ^2. One is the short distance region, where perturbative techniques can be applied.

The other describes the long distance region. Here bound state effects are essential and a perturbative treatment is not possible due to the large coupling involved. By means of the LCE, the two energy scales of the process are associated with two different quantities: the Wilson coefficients $C(z, Q^2/\mu^2)$ and the already mentioned hadronic operator matrix elements or parton densities $f(z, \mu^2)$. The former contain the large scale contributions and can therefore be calculated perturbatively, whereas the latter describe the low scale behavior and are quantities which have to be extracted from experimental data or can be calculated by applying rigorous non-perturbative methods. Schematically, this factorization can be written for a generic structure function $F(x, Q^2)$ to all orders in perturbation theory as

$$F(x, Q^2) = C\left(x, \frac{Q^2}{\mu^2}\right) \otimes f(x, \mu^2) + O\left(\frac{1}{Q^2}\right). \tag{1.2}$$

This is the so called twist-2 approximation, where all power-like contributions $O(1/Q^2)$ are neglected, cf. e.g., [60, 61]. In Eq. 1.2, the Mellin-convolution $\otimes$ is defined by

$$A(x) \otimes B(x) := \int_0^1 \int_0^1 dz_1 dz_2 A(z_1) B(z_2)\, \delta(x - z_1 z_2). \tag{1.3}$$

At this point, it is also convenient to mention a rather technical aspect, which is important for the understanding of this work. Many calculations simplify in Mellin space compared to *x*-space. The *N*th Mellin moment of a function f is defined through the integral

$$M[f](N) := \int_0^1 dx\, x^{N-1} f(x). \tag{1.4}$$

This transform diagonalizes the Mellin-convolution $A \otimes B$ Eq. 1.3 of two functions

$$M[A \otimes B](N) = M[A](N) \times M[B](N), \tag{1.5}$$

and turns it into a product. Hence in many occasions it is more convenient to work in Mellin-space, as will be done throughout this thesis.

Using the LCE, one may derive Feynman's parton model and show the equivalence of the approaches by Feynman and Bjorken in the twist-2 approximation [62]. The LCE also allows to go beyond the naive partonic description, which is formulated in the renormalization group improved parton model. Shortly after the formulation of QCD, logarithmic scaling violations of the deep inelastic cross section where observed [63, 64], which had to be expected since QCD is not an essentially free field theory, neither is it conformally invariant [65]. The theoretical explanation involves the calculation of higher order corrections to the Wilson coefficients as well as to the anomalous dimensions of the composite operators emerging in the LCE [66–68], and predicts the correct logarithmic Q^2 dependence of the structure functions. In fact, the prediction of scaling violations is one of the strongest experimental evidences for QCD.

Thus deeply inelastic scattering played a crucial role in formulating and testing QCD as the theory governing the dynamics of quark systems. Its two most important properties are the confinement postulate—all physical states have to be color singlets—and asymptotic freedom—the strength of the interaction becomes weaker at higher scales, i.e. at shorter distances, cf. e.g., [52, 69–85].

An important step towards completing the Standard Model were the observations of the three heavy quarks charm (c), bottom (b) and top (t). In 1974, two narrow resonances, called Ψ and Ψ', were observed at SLAC in e^+e^- collisions at 3.1 GeV and 3.7 GeV, respectively [86, 87]. At the same time another resonance called J was discovered in proton–proton collisions at BNL [88], which turned out to be the same particle. Its existence could not be explained in terms of the three known quark flavors and was interpreted as a meson consisting of a new quark, the charm quark. This was an important success of the Standard Model since the existence of the charm had been postulated before [89–91]. It is necessary to cancel anomalies for the second family as well as for the GIM-mechanism [92], in order to explain the absence of flavor changing neutral currents. With its mass of $m_c \approx 1.3$ GeV it is much heavier than the light quarks, $m_u \approx 2$ MeV, $m_d \approx 5$ MeV, $m_s \approx 104$ MeV [93], and heavier than the nucleons. In later experiments, two other heavy quarks were detected. In 1977, the Υ $(= b\overline{b})$ resonance was observed at FERMILAB [94], and interpreted as a bound state of the even heavier bottom quark, with $m_b \approx 4.2$ GeV [93]. Ultimately, the quark picture was completed in case of three fermionic families by the discovery of the heaviest quark, the top-quark, in $p\overline{p}$ collisions at the TEVATRON in 1995 [95–98]. Its mass is given by roughly $m_t \approx 171$ GeV [93]. Due to their large masses, heavy quarks cannot be considered as constituents of hadrons at rest or bound in atomic nuclei. They are rather excited in high energy experiments and may form short-lived hadrons, with the exception of the top-quark, which decays before it can form a bound state.

The physics of heavy quarks is a very active area of research, both experimentally and theoretically. From a theory point of view, calculations involving only massless particles are not as challenging as those involving one or more heavy particles. This is due to the larger number of scales involved, leading to more tedious mathematical expressions. Therefore, the level of accuracy reached for processes involving heavy quarks is not yet the same as has been done for massless quarks. However, one finds that the mass-effects are large and can not be neglected.

In this thesis, the theoretical accuracy of scattering processes including heavy quarks is increased. In particular, we study the Wilson coefficients of unpolarized deeply inelastic scattering via single photon exchange. In this case, the double differential scattering cross-section can be expressed in terms of the structure functions $F_2(x, Q^2)$ and $F_L(x, Q^2)$. After the SLAC-MIT experiments forty years ago, many DIS experiments have been performed [63, 64, 99–129]. Recently, the proton was probed to shortest distances at the Hadron–Elektron-Ring-Anlage HERA at DESY in Hamburg [130–134]. In these experiments, a large amount of data has been acquired, and in the case of HERA it is still being processed. This holds especially for the last running period, which was also devoted to the measurement of $F_L(x, Q^2)$ [135]. Up to now, the structure function $F_2(x, Q^2)$ is measured in a wide kinematic region [93], whereas $F_L(x, Q^2)$ was mainly measured in fixed target experiments [136–141], and determined in the region of large ν [142].

In the analysis of DIS data, heavy quarks play an important role, cf. e.g., [143–147]. One finds that the scaling violations of their contributions differ significantly from those of the light partons in a rather wide range starting from lower values of Q^2. It also turns out that the heavy quark contributions to the structure functions of the nucleon may amount up to 25–35%, especially in the small-x region [145, 146, 148, 149]. This demands a more precise theoretical evaluation. The need for considering heavy quarks has additional important aspects. Increasing our knowledge on the perturbative part of deep-inelastic processes allows for a more precise determination of the QCD-scale Λ_{QCD} and the strong coupling constant α_s, as well as of the parton-densities $f_i(z)$ from experimental data. For the former, sufficient knowledge of the next-to-next-to leading order (NNLO)massive corrections in DIS is required to control the theory-errors on the level of the experimental accuracy and below [150–159]. The parton distribution functions are process independent quantities and can be used to describe not only deeply inelastic scattering, but also a large variety of scattering events at (anti-)proton–proton colliders such as the TEVATRON at FERMILAB, and the Large-Hadron–Collider (LHC)at CERN [147]. Heavy quark production is e.g. well suited to extract the gluon density since at leading order (LO) only the photon–gluon fusion process contributes to the cross section [160–164]. It is given by

$$\gamma^* + G \to c + \overline{c}. \tag{1.6}$$

Here, γ^* is the virtual photon exchanged between the electron and the proton, G is the gluon extracted from the proton and $c\overline{c}$ denotes the charm–anticharm pair produced in the final state. Note that we do not consider the heavy quark to be a sea-like constituent of the proton (intrinsic heavy quark production), since data from HERA show that this production mechanism hardly gives any contribution, cf. [165–169]. One also finds that due to the kinematic range of HERA and the previous DIS experiments, charm is produced much more abundantly and gives a higher contribution to the cross section than bottom [146]. Therefore we subsequently limit our discussion to only one heavy quark. Next-to leading order (NLO) calculations, as performed in Refs. [170–172], showed that photon–gluon fusion, Eq. 1.6, is still dominant, although now other processes contribute, too. The gluon density plays a special role, since it carries roughly 50% of the proton momentum, as data from FERMILAB and CERN showed already in the 1970s [173]. Improved knowledge on the gluon distribution $G(x, Q^2)$ is also necessary to describe gluon-initiated processes at the TEVATRON and at the LHC. The study of the gluon density will also help to further understand the small-x behavior of the structure functions, showing a steep rise, which is mainly attributed to its properties.

The perturbatively calculable contributions to the DIS cross section are the Wilson coefficients, Eq. 1.2. In case of light flavors only, these are denoted by $C_{(q,g),(2,L)}(x, Q^2/\mu^2)$ [2] and at present they are known up to the third order in the strong coupling constant [174–193]. Including massive quarks into the analysis, the corresponding terms are known exactly at NLO. The LO terms have been derived in the late seventies [160–164], and the NLO corrections semi-analytically in z-space in the mid-90s [170–172]. A fast numerical implementation was given in [194]. In order to describe DIS at the level of twist $\tau = 2$, also the anomalous dimensions of the local composite operators emerging in the LCE are needed. These have to be combined with the Wilson coefficients and describe, e.g., the scaling violations of the structure functions and parton densities [66–68]. This description is equivalent to the picture in z-space in terms of the splitting functions, [195]. The unpolarized anomalous dimensions are known up to NNLO.[3] At leading [66–68], and at next-to-leading-order level [197–203], they have been known for a long time and were confirmed several times. The NNLO anomalous dimension were calculated by Vermaseren et. al. First, the fixed moments were calculated in Refs. [189, 190, 192] and the complete result was obtained in Refs. [204, 205].

The main parts of this thesis are the extension of the description of the contributions of heavy quark mass effects to the deep-inelastic Wilson coefficients to NNLO.In course of that, we also obtain a first independent calculation of fixed moments of the fermionic parts of the NNLO anomalous dimensions given in Refs. [189, 190] before.

The calculation of the 3-loop heavy flavor Wilson coefficients in the whole Q^2 region is currently not within reach. However, as noticed in Ref. [206], a very precise description of the heavy flavor Wilson coefficients contributing to the structure function $F_2(x, Q^2)$ at NLO is obtained for $Q^2 \gtrsim 10m_Q^2$, disregarding the power corrections $\propto (m_Q^2/Q^2)^k, k \geq 1$. If one considers the charm quark, this covers an important region for deep-inelastic physics at HERA. In this limit, the massive Wilson coefficients factorize into universal massive operator matrix elements (OMEs) $A_{ij}(x, \mu^2/m_Q^2)$ and the light flavor Wilson coefficients $C_{(q,g),(2,L)}(x, Q^2/\mu^2)$. The former are process independent quantities and describe all quark mass effects. They are given by matrix elements of the leading twist local composite operators O_i between partonic states j $(i, j = q, g)$, including quark masses. The process dependence is described by the massless Wilson coefficients. This factorization has been applied in Ref. [207] to obtain the asymptotic limit for $F_L^{c\bar{c}}(x, Q^2)$ at NNLO. However, unlike the case for $F_2^{c\bar{c}}$, the asymptotic result in this case is only valid for much higher values $Q^2 \gtrsim 800m_Q^2$, outside the kinematic domain at HERA for this quantity. An analytic result for the NLO quarkonic massive operator matrix elements A_{qj} needed for the description of the structure functions at this order was derived in Ref. [206] and confirmed in Ref. [208]. A related

---

[2] q = quark, g = gluon

[3] In Ref. [196], the 2nd moment of the 4-loop $\mathsf{N}S^+$ anomalous dimension was calculated.

application of the massive OMEs concerns the formulation of a variable flavor number scheme (VFNS) to describe parton densities of massive quarks at sufficiently high scales. This procedure has been described in detail in Ref. [209], where the remaining gluonic massive OMEs A_{gj} were calculated up to 2-loop order, thereby giving a full NLO description. This calculation was confirmed and extended in [210].

In this work, fixed Mellin-moments, Eq. 1.4, of all contributing massive OMEs at the 3-loop level are calculated and presented, which is a new result [211–214]. The OMEs are then matched with the corresponding known $O(\alpha_s^3)$ light flavor Wilson coefficients to obtain the heavy flavor Wilson coefficients in the limit $Q^2 \gg m^2$, which leads to a precise description for $Q^2/m^2 \gtrsim 10$ in case of $F_2(x, Q^2)$. It is then possible to calculate all logarithmic contributions $\propto \ln(Q^2/m^2)^k$ to the massive Wilson coefficients in the asymptotic region for general values of the Mellin variable N. This applies as well for a large part of the constant term, where also the $O(\varepsilon)$ contributions at the 2-loop level occur (ε being the parameter of dimensional regularization, cf. [215–218]). The first calculation of the latter for all N forms a part of this thesis, too [210, 211, 213, 219–222].Thus only the constant terms of the unrenormalized 3-loop results are at present only known for fixed moments. Since the OMEs are given by the twist $\tau = 2$ composite operators between on-shell partonic states, also fixed moments of the fermionic contributions to the NNLO unpolarized anomalous dimensions are obtained, which are thereby confirmed for the first time in an independent calculation [211–214].

A more technical aspect of this thesis is the study of the mathematical structure of single scale quantities in renormalizable quantum field theories [223–226]. One finds that the known results for a large number of different hard scattering processes are most simply expressed in terms of nested harmonic sums, cf. [227, 228], which are generalizations of the harmonic sum S_1 defined by

$$S_1(N) := \sum_{i}^{N} \frac{1}{i}, \quad N \in \mathbb{N}. \tag{1.7}$$

This holds at least up to 3-loop order for massless Yang–Mills theories, cf. [154, 193, 204, 205, 223, 229–231], including the 3-loop Wilson coefficients and anomalous dimensions. By studying properties of harmonic sums, one may thus obtain significant simplifications [199], because they obey algebraic [232], and structural relations [233, 234]. Performing the calculation in Mellin-space one is naturally led to harmonic sums, which is one of the reasons why we adopt this approach in our calculation. In course of this, new types of infinite sums occur if compared to massless calculations. In the latter case, summation algorithms such as presented in Refs. [228, 235, 236] may be used to calculate the respective sums. The new sums which emerge were calculated using the recent summation package Sigma [237–242], written in MATHEMATICA,which opens up completely new possibilities in symbolic summation and has been tested extensively throughout this work [224].

For fixed values of N, single scale quantities reduce to zero-scale quantities, which can be expressed by rational numbers and certain special numbers as *multiple zeta values* (MZVs) [243, 244], and related objects. Zero scale problems are much easier to calculate than single scale problems. By working in Mellin-space, single scale problems are discrete and one can seek a description in terms of difference equations. One may think of an automated reconstruction of the all N relation out of a *finite number* of Mellin moments given in analytic form. This is possible for recurrent quantities. At least up to 3-loop order, presumably to even higher orders, single scale objects belong to this class. In this work [225, 226], we report on a general algorithm for this purpose, which we applied to a problem being currently one of the most sophisticated ones: the determination of the anomalous dimensions and Wilson coefficients to 3-loop order for unpolarized deeply-inelastic scattering [193, 204, 205].

The thesis is based on the publications Refs. [210, 214, 222, 226, 245], the conference contributions [123, 211–213, 219–221, 224, 225, 246] and the paper in preparation. It is organized as follows. Deeply inelastic scattering within the parton model, the LCE and how one obtains improved results using the renormalization group are described in Chap. 2. Chapter 3 is devoted to the production mechanisms of heavy quarks and their contributions to the cross section. We also discuss the framework of obtaining the heavy flavor Wilson coefficients using massive OMEs in the asymptotic limit $Q^2 \gg m_Q^2$ and comment on the different schemes one may apply to treat heavy quark production [210, 214]. The massive operator matrix elements are considered in Chap. 4 and we describe in detail the renormalization of these objects to 3-loop order, cf. [210–214, 222]. Chapter 5 contains transformation formulas between the different renormalization schemes. We clarify an apparent inconsistency which we find in the renormalization of the massive contributions to the NLO Wilson coefficients given in Refs. [170–172] and the massive OMEs as presented in Refs. [206, 209]. This is due to the renormalization scheme chosen, cf. Ref. [210, 214]. In Chap. 6 the calculation and the results for the 2-loop massive operator matrix elements up to $O(\varepsilon)$ in dimensional regularization are presented. This confirms the results of Ref. [209], cf. [210]. The $O(\varepsilon)$ terms are new results and are needed for renormalization at $O(\alpha_s^3)$, cf. [210, 211, 213, 219–222]. We describe the calculation using hypergeometric functions to set up infinite sums containing the parameter N. These sums are solved using the summation package Sigma, cf. [222, 224]. All sums can then be expressed in terms of nested harmonic sums. The same structure is expected for the 3-loop terms, of which we calculate fixed moments ($N=2,...,10(14)$) using the programs QGRAF [247], FORM, [248, 249], and MATAD [250] in Chap. 7, cf. [211–214]. Thus we confirm the corresponding moments of the fermionic contributions to all unpolarized 3-loop anomalous dimensions which have been calculated before in Refs. [189, 190, 192, 204, 205].

In Chap. 8 we calculate the asymptotic heavy flavor Wilson coefficients for the polarized structure function $g_1(x, Q^2)$ to $O(\alpha_s^2)$ following Ref. [251] and compare them with the results given there. We newly present the terms of $O(\alpha_s^2\varepsilon)$ which contribute to the polarized massive OMEs at $O(\alpha_s^3)$ through renormalization [246]. One may also consider the local flavor non-singlet tensor operator for transversity [252]. This is done in Chap. 9. We derive the corresponding massive OMEs for general values of N up to $O(\alpha_s^2\varepsilon)$ and for the fixed moments $N = 1, \ldots, 13$ at $O(\alpha_s^3)$, [245].

A calculation at 3-loops keeping the full N dependence has not been performed yet. In Chap. 10, we describe several steps which have been undertaken in this direction so far. This involves the calculation of several non-trivial 3-loop scalar integrals for all N and the description of a technique to reconstruct the complete result starting from a fixed number of moments, cf. [225, 226]. Chapter 11 contains the conclusions.

Our conventions are summarized in Appendix A. The set of Feynman-rules used, in particular for the composite operators, is given in Appendix B. In Appendix C, we summarize properties of special functions which frequently occurred in this work. Appendix D contains examples of different types of infinite sums which had to be computed in the present calculation. The main results are shown in Appendices E–G: various anomalous dimensions and the constant contributions of the different massive OMEs for fixed values of N at $O(\alpha_s^3)$. All Figures have been drawn using Axodraw [253].

References

1. M. Gell-Mann, Phys. Lett. **8**, 214 (1964)
2. G. Zweig, An SU(3) model for the strong interaction symmetry and its breaking, CERN-TH-401, **412** (1964)
3. M. Gell-Mann, Y. Neemam, *The Eightfold Way: A Review With a Collection of Reprints*, (Benjamin Press, New York , 1964), pp. 317.
4. J.J.J. Kokkedee, *The Quark Model* (Benjamin Press, New York, 1969), p. 239
5. F.E. Close, *An Introduction to Quarks and Partons* (Academic Press, London, 1979), p. 481
6. V.E. Barnes et al., Phys. Rev. Lett. **12**, 204 (1964)
7. F. Gürsey, L.A. Radicati, Phys. Rev. Lett. **13**, 173 (1964)
8. M.A.B. Beg , B.W. Lee , A. Pais, Phys. Rev. Lett. **13**, 514 (1964)
9. B. Sakita, Phys. Rev. Lett. **13**, 643 (1964)
10. W. Pauli, Phys. Rev. **58**, 716 (1940)
11. O.W. Greenberg, Phys. Rev. Lett. **13**, 598 (1964)
12. Y. Nambu, A Systematics of Hadrons in Subnuclear Physics. In: A. De-Shalit, H. Fehsbach, L. van Hove (eds) *Preludes in Theoretical Physics* (North-Holland, Amsterdam, 1966), p. 133
13. M.Y. Han , Y. Nambu , Phys. Rev. **139**, B1006 (1965)
14. H. Fritzsch, M. Gell-Mann, Current algebra: Quarks and what else? in *Proceedings of 16th International Conference on High-Energy Physics*, ed. by J.D. Jackson, A. Roberts, R. Donaldson. Batavia, Illinois, 6–13 Sept, vol. 2 (1972), p. 135, hep-ph/0208010
15. L.W. Mo, C. Peck, 8-GeV*/cSpectrometer*, SLAC-TN-65-029 (1965)
16. R.E. Taylor Nucleon form–factors above 6-GeV, in *Proceedings of the International Symposium on Electron and Photon Interactions at High Energies*, SLAC, September 5–9 (SLAC, Stanford CA, 1967), SLAC-PUB-0372, p. 78
17. S.D. Drell , J.D. Walecka , Ann. Phys. **28**, 18 (1964)
18. E. Derman, Phys. Rev. **D19**, 133 (1979)
19. J.D. Bjorken, Phys. Rev. **179**, 1547 (1969)
20. D.H. Coward et al., Phys. Rev. Lett. **20**, 292 (1968)
21. W.K.H. Panofsky, Low q^2 electrodynamics, elastic and inelastic electron (and muon) scattering, in *Proceedings of 14th International Conference on High-Energy Physics*, ed. by J. Prentki, J. Steinberger. Vienna (CERN, Geneva, 1968), p. 23
22. E.D. Bloom et al., Phys. Rev. Lett. **23**, 930 (1969)

23. M. Breidenbach et al., Phys. Rev. Lett. **23**, 935 (1969)
24. H.W. Kendall, Rev. Mod. Phys. **63**, 597 (1991)
25. R.E. Taylor, Rev. Mod. Phys. **63**, 573 (1991)
26. J.I. Friedman, Rev. Mod. Phys. **63**, 615 (1991)
27. R.P. Feynman, The behavior of hadron collisions at extreme energies, in *Proceedings of 3rd International Conference on High Energy Collisions*, ed. by C.N. Yang, J.A. Cole, M. Good, R. Hwa, J. Lee-Franzini. Stony Brook, 1969 (Gordon and Breach, New York, 1970), p. 237
28. R.P. Feynman, Phys. Rev. Lett. **23**, 1415 (1969)
29. R.P. Feynman, *Photon-Hadron Interactions* (Benjamin Press, Reading, 1972), p. 282
30. C.G. Callan , D.J. Gross, Phys. Rev. Lett. **22**, 156 (1969)
31. T.D Lee, S. Weinberg , B. Zumino, Phys. Rev. Lett. **18**, 1029 (1967)
32. J. Sakurai, Vector-Meson dominance—present status and future prospects, in *Proceedings 4th International Symposium on Electron and Photon Interactions at High Energies*, ed. by D.W. Braben, R.E. Rand (Daresbury Laboratory, Liverpool, 1969), p. 91
33. J.J. Sakurai, Phys. Rev. Lett. **22**, 981 (1969)
34. Y.-S. Tsai SLAC-PUB-0600 (1969)
35. H. Fraas, D. Schildknecht, Nucl. Phys. **B14**, 543 (1969)
36. J.D. Bjorken, E.A. Paschos, Phys. Rev. **185**, 1975 (1969)
37. S. Weinberg, Phys. Rev. Lett. **19**, 1264 (1967)
38. S.L. Glashow, Nucl. Phys. **22**, 579 (1961)
39. A. Salam, J.C Ward, Phys. Lett. **13**, 168 (1964)
40. A. Salam Weak and Electromagnetic Interactions, in *Proceedings of the 8th Nobel Symposium*, ed. by N. Svartholm. Göteborg, Sweden, 19–25 May (Almqvist and Wiskell, Stockholm, 1968), p. 367
41. G. 't Hooft, M.J.G. Veltman, Nucl. Phys. **B50**, 318 (1972)
42. J.C. Taylor, Nucl. Phys. **B33**, 436 (1971)
43. A.A. Slavnov, Theor. Math. Phys. **10**, 99 (1972)
44. B.W. Lee, J. Zinn-Justin, Phys. Rev. **D5**, 3121 (1972)
45. B.W. Lee, J. Zinn-Justin, Phys. Rev. **D7**, 1049 (1973)
46. J.S. Bell, R. Jackiw, Nuovo Cim. **A60**, 47 (1969)
47. S.L. Adler, Phys. Rev. **177**, 2426 (1969)
48. R.A. Bertlmann, *Anomalies in quantum field theory* (Clarendon, Oxford, 1996), p. 566
49. G. 't Hooft, Nucl. Phys. **B33**, 173 (1971)
50. C.-N. Yang, R.L. Mills, Phys. Rev. **96**, 191 (1954)
51. H. Fritzsch, M. Gell-Mann, H. Leutwyler, Phys. Lett. **B47**, 365 (1973)
52. E. Reya, Phys. Rept. **69**, 195 (1981)
53. T. Muta, *Foundations of Quantum Chromodynamics*, vol. 57, 2nd edn. World Sci Lect. Notes Phys. (World Scientific, Singapore, 1998)
54. D.J. Gross, F. Wilczek, Phys. Rev. Lett. **30**, 1343 (1973)
55. H.D. Politzer, Phys. Rev. Lett. **30**, 1346 (1973)
56. K.G. Wilson, Phys. Rev. **179**, 1499 (1969)
57. W. Zimmermann, *Lectures on Elementary Particle Physics and Quantum Field Theory*, Brandeis Summer Institute vol. 1 (MIT Press, Cambridge, 1970), p. 395
58. Y. Frishman, Ann. Phys. **66**, 373 (1971)
59. R.A. Brandt, G. Preparata, Nucl. Phys. **B27**, 541 (1972)
60. J.C. Collins , D.E. Soper , G. Sterman, Perturbative QCD, ed. by A.H. Mueller. Adv. Ser. Direct. High Energy Phys. **5**, 1 (1988), hep-ph/0409313
61. S.B. Libby, G. Sterman, Phys. Rev. **D18**, 4737 (1978)
62. D.J. Gross, S.B. Treiman, Phys. Rev. **D4**, 1059 (1971)
63. C. Chang et al., Phys. Rev. Lett. **35**, 901 (1975)
64. Y. Watanabe et al., Phys. Rev. Lett. **35**, 898 (1975)
65. S. Ferrara, R. Gatto, A.F. Grillo, Springer Tracts Mod. Phys. **67**, 1 (1973) and references therein

66. D.J. Gross, F. Wilczek, Phys. Rev. **D8**, 3633 (1973)
67. D.J. Gross, F. Wilczek, Phys. Rev. **D9**, 980 (1974)
68. H. Georgi, H.D. Politzer, Phys. Rev. **D9**, 416 (1974)
69. Photon-Hadron Interactions I, International Summer Institute in Theoretical Physics, DESY, July 12–14, Springer Tracts in Modern Physics **62** (1971), p. 147
70. H.D. Politzer, Phys. Rept. **14**, 129 (1974)
71. W.J. Marciano, H. Pagels, Phys. Rept. **36**, 137 (1978)
72. J.R Ellis, C.T Sachrajda, NATO Adv. Study Inst. Ser. B Phys. **59**, 285 (1980)
73. A.J. Buras, Rev. Mod. Phys. **52**, 199 (1980)
74. G. Altarelli, Phys. Rept. **81**, 1 (1982)
75. F. Wilczek, Ann. Rev. Nucl. Part. Sci. **32**, 177 (1982)
76. R.L. Jaffe, *Deep Inelastic Scattering with Application to nuclear targets*, ed. by M.B. Johnson, A. Picklesimer. Lectures presented at the Los Alamos School on Quark Nuclear Physics, Los Alamos, NM, June 10-14, 1985 (Wiley, New York, 1986), p. 82
77. J.C. Collins, D.E. Soper, Ann. Rev. Nucl. Part. Sci. **37**, 383 (1987)
78. R.K. Ellis, *An Introduction to the QCD Parton Model*, ed. by M.B. Johnson, A. Picklesimer. Lectures given at 1987 Theoretical Advanced Study Institute in Elementary Particle Physics, Santa Fe, NM, July 5–Aug 1, 1987 (World Scientific, Singapore, 1988), p. 214
79. A.H. Mueller (eds), *Perturbative Quantum Chromodynamics* (World Scientific, Singapore, 1989)
80. R.G. Roberts, *The Structure of the proton: Deep inelastic scattering* (Cambridge University Press, Cambridge, 1990), p. 182
81. G. Sterman, *An Introduction to quantum field theory* (Cambridge University Press, Cambridge, 1993), p. 572
82. R.K. Ellis, W.J. Stirling, B.R. Webber, *QCD and collider physics*, Camb. Monogr. Part. Phys. Nucl. Phys. Cosmol. 8 (Cambridge University Press, Cambridge, 1996), p. 435
83. CTEQ collaboration, R. Brock et al., Rev. Mod. Phys. **67**, 157 (1995)
84. J. Blümlein, Surveys High Energ. Phys. **7**, 181 (1994)
85. P. Mulders, *Quantum Chromodynamics and Hard Scattering Processes*, Lectures, Dutch Research School for Theoretical Physics, Dalfsen, January 1996, and Dutch Research School for Subatomic Physics, Beekbergen, February, (NIKHEF, Amsterdam, The Netherlands, 1996)
86. SLAC-SP-017 collaboration, J.E. Augustin et al., Phys. Rev. Lett. **33**, 1406 (1974)
87. G.S. Abrams et al., Phys. Rev. Lett. **33**, 1453 (1974)
88. E598 collaboration, J.J. Aubert et al., Phys. Rev. Lett. **33**, 1404 (1974)
89. Z. Maki, M. Nakagawa, Prog. Theor. Phys. **31**, 115 (1964)
90. Y. Hara, Phys. Rev. **134**, B701 (1964)
91. J.D. Bjorken, S.L. Glashow, Phys. Lett. **11**, 255 (1964)
92. S.L. Glashow, J. Iliopoulos, L. Maiani, Phys. Rev. **D2**, 1285 (1970)
93. Particle Data Group, C. Amsler et al., Phys. Lett. **B667**, 1 (2008)
94. S.W. Herb et al., Phys. Rev. Lett. **39**, 252 (1977)
95. CDF collaboration, F. Abe et al., Phys. Rev. Lett. **73**, 225 (1994), hep-ex/9405005
96. CDF collaboration, F. Abe et al., Phys. Rev. **D50**, 2966 (1994)
97. CDF collaboration, F. Abe et al., Phys. Rev. Lett. **74**, 2626 (1995), hep-ex/9503002
98. D0 collaboration, S. Abachi et al., Phys. Rev. Lett. **74**, 2632 (1995), hep-ex/9503003
99. S. Stein et al., Phys. Rev. **D12**, 1884 (1975)
100. W.B. Atwood et al., Phys. Rev. Lett. **23**, 935 (1969)
101. A. Bodek et al., Phys. Rev. **D20**, 1471 (1979)
102. M.D. Mestayer et al., Phys. Rev. **D27**, 285 (1983)
103. EMC collaboration, O. Allkofer et al., Nucl. Instr. Meth. **179**, 445 (1981)
104. EMC collaboration, J.J. Aubert et al., Nucl. Phys. **B259**, 189 (1985)
105. D. Bollini et al., Nucl. Instr. Meth. **204**, 333 (1983)
106. BCDMS collaboration, A.C. Benvenuti et al., Nucl. Instr. Meth. **A226**, 330 (1984)
107. BCDMS collaboration, A.C. Benvenuti et al., Phys. Lett. **B195**, 91 (1987)

108. BCDMS collaboration, A.C. Benvenuti et al., Phys. Lett. **B223**, 485 (1989)
109. BCDMS collaboration, A.C. Benvenuti et al., Phys. Lett. **B237**, 592 (1990)
110. NMC collaboration, P. Amaudruz et al., Nucl. Phys. **B371**, 3 (1992)
111. NMC collaboration, P. Amaudruz et al., Phys. Lett. **B295**, 159 (1992)
112. NMC collaboration, M. Arneodo et al., Phys. Lett. **B364**, 107, hep-ph/9509406 (1995)
113. H.L. Anderson et al., Phys. Rev. **D20**, 2645 (1979)
114. E665 collaboration, M.R. Adams et al., Nucl. Instrum. Meth. **A291**, 533 (1990)
115. E665 collaboration, M.R. Adams et al., Phys. Rev. **D54**, 3006 (1996)
116. CHARM collaboration, M. Jonker et al., Phys. Lett. **B109**, 133 (1982)
117. CHARM collaboration, F. Bergsma et al., Phys. Lett. **B123**, 269 (1983)
118. CHARM collaboration, F. Bergsma et al., Phys. Lett. **B153**, 111 (1985)
119. J.P. Berge et al., Z. Phys. **C49**, 187 (1991)
120. Birmingham-CERN-Imperial College-München(MPI)-Oxford collaboration, G.T. Jones et al., Z. Phys. **C62**, 575 (1994)
121. CCFR collaboration, M.H. Shaevitz et al., Nucl. Phys. Proc. Suppl. **38**, 188 (1995)
122. Aachen-Bonn-CERN-London-Oxford-Saclay collaboration, P.C. Bosetti et al., Nucl. Phys. **B142**, 1 (1978)
123. J.G.H. de Groot et al., Z. Phys. **C1**, 143 (1979)
124. S.M. Heagy et al., Phys. Rev. **D23**, 1045 (1981)
125. Gargamelle SPS collaboration, J.G. Morfin et al., Phys. Lett. **B104**, 235 (1981)
126. Aachen-Bonn-CERN-Democritos-London-Oxford-Saclay collaboration, P.C. Bosetti et al., Nucl. Phys. **B203**, 362 (1982)
127. H. Abramowicz et al., Z. Phys. **C17**, 283 (1983)
128. D. MacFarlane et al., Z. Phys. **C26**, 1 (1984)
129. D. Allasia et al., Z. Phys. **C28**, 321 (1985)
130. HERA—a proposal for a large electron proton colliding beam facility at DESY (Hamburg, DESY), DESY HERA 81-10, 292 (1981)
131. H1 collaboration, I. Abt et al., The H1 detector at HERA, DESY-93-103, 194 (1993)
132. ZEUS collaboration, M. Derrick et al., Phys. Lett. **B303**, 183 (1993)
133. HERMES collaboration, K. Ackerstaff et al., Nucl. Instrum. Meth. **A417**, 230, (1998), hep-ex/9806008
134. E. Hartouni et al., HERA-B: An Experiment to Study CP Violation in the B System Using an Internal Target at the HERA Proton Ring. Design report, DESY-PRC-95-01 (1995), p. 491
135. H1 collaboration, F.D. Aaron et al., Phys. Lett. **B665**, 139 (2008), hep-ex/0805.2809
136. L.W Whitlow, S. Rock, A. Bodek, E.M Riordan, S. Dasu, Phys. Lett. **B250**, 193 (1990)
137. S. Dasu et al., Phys. Rev. **D49**, 5641 (1994)
138. E140X collaboration, L.H. Tao et al., Z. Phys. **C70**, 387 (1996)
139. NMC collaboration, M. Arneodo et al., Nucl. Phys. **B483**, 3 (1997), hep-ph/9610231
140. NMC collaboration, M. Arneodo et al., Nucl. Phys. **B487**, 3 (1997), hep-ex/9611022
141. Y. Liang, M.E Christy, R. Ent, C.E. Keppel, Phys. Rev. **C73**, 065201 (2006) nucl-ex/0410028
142. H1 collaboration, C. Adloff et al., Phys. Lett. **B393**, 452 (1997), hep-ex/9611017
143. M. Klein, On the future measurement of the longitudinal structure function at low x at HERA, in *Proceedings of the 12th International Workshop on Deep Inelastic Scattering*, ed. by D. Bruncko, J. Ferencei, P. Striženec. DIS 2004, Strebske Pleso, Slovakia, 14–18 April (Academic Electronic Press, Bratislava, 2004), p. 309
144. J. Feltesse, Measurement of the longitudinal proton structure function at low x at HERA, in *Proceedings of Ringberg Workshop on New Trends in HERA Physics 2005*, ed. by G. Grindhammer, B.A. Kniehl, G. Kramer, W. Ochs. Ringberg Castle, Tegernsee, Germany, 2–7 Oct (World Scientific, Singapore, 2005), p. 370
145. H1 and ZEUS collaboration, K. Lipka Nucl. Phys. Proc. Suppl. **152**, 128 (2006)
146. P.D. Thompson, J. Phys. **G34**, N177 (2007) hep-ph/0703103

147. H. Jung, A. De Roeck (eds.), in *Proceedings of the workshop: HERA and the LHC workshop series on the implications of HERA for LHC physics*, (2006–2008, Hamburg, Geneve), DESY-PROC-2009-02, March 2009, 794, hep-ph/0903.3861
148. ZEUS, S. Chekanov et al., Eur. Phys. J. **C63**, 171, hep-ex/0812.3775 (2009)
149. J. Blümlein, S. Riemersma, *QCD corrections to* $F_L(x, Q^2)$ (1996), hep-ph/9609394
150. S. Bethke, J. Phys. **G26**, R27 (2000) hep-ex/0004021
151. S. Bethke, Nucl. Phys. Proc. Suppl. **135**, 345 (2004) hep-ex/0407021
152. J. Blümlein, H. Böttcher, A. Guffanti, Nucl. Phys. Proc. Suppl. **135**, 152 (2004) hep-ph/0407089
153. S. Alekhin et al., in *HERA and the LHC—A workshop on the implications of HERA for LHC physics: Proceedings Part A, B* (2005), hep-ph/0601012, hep-ph/0601013
154. M. Dittmar et al., *Parton distributions: Summary report for the HERA—LHC workshop* (2005), hep-ph/0511119
155. M. Glück, E. Reya, C. Schuck, Nucl. Phys. **B754**, 178 (2006) hep-ph/0604116
156. S. Alekhin, K. Melnikov, F. Petriello, Phys. Rev. **D74**, 054033 (2006) hep-ph/0606237
157. J. Blümlein, H. Böttcher , A. Guffanti, Nucl. Phys. **B774**, 182 (2007) hep-ph/0607200
158. J. Blümlein, Λ_{QCD} and $\alpha_s(M_Z^2)$ *from DIS Structure Functions* (2007), hep-ph/0706.2430
159. H. Jung et al., *What HERA may provide?* (2008), hep-ph/0809.0549
160. E. Witten, Nucl. Phys. **B104**, 445 (1976)
161. J. Babcock, D.W. Sivers, S. Wolfram, Phys. Rev. **D18**, 162 (1978)
162. M.A. Shifman, A.I. Vainshtein, V.I. Zakharov, Nucl. Phys. **B136**, 157 (1978)
163. J.P. Leveille, T.J. Weiler, Nucl. Phys. **B147**, 147 (1979)
164. M. Glück, E. Hoffmann, E. Reya, Z. Phys. **C13**, 119 (1982)
165. S.J. Brodsky, P. Hoyer, C. Peterson, N. Sakai, Phys. Lett. **B93**, 451 (1980)
166. E. Hoffmann, R. Moore, Z. Phys. **C20**, 71 (1983)
167. ZEUS collaboration, M. Derrick et al., Phys. Lett. **B349**, 225 (1995), hep-ex/9502002
168. B.W Harris, J. Smith, R. Vogt, Nucl. Phys. **B461**, 181 (1996) hep-ph/9508403
169. H1 collaboration, C. Adloff et al., Z. Phys. **C72**, 593 (1996), hep-ex/9607012
170. E. Laenen, S. Riemersma, J. Smith, W.L. van Neerven, Nucl. Phys. **B392**, 162 (1993)
171. E. Laenen, S. Riemersma, J. Smith, W.L. van Neerven, Nucl. Phys. **B392**, 229 (1993)
172. S. Riemersma, J. Smith, W.L. van Neerven, Phys. Lett. **B347**, 143 (1995) hep-ph/9411431
173. R.E. Taylor, Inelastic Electron-Nucleon Scattering Experiments, ed. by W.T. Kirk. Invited paper presented at International Symposium on Lepton and Photon Interactions, Stanford University, California, Aug 21–27, 1975, (SLAC, Stanford CA, 1976), SLAC–PUB–1729, p. 29
174. A. Zee, F. Wilczek, S.B. Treiman, Phys. Rev. **D10**, 2881 (1974)
175. W.A. Bardeen, A.J. Buras, D.W. Duke, T. Muta, Phys. Rev. **D18**, 3998 (1978)
176. W. Furmanski, R. Petronzio, Z. Phys. **C11**, 293 (1982) and references therein
177. D.W. Duke , J.D. Kimel , G.A. Sowell, Phys. Rev. **D25**, 71 (1982)
178. A. , D.W. Duke, J.D. Kimel, G.A. Sowell, Phys. Rev. **D30**, 541 (1984)
179. D.I. Kazakov, A.V. Kotikov, Nucl. Phys. **B307**, 721 (1988)
180. D.I. Kazakov, A.V. Kotikov, G. Parente, O.A. Sampayo, J. Sanchez Guillen, Phys. Rev. Lett. **65**, 1535 (1990)
181. J. Sanchez Guillen, J. Miramontes, M. Miramontes, G. Parente, O.A. Sampayo, Nucl. Phys. **B353**, 337 (1991)
182. W.L. van Neerven, E.B. Zijlstra, Phys. Lett. **B272**, 127 (1991)
183. W.L. van Neerven, E.B. Zijlstra, Phys. Lett. **B273**, 476 (1991)
184. W.L. van Neerven, E.B. Zijlstra, Nucl. Phys. **B383**, 525 (1992)
185. D.I. Kazakov, A.V. Kotikov, Phys. Lett. **B291**, 171 (1992)
186. S.A. Larin, J.A.M. Vermaseren, Z. Phys. **C57**, 93 (1993)
187. S. Moch, J.A.M. Vermaseren , Nucl. Phys. **B573**, 853 (2000) hep-ph/9912355
188. S.A. Larin, T. van Ritbergen, J.A.M. Vermaseren, Nucl. Phys. **B427**, 41 (1994)

189. S.A. Larin, P. Nogueira, T. van Ritbergen, J.A.M. Vermaseren, Nucl. Phys. **B492**, 338 (1997) hep-ph/9605317
190. A. Retey, J.A.M. Vermaseren, Nucl. Phys. **B604**, 281 (2001) hep-ph/0007294
191. S. Moch, J.A.M. Vermaseren, A. Vogt, Phys. Lett. **B606**, 123 (2005) hep-ph/0411112
192. J. Blümlein, J.A.M. Vermaseren, Phys. Lett. **B606**, 130 (2005) hep-ph/0411111
193. J.A.M. Vermaseren, A. Vogt, S. Moch, Nucl. Phys. **B724**, 3 (2005) hep-ph/0504242
194. S.I. Alekhin, J. Blümlein, Phys. Lett. **B594**, 299 (2004) hep-ph/0404034
195. G. Altarelli, G. Parisi, Nucl. Phys. **B126**, 298 (1977)
196. P.A. Baikov, K.G. Chetyrkin, Nucl. Phys. Proc. Suppl. **160**, 76 (2006)
197. E.G. Floratos , D.A. Ross , C.T. Sachrajda, Nucl. Phys. **B129**, 66 (1977); [Erratum-ibid.] **B139**, 545 (1978)
198. E.G Floratos, D.A. Ross, C.T. Sachrajda, Nucl. Phys. **B152**, 493 (1979)
199. A. Gonzalez-Arroyo, C. Lopez, F.J. Yndurain, Nucl. Phys. **B153**, 161 (1979)
200. A. Gonzalez-Arroyo, C. Lopez, Nucl. Phys. **B166**, 429 (1980)
201. G. Curci, W. Furmanski, R. Petronzio, Nucl. Phys. **B175**, 27 (1980)
202. W. Furmanski, R. Petronzio, Phys. Lett. **B97**, 437 (1980)
203. R. Hamberg, W.L. van Neerven, Nucl. Phys. **B379**, 143 (1992)
204. S. Moch, J.A.M. Vermaseren, A. Vogt, Nucl. Phys. **B688**, 101 (2004) hep-ph/0403192
205. A. Vogt, S. Moch, J.A.M. Vermaseren, Nucl. Phys. **B691**, 129 (2004) hep-ph/0404111
206. M. Buza, Y. Matiounine, J. Smith, R. Migneron, W.L. van Neerven, Nucl. Phys. **B472**, 611 (1996) hep-ph/9601302
207. J. Blümlein, A. De Freitas, W.L. van Neerven, S. Klein, Nucl. Phys. **B755**, 272 (2006) hep-ph/0608024
208. I. Bierenbaum, J. Blümlein, S. Klein, Nucl. Phys. **B780**, 40 (2007) hep-ph/0703285
209. M. Buza, Y. Matiounine, J. Smith, W.L. van Neerven, Eur. Phys. J. **C1**, 301 (1998) hep-ph/9612398
210. I. Bierenbaum, J. Blümlein, S. Klein, Phys. Lett. **B672**, 401 (2009) hep-ph/0901.0669
211. I. Bierenbaum, J. Blümlein, S. Klein, Nucl. Phys. Proc. Suppl. **183**, 162 (2008) hep-ph/0806.4613
212. I. Bierenbaum , J. Blümlein , S. Klein, PoS Confinement 8, 185 (2008), hep-ph/0812.2427
213. I. Bierenbaum , J. Blümlein , S. Klein, in *2- and 3-loop heavy flavor contributions to* $F_2(x, Q^2)$, $F_L(x, Q^2)$ *and* $g_{1,2}(x, Q^2)$ vol. 216, p. 363
214. I. Bierenbaum, J. Blümlein, S. Klein, Nucl.Phys. **B820**, 417 (2009) hep-ph/0904.3563
215. G. 't Hooft, M.J.G. Veltman, Nucl. Phys. **B44**, 189 (1972)
216. J.F. Ashmore, Lett. Nuovo Cim. **4**, 289 (1972)
217. G.M. Cicuta, E. Montaldi, Nuovo Cim. Lett. **4**, 329 (1972)
218. C.G. Bollini, J.J. Giambiagi, Nuovo Cim. **B12**, 20 (1972)
219. I. Bierenbaum, J. Blümlein, S. Klein, Acta Phys. Polon. **B38**, 3543 (2007)
220. I. Bierenbaum, J. Blümlein, S. Klein, Pos RADCOR 2007, 034 (2007), hep-ph/0710.3348
221. I. Bierenbaum, J. Blümlein, S. Klein, Acta Phys. Polon. **B39**, 1531 (2008) hep-ph/0806.0451
222. I. Bierenbaum, J. Blümlein, S. Klein, C. Schneider, Nucl. Phys. **B803**, 1 (2008) hep-ph/0803.0273
223. J. Blümlein, S. Klein, PoS ACAT 2007, 084 (2007), hep-ph/0706.2426
224. I. Bierenbaum, J. Blümlein, S. Klein, C. Schneider, PoS ACAT 2007, 082 (2007), math-ph/0707.4659
225. J. Blümlein, M. Kauers, S. Klein, C. Schneider, PoS ACAT 2008, 106 (2008), hep-ph/0902.4095
226. J. Blümlein, M. Kauers, S. Klein, C. Schneider, Comput.Phys.Commun. **180**, 2143 (2009) hep-ph/0902.4091
227. J. Blümlein, S. Kurth, Phys. Rev. **D60**, 014018 (1999) hep-ph/9810241
228. J.A.M. Vermaseren, Int. J. Mod. Phys. **A14**, 2037 (1999) hep-ph/9806280
229. J. Blümlein, Nucl. Phys. Proc. Suppl. **135**, 225 (2004) hep-ph/0407044
230. J. Blümlein, V. Ravindran, Nucl. Phys. **B716**, 128 (2005) hep-ph/0501178

231. J. Blümlein, V. Ravindran, Nucl. Phys. B **749**, 1 (2006) hep-ph/0604019
232. J. Blümlein, Comput. Phys. Commun. **159**, 19 (2004) hep-ph/0311046
233. J. Blümlein, Comput. Phys. Commun. **180**, 2218 (2009) hep-ph/0901.3106
234. J. Blümlein, Structural Relations of Harmonic Sums and Mellin Transforms at Weight w=6, in *Proceedings of the Workshop Motives, Quantum Field Theory, and Pseudodifferential Operators*, June 2008 (Clay Institute, Boston University, 2009), math-ph/0901.0837
235. S. Weinzierl, Comput. Phys. Commun. **145**, 357 (2002) math-ph/0201011
236. S. Moch, P. Uwer, Comput. Phys. Commun. **174**, 759 (2006) math-ph/0508008
237. C. Schneider, Ann.Comb. **9**(1), 75 (2005)
238. C. Schneider, *Proc ISSAC'05.*, (ACM Press, , 2005) pp. 285.
239. C. Schneider, in *Proceedings of FPSAC'07* (2007), p. 1
240. C. Schneider, J. Algebra Appl. **6**(3), 415 (2007)
241. C. Schneider, J. Diffr. Equations Appl. **11**(9), 799 (2005)
242. C. Schneider, Sém. Lothar. Combin. **56** (2007) Article B56b and Habilitation Thesis, JKU Linz (2007)
243. J.M. Borwein, Bradley D.M. , D.J. Broadhurst, P. Lisonek, Trans. Am. Math. Soc. **353**, 907 (2001) math/9910045
244. J. Blümlein, D. Broadhurst, J. Vermaseren, Comput.Phys.Commun. **181**, 582 (2010) math-ph/0907.2557
245. J. Blümlein, S. Klein, B. Tödtli, Phys.Rev. **D80**, 094010 (2009) hep-ph/0909.1547
246. I. Bierenbaum, J. Blümlein , S. Klein, Two-Loop Massive Operator Matrix Elements for Polarized and Unpolarized Deep-Inelastic Scattering, in *Proceedings of 15th International Workshop On Deep-Inelastic Scattering and Related Subjects*, ed. by G. Grindhammer, K. Sachs. (DIS2007; 16–20 April 2007, Munich), vol. 2, p. 821, hep-ph/0706.2738
247. P. Nogueira, J. Comput. Phys. **105**, 279 (1993)
248. J.A.M. Vermaseren, New features of FORM (2000), math-ph/0010025
249. T. van Ritbergen , A.N. Schellekens, J.A.M. Vermaseren, Int. J. Mod. Phys. **A14**, 41 (1999) hep-ph/9802376
250. M. Steinhauser, Comput. Phys. Commun. **134**, 335 (2001) hep-ph/0009029
251. M. Buza, Y. Matiounine, J. Smith, W.L. van Neerven, Nucl. Phys. B **B485**, 420 (1997) hep-ph/9608342
252. V. Barone, A. Drago, P.G. Ratcliffe, Phys. Rept. **359**, 1 (2002) hep-ph/0104283
253. J.A.M. Vermaseren, Comput. Phys. Commun. **83**, 45 (1994)

Chapter 2
Deeply Inelastic Scattering

Deep—inelastic scattering experiments provide one of the cleanest possibilities to probe the space—like short distance structure of hadrons through the reactions

$$l^{\pm}N \to l^{\pm} + X \tag{2.1}$$

$$\nu_l(\overline{\nu}_l)N \to l^{\mp} + X \tag{2.2}$$

$$l^{\mp}N \to \nu_l(\overline{\nu}_l) + X, \tag{2.3}$$

with $l = e, \mu,\ \nu_l = \nu_{e,\mu,\tau},\ N = p, d$ or a nucleus, and X the inclusive hadronic final state. The 4-momentum transfers $q^2 = -Q^2$ involved are at least of the order of $Q^2 \geq 4\,\text{GeV}^2$ and one may resolve spatial scales of approximately $1/\sqrt{Q^2}$. The different deep inelastic charged—and neutral current reactions offer complementary sensitivity to unfold the quark flavor and gluonic structure of the nucleons. Furthermore, polarized lepton scattering off polarized targets is studied in order to investigate the spin structure of the nucleons.

The electron–proton experiments performed at SLAC in 1968 [1–6] cf. also [7–9] and at DESY [10–12] found the famous scaling behavior of the structure functions which had been predicted by Bjorken before [13]. These measurements led to the creation of the parton model [14–16]. Several years later, after a series of experiments had confirmed its main predictions, the partons were identified with the quarks, antiquarks and gluons as real quantum fields, which are confined inside hadrons. Being formerly merely mathematical objects [17, 18] they became essential building blocks of the Standard Model of elementary particle physics, besides the leptons and the electroweak gauge fields, thereby solving the anomaly-problem [19–21].

In the following years, more studies were undertaken at higher energies, such as the electron–proton/neutron scattering experiments at SLAC [22–25]. Muons were used as probes of the nucleons by EMC [26, 27] BCDMS [28–32] and NMC [33–35] at the SPS [36] at CERN, as well as by the E26—[37–39] CHIO—[40]

S. Klein, *Charm Production in Deep Inelastic Scattering*, Springer Theses,
DOI: 10.1007/978-3-642-23286-2_2, © Springer-Verlag Berlin Heidelberg 2012

and E665—[41, 42] collaborations at FERMILAB. For a general review of $\mu^{\pm}N$–scattering, see [43]. The latter experiments were augmented by several high energy neutrino scattering experiments by the CHARM– and CDHSW–collaborations [44–49] and the WA21/25–experiments [50, 51] at the SPS, and by the CCFR–collaboration [52–54] at FERMILAB. Further results on neutrinos were reported in Refs. [55–62] cf. also [63–67]. The data of these experiments confirmed QCD as the theory describing the strong interactions within hadrons, most notably by the observation of logarithmic scaling violations of the structure functions at higher energies and lower values of x, which had been precisely predicted by theoretical calculations [68–70].

All these experiments had in common that they were fixed target experiments and therefore could only probe a limited region of phase space, up to $x \geq 10^{-3}$, $Q^2 \leq 500\,\mathrm{GeV}^2$. The first electron–proton collider experiments became possible with the advent of the HERA facility, which began operating in the beginning of the 1990ies at DESY [71]. This allowed measurements at much larger values of Q^2 and at far smaller values of x than before, $x \geq 10^{-4}$, $Q^2 \leq 20.000\,\mathrm{GeV}^2$. The physics potential for the deep–inelastic experiments at HERA was studied during a series of workshops, see [72–80]. HERA collected a vast amount of data until its shutdown in 2007, a part of which is still being analyzed, reaching unprecedented experimental precisions below the level of 1%. Two general purpose experiments to study inclusive and various semi-inclusive unpolarized deep–inelastic reactions, H1 [81] and ZEUS [82] were performed. Both experiments measured the structure functions $F_{2,L}(x, Q^2)$ as well as the heavy quark contributions to these structure functions to high precision. The theoretical calculations in this thesis are important for the analysis and understanding of the latter, as will be outlined in Chap. 3 . The HERMES—experiment [83] studied scattering of polarized electrons and positrons off polarized gas—targets. HERA-B [84] was dedicated to the study of CP—violations in the B–sector.

In the following, we give a brief introduction into the theory of DIS and the theoretical tools which are used to predict the properties of structure functions, such as asymptotic scaling and scaling violations. In Sect. 2.1, we discuss the kinematics of the DIS process and derive the cross section for unpolarized electromagnetic electron-proton scattering. In Sect. 2.2, we give a description of the naive parton model, which was employed to explain the results obtained at SLAC and gave a first correct qualitative prediction of the observed experimental data. A rigorous treatment of DIS can be obtained by applying the light-cone expansion to the forward Compton amplitude [85–88] which is described in Sect. 2.3 This is equivalent to the QCD–improved parton model at the level of twist $\tau = 2$, cf. e.g. [89–93]. One obtains evolution equations for the structure functions and the parton densities with respect to the mass scales considered. The evolution is governed by the splitting functions [94] or the anomalous dimensions [68–70] cf. Sect. 2.4.

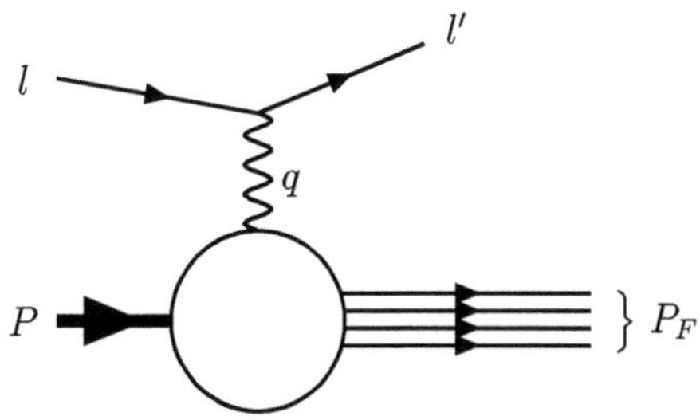

Fig. 2.1 Schematic graph of deeply inelastic scattering for single boson exchange

2.1 Kinematics and Cross Section

The schematic diagram for the Born cross section of DIS is shown in Fig. 2.1 for single gauge boson exchange. A lepton with momentum l scatters off a nucleon of mass M and momentum P via the exchange of a virtual vector boson with momentum q. The momenta of the outgoing lepton and the set of hadrons are given by l' and P_F, respectively. Here F can consist of any combination of hadronic final states allowed by quantum number conservation. We consider inclusive final states and thus all the hadronic states contributing to F are summed over. The kinematics of the process can be measured from the scattered lepton or the hadronic final states, cf. e.g. [95–97] depending on the respective experiment. The virtual vector boson has space-like momentum with a virtuality Q^2

$$Q^2 \equiv -q^2\,, \quad q = l - l'. \tag{2.4}$$

There are two additional independent kinematic variables for which we choose

$$s \equiv (P + l)^2, \tag{2.5}$$

$$W^2 \equiv (P + q)^2 = P_F^2. \tag{2.6}$$

Here, s is the total cms energy squared and W denotes the invariant mass of the hadronic final state. In order to describe the process, one usually refers to Bjorken's scaling variable x, the inelasticity y, and the total energy transfer ν of the lepton to the nucleon in the nucleon's rest frame [98]. They are defined by

$$\nu \equiv \frac{P \cdot q}{M} = \frac{W^2 + Q^2 - M^2}{2M}, \tag{2.7}$$

$$x \equiv \frac{-q^2}{2P \cdot q} = \frac{Q^2}{2M\nu} = \frac{Q^2}{W^2 + Q^2 - M^2}, \tag{2.8}$$

$$y \equiv \frac{P \cdot q}{P \cdot l} = \frac{2M\nu}{s - M^2} = \frac{W^2 + Q^2 - M^2}{s - M^2}, \tag{2.9}$$

where lepton masses are disregarded. In general, the virtual vector boson exchanged can be a γ, Z or $W^{\pm}$–boson with the in—and outgoing lepton, respectively, being an electron, muon or neutrino. In the following, we consider only unpolarized neutral current charged lepton–nucleon scattering. In addition, we will disregard weak gauge boson effects caused by the exchange of a Z–boson. This is justified as long as the virtuality is not too large, i.e. $Q^2 < 500\,\text{GeV}^2$, cf. [99]. We assume the QED- and electroweak radiative corrections to have been carried out [96, 97, 100].

The kinematic region of DIS is limited by a series of conditions. The hadronic mass obeys

$$W^2 \geq M^2. \tag{2.10}$$

Furthermore,

$$\nu \geq 0, \quad 0 \leq y \leq 1, \quad s \geq M^2. \tag{2.11}$$

From (2.10) follows the kinematic region for Bjorken-x via

$$W^2 = (P+q)^2 = M^2 - Q^2\left(1 - \frac{1}{x}\right) \geq M^2 \quad \Longrightarrow 0 \leq x \leq 1. \tag{2.12}$$

Note that $x = 1$ describes the elastic process, while the inelastic region is defined by $x < 1$. Additional kinematic constraints follow from the design parameters of the accelerator [101, 102]. In the case of HERA, these were 820(920) GeV for the proton beam and 27.5 GeV for the electron beam, resulting in a cms–energy $\sqrt{s}$ of 300.3(319) GeV.[1] This additionally imposes kinematic constraints which follow from

$$Q^2 = xy(s - M^2), \tag{2.13}$$

correlating s and Q^2. For the kinematics at HERA, this implies

$$Q^2 \leq sx \approx 10^5 x. \tag{2.14}$$

In order to calculate the cross section of deeply inelastic ep–scattering, one considers the tree–level transition matrix element for the electromagnetic current. It is given by, cf. e.g. [89–91]

$$M_{fi} = e^2 \overline{u}(l', \eta')\gamma^\mu u(l, \eta) \frac{1}{q^2} \langle P_F | J_\mu^{em}(0) | P, \sigma \rangle. \tag{2.15}$$

Here, the spin of the charged lepton or nucleon is denoted by $\eta(\eta')$ and σ, respectively. The state vectors of the initial-state nucleons and the hadronic final state are

---

[1] During the final running period of HERA, low–energy measurements were carried out with $E_p = 460(575)$ GeV in order to extract the longitudinal structure function $F_L(x, Q^2)$, [103].

$|P, \sigma\rangle$ and $|P_F\rangle$. The Dirac-matrices are denoted by γ_μ and bi–spinors by u, see Appendix A. Further e is the electric unit charge and $J_\mu^{em}(\xi)$ the quarkonic part of the electromagnetic current operator, which is self-adjoint :

$$J_\mu^\dagger(\xi) = J_\mu(\xi). \tag{2.16}$$

In QCD, it is given by

$$J_\mu^{em}(\xi) = \sum_{f,f'} \overline{\Psi}_f(\xi)\gamma_\mu \lambda_{ff'}^{em} \Psi_{f'}(\xi), \tag{2.17}$$

where $\Psi_f(\xi)$ denotes the quark field of flavor f. For three light flavors, λ^{em} is given by the following combination of Gell–Mann matrices of the flavor group $SU(3)_{flavor}$, cf. [104, 105]

$$\lambda^{em} = \frac{1}{2}\left(\lambda^3_{flavor} + \frac{1}{\sqrt{3}}\lambda^8_{flavor}\right). \tag{2.18}$$

According to standard definitions [89–91, 106] the differential inclusive cross section is then given by

$$l_0' \frac{d\sigma}{d^3 l'} = \frac{1}{32(2\pi)^3 (l \cdot P)} \sum_{\eta',\eta,\sigma,F} (2\pi)^4 \delta^4(P_F + l' - P - l)|M_{fi}|^2. \tag{2.19}$$

Inserting the transition matrix element (2.15) into the relation for the scattering cross section (2.19), one notices that the trace over the leptonic states forms a separate tensor, $L^{\mu\nu}$. Similarly, the hadronic tensor $W_{\mu\nu}$ is obtained,

$$L_{\mu\nu}(l, l') = \sum_{\eta',\eta} \left[\overline{u}(l', \eta')\gamma^\mu u(l, \eta)\right]^* \left[\overline{u}(l', \eta')\gamma^\nu u(l, \eta)\right], \tag{2.20}$$

$$W_{\mu\nu}(q, P) = \frac{1}{4\pi} \sum_{\sigma,F} (2\pi)^4 \delta^4(P_F - q - P)\langle P, \sigma|J_\mu^{em}(0)|P_F\rangle\langle P_F|J_\nu^{em}(0)|P, \sigma\rangle. \tag{2.21}$$

Thus one arrives at the following relation for the cross section

$$l_0' \frac{d\sigma}{d^3 l'} = \frac{1}{4P \cdot l} \frac{\alpha^2}{Q^4} L^{\mu\nu} W_{\mu\nu} = \frac{1}{2(s - M^2)} \frac{\alpha^2}{Q^4} L^{\mu\nu} W_{\mu\nu}, \tag{2.22}$$

where α denotes the fine-structure constant, see Appendix A. The leptonic tensor in (2.22) can be easily computed in the context of the Standard Model,

$$L_{\mu\nu}(l, l') = Tr[\not{l}\gamma^\mu \not{l}'\gamma^\nu] = 4\left(l_\mu l'_\nu + l'_\mu l_\nu - \frac{Q^2}{2} g_{\mu\nu}\right). \tag{2.23}$$

This is not the case for the hadronic tensor, which contains non–perturbative hadronic contributions due to long-distance effects. To calculate these effects a priori, non-perturbative QCD calculations have to be performed, as in QCD lattice simulations. During the last years these calculations were performed with increasing systematic and numerical accuracy, cf. e.g. [107–111].

The general structure of the hadronic tensor can be fixed using S–matrix theory and the global symmetries of the process. In order to obtain a form suitable for the subsequent calculations, one rewrites Eq. 2.21 as, cf. [90, 112]

$$\begin{aligned} W_{\mu\nu}(q,P) &= \frac{1}{4\pi}\sum_{\sigma}\int d^4\xi \exp(iq\xi)\langle P|[J_\mu^{em}(\xi), J_\nu^{em}(0)]|P\rangle \\ &= \frac{1}{2\pi}\int d^4\xi \exp(iq\xi)\langle P|[J_\mu^{em}(\xi), J_\nu^{em}(0)]|P\rangle. \end{aligned} \tag{2.24}$$

Here, the following notation for the spin-average is introduced in Eq. 2.24

$$\frac{1}{2}\sum_{\sigma}\langle P,\sigma|X|P,\sigma\rangle \equiv \langle P|X|P\rangle. \tag{2.25}$$

Further, $[a,b]$ denotes the commutator of a and b. Using symmetry and conservation laws, the hadronic tensor can be decomposed into different scalar structure functions and thus be stripped of its Lorentz–structure. In the most general case, including polarization, there are 14 independent structure functions [113, 114] which contain all information on the structure of the proton. However, in the case considered here, only two structure functions contribute. One uses Lorentz–and time–reversal invariance [85–88] and additionally the fact that the electromagnetic current is conserved. This enforces electromagnetic gauge invariance for the hadronic tensor,

$$q_\mu W^{\mu\nu} = 0. \tag{2.26}$$

The leptonic tensor (2.23) is symmetric and thus $W_{\mu\nu}$ can be taken to be symmetric as well, since all antisymmetric parts are canceled in the contraction. By making a general ansatz for the hadronic tensor using these properties, one obtains

$$\begin{aligned} W_{\mu\nu}(q,P) = &\frac{1}{2x}\left(g_{\mu\nu}+\frac{q_\mu q_\nu}{Q^2}\right)F_L(x,Q^2) \\ &+\frac{2x}{Q^2}\left(P_\mu P_\nu + \frac{q_\mu P_\nu + q_\nu P_\mu}{2x} - \frac{Q^2}{4x^2}g_{\mu\nu}\right)F_2(x,Q^2). \end{aligned} \tag{2.27}$$

The dimensionless structure functions $F_2(x,Q^2)$ and $F_L(x,Q^2)$ depend on two variables, Bjorken-x and Q^2, contrary to the case of elastic scattering, in which only one variable, e.g. Q^2, determines the cross section. Due to hermiticity of the hadronic tensor, the structure functions are real. The decomposition (2.27) of the hadronic tensor leads to the differential cross section of unpolarized DIS in case of single photon exchange

$$\frac{d\sigma}{dxdy} = \frac{2\pi\alpha^2}{xyQ^2}\left\{\left[1+(1-y)^2\right]F_2(x,Q^2) - y^2 F_L(x,Q^2)\right\}. \tag{2.28}$$

A third structure function, $F_1(x, Q^2)$,

$$F_1(x,Q^2) = \frac{1}{2x}\left[F_2(x,Q^2) - F_L(x,Q^2)\right], \tag{2.29}$$

which is often found in the literature, is not independent of the previous ones.

For completeness, we finally give the full Born cross section for the neutral current, including the exchange of Z–bosons, cf. [97]. Not neglecting the lepton mass m, it is given by

$$\begin{aligned}\frac{d^2\sigma_{\rm NC}}{dxdy} = \frac{2\pi\alpha^2}{xyQ^2}&\left\{\left[2(1-y) - 2xy\frac{M^2}{s} + \left(1-2\frac{m^2}{Q^2}\right)\left(1+4x^2\frac{M^2}{Q^2}\right)\right.\right.\\ &\left.\left.\times\frac{y^2}{1+R(x,Q^2)}\right]\mathcal{F}_2(x,Q^2) + xy(2-y)\mathcal{F}_3(x,Q^2)\right\}.\end{aligned} \tag{2.30}$$

Here, $R(x, Q^2)$ denotes the ratio

$$R(x,Q^2) = \frac{\sigma_L}{\sigma_T} = \left(1+4x^2\frac{M^2}{Q^2}\right)\frac{\mathcal{F}_2(x,Q^2)}{2x\mathcal{F}_1(x,Q^2)} - 1, \tag{2.31}$$

and the effective structure functions $\mathcal{F}_l(x, Q^2), l = 1, ..., 3$ are represented by the structure functions F_l, G_l and H_l via

$$\begin{aligned}\mathcal{F}_{1,2}(x,Q^2) = F_{1,2}(x,Q^2) &+ 2|Q_e|\,(v_e+\lambda a_e)\,\chi(Q^2)G_{1,2}(x,Q^2)\\ &+ 4\left(v_e^2+a_e^2+2\lambda v_e a_e\right)\chi^2(Q^2)H_{1,2}(x,Q^2),\end{aligned} \tag{2.32}$$

$$\begin{aligned}x\mathcal{F}_3(x,Q^2) = -2\,\mathrm{sign}(Q_e)&\left\{|Q_e|\,(a_e+\lambda v_e)\,\chi(Q^2)xG_3(x,Q^2)\right.\\ &\left.+\left[2v_e a_e + \lambda\left(v_e^2+a_e^2\right)\right]\chi^2(Q^2)xH_3(x,Q^2)\right\}.\end{aligned} \tag{2.33}$$

Here, $Q_e = -1,\ a_e = 1$ in case of electrons and

$$\lambda = \xi\ \mathrm{sign}(Q_e), \tag{2.34}$$

$$v_e = 1 - 4\sin^2\theta_W^{\rm eff}, \tag{2.35}$$

$$\chi(Q^2) = \frac{G_\mu}{\sqrt{2}}\frac{M_Z^2}{8\pi\alpha(Q^2)}\frac{Q^2}{Q^2+M_Z^2}, \tag{2.36}$$

with ξ the electron polarization, $\theta_W^{\rm eff}$ the effective weak mixing angle, G_μ the Fermi constant and M_Z the Z–boson mass.

2.2 The Parton Model

The structure functions (2.27) depend on two kinematic variables, x and Q^2. Based on an analysis using current algebra, Bjorken predicted scaling of the structure functions, cf. [13]

$$\lim_{\{Q^2,\, \nu\} \to \infty,\, x=const.} F_{(2,L)}\left(x, Q^2\right) = F_{(2,L)}(x)\,. \tag{2.37}$$

This means that in the Bjorken limit $\{Q^2, \;\nu\;\} \to \;\infty$, with x fixed, the structure functions depend on the ratio Q^2/ν only. Soon after this prediction, approximate scaling was observed experimentally in electron-proton collisions at SLAC (1968) [3–6] cf. also [7–9].[2] Similar to the $\alpha-$ particle scattering experiments by Rutherford in 1911 [115] the cross section remained large at high momentum transfer Q^2, a behavior which is known from point–like targets. This was found in contradiction to the expectation that the cross section should decrease rapidly with increasing Q^2, since the size of the proton had been determined to be about 10^{-13} cm with a smooth charge distribution [116–118]. However, only in rare cases a single proton was detected in the final state, instead it consisted of a large number of hadrons. A proposal by Feynman contained the correct ansatz. To account for the observations, he introduced the parton model [14, 15] cf. also [16, 89, 91–93, 119]. He assumed the proton as an extended object, consisting of several point-like particles, the partons. They are bound together by their mutual interaction and behave like free particles during the interaction with the highly virtual photon in the Bjorken-limit.[3] One arrives at the picture of the proton being "frozen" while the scattering takes place. The electron scatters elastically off the partons and this process does not interfere with the other partonic states, the "spectators". The DIS cross section is then given by the incoherent sum over the individual virtual electron–parton cross sections. Since no information on the particular proton structure is known, Feynman described parton i by the parton distribution function (PDF) $f_i(z)$. It gives the probability to find parton i in the "frozen" proton, carrying the fraction z of its momentum. Fig 2.2 shows a schematic picture of the parton model in Born approximation. The in- and out-going parton momenta are denoted by p and p', respectively.

Similar to the scaling variable x, one defines the partonic scaling variable τ,

$$\tau \equiv \frac{Q^2}{2p \cdot q}\,. \tag{2.38}$$

It plays the same role as the Bjorken-variable, but for the partonic sub-process. In the collinear parton model,[4] which is applied throughout this thesis, $p = zP$ holds,

---

2 The results obtained at DESY [10–12] pointed in the same direction, but were less decisive, because not as large values of Q^2 as at SLAC could be reached.

3 Asymptotic freedom, which was discovered later, is instrumental for this property.

4 For other parton models, as the covariant parton model, cf. [120–125].

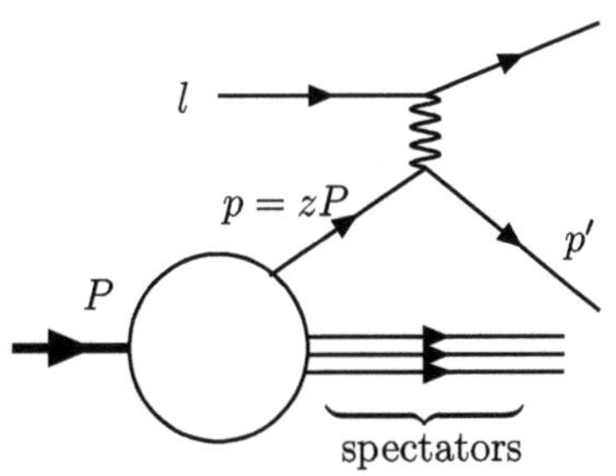

Fig. 2.2 Deeply inelastic electron–proton scattering in the parton model

i.e., the momentum of the partons is taken to be collinear to the proton momentum. From (2.38) one obtains

$$\tau z = x \ . \tag{2.39}$$

Feynman's original parton model, referred to as the naive parton model, neglects the mass of the partons and enforces the strict correlation

$$\delta\left(\frac{q.p}{M} - \frac{Q^2}{2M}\right), \tag{2.40}$$

due to the experimentally observed scaling behavior, which leads to $z = x$. The naive parton model then assumes, in accordance with the quark hypothesis [16–18] that the proton is made up of three valence quarks, two up and one down type, cf. e.g. [126]. This conclusion was generally accepted only several years after the introduction of the parton model, when various experiments had verified its predictions.

Let us consider a simple example, which reproduces the naive parton model at LO and incorporates already some aspects of the improved parton model. The latter allows virtual quark states (sea-quarks) and gluons as partons as well. In the QCD–improved parton model, cf. [89, 91,–93] besides the δ-distribution, (2.40), a function $\mathcal{W}^i_{\mu\nu}(\tau, Q^2)$ contributes to the hadronic tensor. It is called partonic tensor and given by the hadronic tensor, Eq. 2.24, replacing the hadronic states by partonic states i. The basic assumption is that the hadronic tensor can be factorized into the PDFs and the partonic tensor, cf. e.g. [127–134]. The PDFs are non-perturbative quantities and have to be extracted from experiment, whereas the partonic tensors are calculable perturbatively. A more detailed discussion of this using the LCE is given in Sect. 2.3 The hadronic tensor reads, cf. [135]

$$W_{\mu\nu}(x, Q^2) = \frac{1}{4\pi}\sum_i \int_0^1 dz \int_0^1 d\tau \left(f_i(z) + f_{\bar{i}}(z)\right) \mathcal{W}^i_{\mu\nu}(\tau, Q^2)\delta(x - z\tau) \ . \tag{2.41}$$

Here, the number of partons and their respective type are not yet specified and we have included the corresponding PDF of the respective anti-parton, denoted by $f_{\bar{i}}(z)$. Let us assume that the electromagnetic parton current takes the simple form

$$\langle i | j^i_\mu(\tau) | i \rangle = -i e_i \overline{u}^i \gamma_\mu u^i , \qquad (2.42)$$

similar to the leptonic current (2.15). Here e_i is the electric charge of parton i. At LO one finds

$$\mathcal{W}^i_{\mu\nu}(\tau, Q^2) = \frac{2\pi e_i^2}{q.p^i} \delta(1-\tau) \left[2 p^i_\mu p^i_\nu + p^i_\mu q_\nu + p^i_\nu q_\mu - g_{\mu\nu} q.p^i \right] . \qquad (2.43)$$

The δ-distribution in (2.43), together with the δ-distribution in (2.41), just reproduces Feynman's assumption of the naive parton model, $z = x$. Substitution of (2.43) into the general expression for the hadronic tensor (2.27) and projecting onto the structure functions yields

$$\begin{aligned} F_L(x, Q^2) &= 0 , \\ F_2(x, Q^2) &= x \sum_i e_i^2 \left(f_i(x) + f_{\bar{i}}(x) \right) . \end{aligned} \qquad (2.44)$$

This result, at LO, is the same as in the naive parton model. It predicts

- the Callan–Gross relation, cf. [136]

$$F_L(x, Q^2) = F_2(x, Q^2) - 2x F_1(x, Q^2) = 0 . \qquad (2.45)$$

- the structure functions are scale-independent.

These findings were a success of the parton model, since they reproduced the general behavior of the data as observed by the MIT/SLAC experiments.

Finally, we present for completeness the remaining structure functions $G_{2,3}$ and $H_{2,3}$ at the Born level for the complete neutral current, cf. Eq. (2.30),

$$G_2(x, Q^2) = x \sum_i |e_i| v_i \left(f_i(x) + f_{\bar{i}}(x) \right) , \qquad (2.46)$$

$$H_2(x, Q^2) = x \sum_i \frac{1}{4} \left(v_i^2 + a_i^2 \right) \left(f_i(x) + f_{\bar{i}}(x) \right) , \qquad (2.47)$$

$$x G_3(x, Q^2) = x \sum_i |e_i| a_i \left(f_i(x) - f_{\bar{i}}(x) \right) , \qquad (2.48)$$

$$x H_3(x, Q^2) = x \sum_i \frac{1}{2} v_i a_i \left(f_i(x) - f_{\bar{i}}(x) \right) , \qquad (2.49)$$

with $a_i = 1$ and

$$v_i = 1 - 4|e_i| \sin^2 \theta_W^{\text{eff}} . \qquad (2.50)$$

2.2.1 Validity of the Parton Model

The validity of the parton picture can be justified by considering an impulse approximation of the scattering process as seen from a certain class of reference frames, in which the proton momentum is taken to be very large (P_∞-frames). Two things happen to the proton when combining this limit with the Bjorken–limit: The internal interactions of its partons are time dilated, and it is Lorentz contracted in the direction of the collision. As the cms energy increases, the parton lifetimes are lengthened and the time it takes the electron to interact with the proton is shortened. Therefore the condition for the validity of the parton model is given by, cf. [16, 137]

$$\frac{\tau_{\text{int}}}{\tau_{\text{life}}} \ll 1\,. \tag{2.51}$$

Here τ_{int} denotes the interaction time and τ_{life} the average life time of a parton. If (2.51) holds, the proton will be in a single virtual state characterized by a certain number of partons during the entire interaction time. This justifies the assumption that parton i carries a definite momentum fraction z_i, $0 \le z_i \le 1$, of the proton in the cms. This parton model is also referred to as collinear parton model, since the proton is assumed to consist out of a stream of partons with parallel momenta. Further $\sum_i z_i = 1$ holds. In order to derive the fraction of times in (2.51), one aligns the coordinate system parallel to the proton's momentum. Thus one obtains in the limit $P_3^2 \gg M^2$, [138]

$$P = \left(\sqrt{P_3^2 + M^2}; 0, 0, P_3\right) \approx \left(P_3 + \frac{M^2}{2 \cdot P_3}; 0, 0, P_3\right)\,. \tag{2.52}$$

The photon momentum can be parametrized by

$$q = (q_0; q_3, \vec{q}_\perp), \tag{2.53}$$

where $\vec{q}_\perp$ denotes its transverse momentum with respect to the proton. By choosing the cms of the initial states as reference and requiring that νM and q^2 approach a limit independent of P_3 as $P_3 \to \infty$, one finds for the characteristic interaction time scale, using an (approximate) time–energy uncertainty relation,

$$\tau_{\text{int}} \simeq \frac{1}{q_0} = \frac{4P_3 x}{Q^2(1-x)}\,. \tag{2.54}$$

The life time of the individual partons is estimated accordingly to be inversely proportional to the energy fluctuations of the partons around the average energy E

$$\tau_{\text{life}} \simeq \frac{1}{\sum_i E_i - E}\,. \tag{2.55}$$

Here E_i denote the energies of the individual partons. After introducing the two-momentum $\vec{k}_{\perp i}$ of the partons perpendicular to the direction of motion of the proton as given in (2.52), a simple calculation yields, cf. [138]

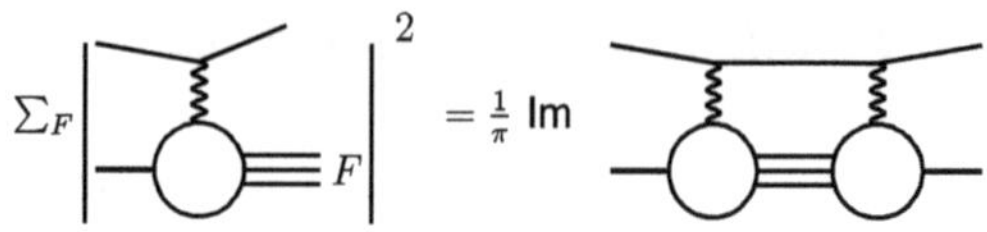

Fig. 2.3 Schematic picture of the optical theorem

$$\frac{\tau_{\rm int}}{\tau_{\rm life}} = \frac{2x}{Q^2(1-x)}\left(\sum_i \frac{(m_i^2+k_{\perp i}^2)}{z_i} - M^2\right), \tag{2.56}$$

where m_i denotes the mass of the ith parton. This expression is independent of P_3. The above procedure allows therefore to estimate the probability of deeply inelastic scattering to occur independently of the large momentum of the proton. Accordingly, we consider now the case of two partons with momentum fractions x and $1-x$ and equal perpendicular momentum, neglecting all masses. One obtains

$$\frac{\tau_{\rm int}}{\tau_{\rm life}} \approx \frac{2k_\perp^2}{Q^2(1-x)^2} . \tag{2.57}$$

This example leads to the conclusion, that deeply inelastic scattering probes single partons if the virtuality of the photon is much larger than the transverse momenta squared of the partons and Bjorken-x is neither close to one nor zero. In the latter case, xP_3 would be the large momentum to be considered. If one does not neglect the quark masses, one has to adjust this picture, as will be described in Sect. 3.3.

2.3 The Light–Cone Expansion

In quantum field theory one usually considers time-ordered products, denoted by T, rather than a commutator as it appears in the hadronic tensor in Eq. 2.24. The hadronic tensor can be expressed as the imaginary part of the forward Compton amplitude for virtual gauge boson–nucleon scattering, $T_{\mu\nu}(q,P)$. The optical theorem, depicted graphically in Fig. 2.3, yields

$$W_{\mu\nu}(q,P) = \frac{1}{\pi}\mathrm{Im}\, T_{\mu\nu}(q,P), \tag{2.58}$$

where the Compton amplitude is given by cf. [104]

$$T_{\mu\nu}(q,P) = i\int d^4\xi\, \exp(iq\xi)\langle P|\mathsf{T}J_\mu(\xi)J_\nu(0)|P\rangle. \tag{2.59}$$

By applying the same invariance and conservation conditions as for the hadronic tensor, the Compton amplitude can be expressed in the unpolarized case by two amplitudes $T_L(x,Q^2)$ and $T_2(x,Q^2)$. It is then given by

$$
\begin{aligned}
T_{\mu\nu}(q,P) = {} & \frac{1}{2x}\left(g_{\mu\nu} + \frac{q_\mu q_\nu}{Q^2}\right) T_L(x,Q^2) \\
& + \frac{2x}{Q^2}\left(P_\mu P_\nu + \frac{q_\mu P_\nu + q_\nu P_\mu}{2x} - \frac{Q^2}{4x^2} g_{\mu\nu}\right) T_2(x,Q^2)\,.
\end{aligned}
\tag{2.60}
$$

Using translation invariance, one can show that (2.59) is crossing symmetric under $q \rightarrow -q$, cf. [113, 139]

$$
T_{\mu\nu}(q,P) = T_{\mu\nu}(-q,P), \tag{2.61}
$$

with $q \rightarrow -q$ being equivalent to $\nu, x \rightarrow (-\nu), (-x)$. The corresponding relations for the amplitudes are then obtained by considering (2.60)

$$
T_{(2,L)}(x,Q^2) = T_{(2,L)}(-x,Q^2)\,. \tag{2.62}
$$

By (2.58) these amplitudes relate to the structure functions F_L and F_2 as

$$
F_{(2,L)}(x,Q^2) = \frac{1}{\pi}\mathrm{Im}\, T_{(2,L)}(x,Q^2)\,. \tag{2.63}
$$

Another general property of the Compton amplitude is that T_L and T_2 are real analytic functions of x at fixed Q^2, cf. [140] i.e.

$$
T_{(2,L)}(x^*,Q^2) = T^*_{(2,L)}(x,Q^2)\,. \tag{2.64}
$$

Using this description one can perform the LCE [85–88] or the cut—vertex method in the time—like case [141–143] respectively, and derive general properties of the moments of the structure functions as will be shown in the subsequent Section. A technical aspect which has been proved very useful is to work in Mellin space rather than in x-space. The Nth Mellin moment of a function f is defined through the integral

$$
\mathbf{M}[f](N) \equiv \int_0^1 dz\, z^{N-1} f(z)\,. \tag{2.65}
$$

This transform diagonalizes the Mellin–convolution $f \otimes g$ of two functions f, g

$$
[f \otimes g](z) = \int_0^1 dz_1 \int_0^1 dz_2\, \delta(z - z_1 z_2)\, f(z_1) g(z_2)\,. \tag{2.66}
$$

The convolution (2.66) decomposes into a simple product of the Mellin-transforms of the two functions,

$$
\mathbf{M}[f \otimes g](N) = \mathbf{M}[f](N)\mathbf{M}[g](N)\,. \tag{2.67}
$$

In Eqs. 2.65, 2.67 N is taken to be an integer. However, later on one may perform an analytic continuation to arbitrary complex values of N [144, 145]. Note that it is enough to know all even or odd integer moments—as is the case for inclusive DIS—of the functions $f,\ g$ to perform an analytic continuation to arbitrary complex values $N \in \mathbb{C}$, [146, 147]. Then Eq. 2.66 can be obtained from the relation for the moments (2.67) by an inverse Mellin–transform. Hence in this case the z- and N-space description are equivalent, which we will frequently use later on.

2.3.1 Light–Cone Dominance

It can be shown that in the Bjorken limit, $Q^2 \rightarrow \infty,\ \nu \rightarrow \infty$, x fixed, the hadronic tensor is dominated by its contribution near the light–cone, i.e. by the values of the integrand in (2.24) at $\xi^2 \approx 0$, cf. [85–88]. This can be understood by considering the infinite momentum frame, see Sect. 2.2.1,

$$P = (P_3; 0, 0, P_3), \tag{2.68}$$

$$q = \left(\frac{\nu}{2P_3}; \sqrt{Q^2}, 0, \frac{-\nu}{2P_3}\right), \tag{2.69}$$

$$P_3 \approx \sqrt{\nu} \rightarrow \infty\ . \tag{2.70}$$

According to the Riemann–Lebesgue theorem, the integral in (2.24) is dominated by the region where $q.\xi \approx 0$ due to the rapidly oscillating exponential $\exp(iq.\xi)$, [89]. One can now rewrite the dot product as, cf. [105]

$$q.\xi = \frac{1}{2}(q^0 - q^3)(\xi^0 + \xi^3) + \frac{1}{2}(q^0 + q^3)(\xi^0 - \xi^3) - q^1\xi^1, \tag{2.71}$$

and infer that the condition $q.\xi \approx 0$ in the Bjorken-limit is equivalent to

$$\xi^0 \pm \xi^3 \propto \frac{1}{\sqrt{\nu}}\ , \quad \xi^1 \propto \frac{1}{\sqrt{\nu}}, \tag{2.72}$$

which results in

$$\xi^2 \approx 0, \tag{2.73}$$

called light–cone dominance: for DIS in the Bjorken-limit the dominant contribution to the hadronic tensor $W_{\mu\nu}(q, P)$ and the Compton Amplitude comes from the region where $\xi^2 \approx 0$.

This property allows to apply the LCE of the current–current correlation in Eq. 2.24 and for the time ordered product in Eq. 2.59, respectively. In the latter case it reads for scalar currents, cf. [85–88]

$$\lim_{\xi^2\to 0} \mathsf{T}J(\xi), J(0) \propto \sum_{i,N,\tau} \overline{C}_{i,\tau}^N(\xi^2,\mu^2)\xi_{\mu_1}...\xi_{\mu_N} O_{i,\tau}^{\mu_1...\mu_N}(0,\mu^2) . \tag{2.74}$$

The $O_{i,\tau}(\xi,\mu^2)$ are local operators which are finite as $\xi^2 \to 0$. The singularities which appear for the product of two operators as their arguments become equal are shifted to the c-number coefficients $\overline{C}_{i,\tau}^N(\xi^2,\mu^2)$, the Wilson coefficients, and can therefore be treated separately. In Eq. 2.74, μ^2 is the factorization scale describing at which point the separation between the perturbative and non–perturbative contributions takes place. The summation index i runs over the set of allowed operators in the model, while the sum over N extends to infinity. Dimensional analysis shows that the degree of divergence of the functions $\overline{C}_{i,\tau}^N$ as $\xi^2 \to 0$ is given by

$$\overline{C}_{i,\tau}^N(\xi^2,\mu^2) \propto \left(\frac{1}{\xi^2}\right)^{-\tau/2+d_J} . \tag{2.75}$$

Here, d_J denotes the canonical dimension of the current $J(\xi)$ and τ is the twist of the local operator $O_{i,\tau}^{\mu_1..\mu_N}(\xi,\mu^2)$, which is defined by, cf. [148]

$$\tau \equiv D_O - N . \tag{2.76}$$

D_O is the canonical (mass) dimension of $O_{i,\tau}^{\mu_1..\mu_N}(\xi,\mu^2)$ and N is called its spin. From (2.75) one can infer that the most singular coefficients are those related to the operators of lowest twist, i.e. in the case of the LCE of the electromagnetic current (2.17), twist $\tau = 2$. The contributions due to higher twist operators are suppressed by factors of $(\overline{\mu}^2/Q^2)^k$, with $\overline{\mu}$ a typical hadronic mass scale of $O(1 \text{ GeV})$. In a wide range of phase–space it is thus sufficient to consider the leading twist contributions only, which we will do in the following and omit the index τ.

2.3.2 A Simple Example

In this Section, we consider a simple example of the LCE applied to the Compton amplitude and its relation to the hadronic tensor, neglecting all Lorentz-indices and model dependence, cf. Refs. [90, 149]. The generalization to the case of QCD is straightforward and hence we will already make some physical arguments which apply in both cases. The scalar expressions corresponding to the hadronic tensor and the Compton amplitude are given by

$$W(x,Q^2) = \frac{1}{2\pi}\int d^4\xi \exp(iq\xi)\langle P|[J(\xi), J(0)]|P\rangle, \tag{2.77}$$

$$T(x,Q^2) = i\int d^4\xi \ \exp(iq\xi)\langle P|\mathsf{T}J(\xi)J(0)|P\rangle. \tag{2.78}$$

Eq. 2.78 can be evaluated in the limit $\xi^2 \to 0$ for twist $\tau = 2$ by using the LCE given in Eq. 2.74, where for brevity only one local operator is considered. The coefficient functions in momentum space are defined as

$$\int \exp(iq.\xi)\xi_{\mu_1}..\xi_{\mu_N}\overline{C}^N(\xi^2,\mu^2) \equiv -i\left(\frac{2}{-q^2}\right)^N q_{\mu_1}...q_{\mu_N}C^N\left(\frac{Q^2}{\mu^2}\right) . \qquad (2.79)$$

The nucleon states act on the composite operators only and the corresponding matrix elements can be expressed as

$$\langle P|O^{\mu_1...\mu_N}(0,\mu^2)|P\rangle = A^N\left(\frac{P^2}{\mu^2}\right)P^{\mu_1}...P^{\mu_N} + \text{trace terms}. \qquad (2.80)$$

The trace terms in the above equation can be neglected, because due to dimensional counting they would give contributions of the order $1/Q^2$, $1/\nu$ and hence are irrelevant in the Bjorken–limit. Thus the Compton amplitude reads cf. e.g. [90, 91]

$$T(x',Q^2) = 2\sum_{N=0,2,4,..} C^N\left(\frac{Q^2}{\mu^2}\right)A^N\left(\frac{P^2}{\mu^2}\right)x'^N, \quad x' = \frac{1}{x} \qquad (2.81)$$

In (2.81) only the even moments contribute. This is a consequence of crossing symmetry, Eq. 2.62, and holds as well in the general case of unpolarized DIS for single photon exchange. In other cases the projection is onto the odd moments. Depending on the type of the observable the series may start at different initial values, cf. e.g. [113, 114]. The sum in Eq. 2.81 is convergent in the unphysical region $x \geq 1$ and an analytic continuation to the physical region $0 \leq x \leq 1$ has to be performed. Here, one of the assumptions is that scattering amplitudes are analytic in the complex plane except at values of kinematic variables allowing intermediate states to be on mass–shell. This general feature has been proved to all orders in perturbation theory [150, 151]. In QCD, it is justified on grounds of the parton model. When $\nu \geq Q^2/2M$, i.e. $0 \leq x \leq 1$, the virtual photon–proton system can produce a physical hadronic intermediate state, so the $T_{(2,L)}(x,Q^2)$ and $T(x,Q^2)$, respectively, have cuts along the positive (negative) real x-axis starting from $1(-1)$ and poles at $\nu = Q^2/2M$ $(x = 1,-1)$. The discontinuity along the cut is then just given by (2.58). The Compton amplitude can be further analyzed by applying (subtracted) dispersion relations, cf. [113, 114]. Equivalently, one can divide both sides of Eq. 2.81 by x'^m and integrate along the path shown in Fig. 2.4, cf. [90, 135].

For the left–hand side of (2.81) one obtains

$$\frac{1}{2\pi i}\oint dx'\frac{T(x',Q^2)}{x'^m} = \frac{2}{\pi}\int_1^\infty \frac{dx'}{x'^m}\text{Im}\,T(x',Q^2) = 2\int_0^1 dx\, x^{m-2}W(x,Q^2), \qquad (2.82)$$

where the optical theorem (2.58) and crossing symmetry (2.62) have been used. The right-hand side of (2.81) yields

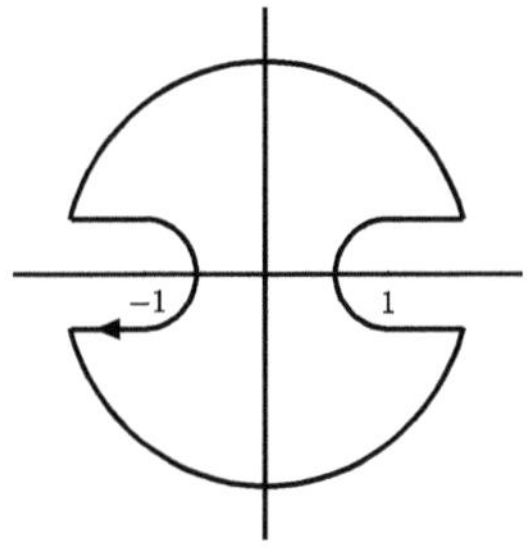

Fig. 2.4 Integration contour in the complex x'-plane

$$\frac{1}{\pi i}\sum_{N=0,2,4,..} C^N\left(\frac{Q^2}{\mu^2}\right)A^N\left(\frac{P^2}{\mu^2}\right)\oint dx'\, x'^{N-m} = 2C^{m-1}\left(\frac{Q^2}{\mu^2}\right)A^{m-1}\left(\frac{P^2}{\mu^2}\right). \tag{2.83}$$

Thus from Eqs. 2.82 and 2.83 one obtains for the moments of the scalar hadronic tensor defined in Eq. 2.77

$$\int_0^1 dx\, x^{N-1}W(x,Q^2) = C^N\left(\frac{Q^2}{\mu^2}\right)A^N\left(\frac{P^2}{\mu^2}\right). \tag{2.84}$$

2.3.3 The Light–Cone Expansion Applied to DIS

In order to derive the moment–decomposition of the structure functions one essentially has to go through the same steps as in the previous section. The LCE of the physical forward Compton amplitude (2.59) at the level of twist $\tau = 2$ in the Bjorken–limit is given by, cf. [149, 152]

$$\begin{aligned} T_{\mu\nu}(q,P) \to \sum_{i,N}\Bigg\{&\Big[Q^2 g_{\mu\mu_1}g_{\nu\mu_2} + g_{\mu\mu_1}q_\nu q_{\mu_2} + g_{\nu\mu_2}q_\mu q_{\mu_1} - g_{\mu\nu}q_{\mu_1}q_{\mu_2}\Big] \\ &\times C_{i,2}\left(N,\frac{Q^2}{\mu^2}\right) + \left[g_{\mu\nu} + \frac{q_\mu q_\mu}{Q^2}\right]q_{\mu_1}q_{\mu_2}C_{i,L}\left(N,\frac{Q^2}{\mu^2}\right)\Bigg\} \\ &\times q_{\mu_3}\cdots q_{\mu_N}\left(\frac{2}{Q^2}\right)^N \langle P|O_i^{\mu_1\ldots\mu_N}(\mu^2)|P\rangle. \end{aligned} \tag{2.85}$$

Additionally to Sect. 2.3.2, the index i runs over the allowed operators which emerge from the expansion of the product of two electromagnetic currents, Eq. 2.17. The possible twist–2 operators are given by[5] [142]

[5] Here we consider only the spin–averaged case for single photon exchange. Other operators contribute for parity–violating processes, in the polarized case and for transversity, cf. Chaps. 8 and 9.

$$O^{\mathsf{NS}}_{q,r;\mu_1,\ldots,\mu_N} = i^{N-1}\mathbf{S}\left[\overline{\psi}\gamma_{\mu_1}D_{\mu_2}\ldots D_{\mu_N}\frac{\lambda_r}{2}\psi\right]\text{—trace terms,} \tag{2.86}$$

$$O^{\mathsf{S}}_{q;\mu_1,\ldots,\mu_N} = i^{N-1}\mathbf{S}\left[\overline{\psi}\gamma_{\mu_1}D_{\mu_2}\ldots D_{\mu_N}\psi\right]\text{—trace terms,} \tag{2.87}$$

$$O^{\mathsf{S}}_{g;\mu_1,\ldots,\mu_N} = 2i^{N-2}\mathbf{SSp}\left[F^a_{\mu_1\alpha}D_{\mu_2}\ldots D_{\mu_{N-1}}F^{\alpha,a}_{\mu_N}\right]\text{—trace terms.} \tag{2.88}$$

Here, **S** denotes the symmetrization operator of the Lorentz indices $\mu_1, \ldots, \mu_N$. λ_r is the flavor matrix of $SU(n_f)$ with n_f light flavors, ψ denotes the quark field, $F^a_{\mu\nu}$ the gluon field–strength tensor, and D_μ the covariant derivative. The indices $q,\ g$ represent the quark- and gluon-operator, respectively. **Sp** in (2.88) is the color-trace and a the color index in the adjoint representation, cf. Appendix A. The quark-fields carry color indices in the fundamental representation, which have been suppressed. The classification of the composite operators (2.86–2.88) in terms of flavor singlet (S) and non-singlet (NS) refers to their symmetry properties with respect to the flavor group $SU(n_f)$. The operator in Eq. 2.86 belongs to the adjoint representation of $SU(n_f)$, whereas the operators in Eqs. 2.87, 2.88 are singlets under $SU(n_f)$. Neglecting the trace terms, one rewrites the matrix element of the composite operators in terms of its Lorentz structure and the scalar operator matrix elements, cf. [91, 105]

$$\langle P|O_i^{\mu_1\ldots\mu_N}|P\rangle = A_i\left(N, \frac{P^2}{\mu^2}\right)P^{\mu_1}\ldots P^{\mu_N}. \tag{2.89}$$

Eq. 2.85 then becomes

$$\begin{aligned} T_{\mu\nu}(q,P) = 2\sum_{i,N}&\left\{\frac{2x}{Q^2}\left[P_\mu P_\nu + \frac{P_\mu q_\nu + P_\nu q_\mu}{2x} - \frac{Q^2}{4x^2}g_{\mu\nu}\right]C_{i,2}\left(N, \frac{Q^2}{\mu^2}\right)\right.\\ &\left. + \frac{1}{2x}\left[g_{\mu\nu} + \frac{q_\mu q_\mu}{Q^2}\right]C_{i,L}\left(N, \frac{Q^2}{\mu^2}\right)\right\}\\ &\times\frac{1}{x^{N-1}}A_i\left(N, \frac{P^2}{\mu^2}\right). \end{aligned} \tag{2.90}$$

Comparing Eq. 2.90 with the general Lorentz structure expected for the Compton amplitude Eq. 2.60, the relations of the scalar forward amplitudes to the Wilson coefficients and nucleon matrix elements can be read off

$$T_{(2,L)}(x, Q^2) = 2\sum_{i,N}\frac{1}{x^{N-1}}C_{i,(2,L)}\left(N, \frac{Q^2}{\mu^2}\right)A_i\left(N, \frac{P^2}{\mu^2}\right). \tag{2.91}$$

Eq. 2.91 is of the same type as Eq. 2.81 and one thus obtains for the moments of the structure functions

$$F_{(2,L)}(N, Q^2) = \mathbf{M}[F_{(2,L)}(x, Q^2)](N) \tag{2.92}$$

$$= \sum_i C_{i,(2,L)}\left(\frac{Q^2}{\mu^2}, N\right) A_i\left(\frac{P^2}{\mu^2}, N\right). \tag{2.93}$$

The above equations have already been written in Mellin space, which we will always do from now on, if not indicated otherwise. Eqs. 2.91, 2.93 together with the general structure of the Compton amplitude, Eqs. 2.60, 2.90 and the hadronic tensor, Eq. 2.27 are the basic equations for theoretical or phenomenological analysis of DIS in the kinematic regions where higher twist effects can be safely disregarded. Note that the generalization of these equations to electroweak or polarized interactions is straightforward by including additional operators and Wilson coefficients. In order to interpret Eqs. 2.91, 2.93 one uses the fact that the Wilson coefficients $C_{i,(2,L)}$ are independent of the proton state. This is obvious since the wave function of the proton only enters into the definition of the operator matrix elements, cf. Eq. 2.89. In order to calculate the Wilson coefficients, the proton state has therefore to be replaced by a suitably chosen quark or gluon state i with momentum p. The corresponding partonic tensor is denoted by $\mathcal{W}^i_{\mu\nu}(q, p)$, cf. below Eq. 2.40, with scalar amplitudes $\mathcal{F}^i_{(2,L)}(\tau, Q^2)$. Here τ is the partonic scaling variable defined in Eq. 2.38. The LCE of the electromagnetic current does not change and the replacement only affects the operator matrix elements. The forward Compton amplitude for photon–quark (gluon) scattering corresponding to $\mathcal{W}^i_{\mu\nu}(q, p)$ can be calculated order by order in perturbation theory, provided the scale Q^2 is large enough for the strong coupling constant to be small. In the same manner, the contributing operator matrix elements with external partons may be evaluated. Finally, one can read off the Wilson coefficients from the partonic equivalent of Eq. 2.91.[6] By identifying the nucleon OMEs (2.89) with the PDFs, one obtains the QCD improved parton model. At LO it coincides with the naive parton model, which we described in Sect. 2.2, as can be inferred from the discussion below Eq. 2.41. The improved parton model states that in the Bjorken limit at the level of twist $\tau = 2$ the unpolarized nucleon structure functions $F_i(x, Q^2)$ are obtained in Mellin space as products of the universal parton densities $f_i(N, \mu^2)$ with process–dependent Wilson coefficients $C_{i,(2,L)}(N, Q^2/\mu^2)$

$$F_{(2,L)}(N, Q^2) = \sum_i C_{i,(2,L)}\left(N, \frac{Q^2}{\mu^2}\right) f_i(N, \mu^2) \tag{2.94}$$

to all orders in perturbation theory. This property is also formulated in the factorization theorems [127, 129–134] cf. also [128] where it is essential that an inclusive, infrared–safe cross section is considered [153]. We have not yet dealt with the question of how renormalization is being performed. However, we have already intro-

[6] Due to the optical theorem, one may also obtain the Wilson coefficients by calculating the inclusive hard scattering cross sections of a virtual photon with a quark(gluon) using the standard Feynman–rules and phase–space kinematics.

duced the scale μ^2 into the right–hand side of Eq. 2.94. This scale is called factorization scale. It describes a mass scale at which the separation of the structure functions into the perturbative hard scattering coefficients $C_{i,(2,L)}$ and the non–perturbative parton densities f_i can be performed. This choice is arbitrary at large enough scales and the physical structure functions do not depend on it. This independence is used in turn to establish the corresponding renormalization group equation [154–158] which describes the scale–evolution of the Wilson coefficients, parton densities and structure functions w.r.t. to μ^2 and Q^2, cf. Refs. [89, 91, 152, 159–161] and Sect. 2.4

These evolution equations then predict scaling violations and are used to analyze experimental data in order to unfold the twist–2 parton distributions at some scale Q_0^2, together with the QCD–scale $\Lambda_{\rm QCD}$, cf. [159, 162, 163].

Before finishing this Section, we describe the quantities appearing in Eq. 2.94 in detail. Starting from the operators defined in Eqs. 2.86–2.88, three types of parton densities are expected. Since the question how heavy quarks are treated in this framework will be discussed in Chap. 3, we write the following equations for n_f light flavors in massless QCD. The gluon density is denoted by $G(n_f, N, \mu^2)$ and multiplies the gluonic Wilson coefficients $C_{g,(2,L)}(n_f, N, Q^2/\mu^2)$, which describe the interaction of a gluon with a photon and emerge for the first time at $O(\alpha_s)$. Each quark and its anti–quark have a parton density, denoted by $f_{k(\bar{k})}(n_f, N, \mu^2)$. These are grouped together into the flavor singlet combination $\Sigma(n_f, N, \mu^2)$ and a non–singlet combination $\Delta_k(n_f, N, \mu^2)$ as follows

$$\Sigma(n_f, N, \mu^2) = \sum_{l=1}^{n_f} \Big[f_l(n_f, N, \mu^2) + f_{\bar{l}}(n_f, N, \mu^2) \Big], \tag{2.95}$$

$$\Delta_k(n_f, N, \mu^2) = f_k(n_f, N, \mu^2) + f_{\bar{k}}(n_f, N, \mu^2) - \frac{1}{n_f}\Sigma(n_f, N, \mu^2)\,. \tag{2.96}$$

The distributions multiply the quarkonic Wilson coefficients $C_{q,(2,L)}^{\mathsf{S},NS}(n_f, N, Q^2/\mu^2)$, which describe the hard scattering of a photon with a light quark. The complete factorization formula for the structure functions is then given by

$$\begin{aligned} F_{(2,L)}(n_f, N, Q^2) = \frac{1}{n_f}\sum_{k=1}^{n_f} e_k^2 \Bigg[&\Sigma(n_f, N, \mu^2) C_{q,(2,L)}^{\mathsf{S}}\left(n_f, N, \frac{Q^2}{\mu^2}\right) + G(n_f, N, \mu^2) \\ &\times C_{g,(2,L)}^{\mathsf{S}}\left(n_f, N, \frac{Q^2}{\mu^2}\right) + n_f \Delta_k(n_f, N, \mu^2) C_{q,(2,L)}^{\mathsf{NS}}\left(n_f, N, \frac{Q^2}{\mu^2}\right) \Bigg]. \end{aligned} \tag{2.97}$$

Note, that one usually splits the quarkonic S contributions into a NS and pure-singlet (PS) part via $\mathsf{S} = \mathsf{PS} + \mathsf{NS}$. The perturbative expansions of the Wilson coefficients read

$$C_{g,(2,L)}^{\mathsf{S}}\left(n_f, N, \frac{Q^2}{\mu^2}\right) = \sum_{i=1}^{\infty} a_s^i C_{g,(2,L)}^{(i),\mathsf{S}}\left(n_f, N, \frac{Q^2}{\mu^2}\right), \tag{2.98}$$

$$C_{q,(2,L)}^{\mathsf{PS}}\left(n_f, N, \frac{Q^2}{\mu^2}\right) = \sum_{i=2}^{\infty} a_s^i C_{q,(2,L)}^{(i),\mathsf{PS}}\left(n_f, N, \frac{Q^2}{\mu^2}\right), \tag{2.99}$$

$$C_{q,(2,L)}^{\mathsf{NS}}\left(n_f, N, \frac{Q^2}{\mu^2}\right) = \delta_2 + \sum_{i=1}^{\infty} a_s^i C_{q,(2,L)}^{(i),\mathsf{NS}}\left(n_f, N, \frac{Q^2}{\mu^2}\right), \tag{2.100}$$

where $a_s \equiv \alpha_s/(4\pi)$ and

$$\delta_2 = 1 \text{ for } F_2 \text{ and } \delta_2 = 0 \text{ for } F_L\ . \tag{2.101}$$

These terms are at present known up to $O(a_s^3)$. The $O(a_s)$ terms have been calculated in Refs. [149, 164, 165] and the $O(a_s^2)$ contributions by various groups in Refs. [166–176]. The $O(a_s^3)$ terms have first been calculated for fixed moments in Refs. [177–180] and the complete result for all N has been obtained in Ref. [181].[7]

2.4 RGE–Improved Parton Model and Anomalous Dimensions

In the following, we present a derivation of the RGEs for the Wilson coefficients, and subsequently, the evolution equations for the parton densities. When calculating scattering cross sections in quantum field theories, they usually contain divergences of different origin. The infrared and collinear singularities are connected to the limit of soft—and collinear radiation, respectively. Due to the Bloch–Nordsieck theorem [183, 184] it is known that the infrared divergences cancel between virtual and bremsstrahlung contributions. The structure functions are inclusive quantities. Therefore, all final state collinear (mass) singularities cancel as well, which is formulated in the Lee–Kinoshita–Nauenberg theorem [185, 186]. Thus in case of the Wilson coefficients, only the initial state collinear divergences of the external light partons and the ultraviolet divergences remain. The latter are connected to the large scale behavior and are renormalized by a redefinition of the parameters of the theory, as the coupling constant, the masses, the fields, and the composite operators [187, 188]. This introduces a renormalization scale μ_r, which forms the subtraction point for renormalization. The scale which appears in the factorization formulas (2.94, 2.97) is denoted by μ_f and called factorization scale, cf. [127–133]. Its origin lies in the arbitrariness of the point at which short- and long-distance effects are separated and is connected to the redefinition of the bare parton densities by absorbing the initial state collinear singularities of the Wilson coefficients into them. Note, that one usually adopts dimensional regularization to regularize the infinities in perturbative calculations, cf. Chap. 4, which causes another scale μ to appear. It is associated to the mass dimension of the coupling constant in $D \neq 4$ dimensions. In principle all

[7] Recently, the $O(a_s^3)$ Wilson coefficient for the structure function $xF_3(x, Q^2)$ was calculated in Ref. [182].

these three scales have to be treated separately, but we will set them equal in the subsequent analysis, $\mu = \mu_r = \mu_f$.

The renormalization group equations are obtained using the argument that all these scales are arbitrary and therefore physical quantities do not alter when changing these scales [154–157, 187, 188]. One therefore defines the total derivative w.r.t. to μ^2

$$\mathcal{D}(\mu^2) \equiv \mu^2 \frac{\partial}{\partial \mu^2} + \beta(a_s(\mu^2)) \frac{\partial}{\partial a_s(\mu^2)} - \gamma_m(a_s(\mu^2)) m(\mu^2) \frac{\partial}{\partial m(\mu^2)}. \tag{2.102}$$

Here the β–function and the anomalous dimension of the mass, γ_m, are given by

$$\beta(a_s(\mu^2)) \equiv \mu^2 \frac{\partial a_s(\mu^2)}{\partial \mu^2}, \tag{2.103}$$

$$\gamma_m(a_s(\mu^2)) \equiv -\frac{\mu^2}{m(\mu^2)} \frac{\partial m(\mu^2)}{\partial \mu^2}, \tag{2.104}$$

cf. Sects. 4.3 and 4.4. The derivatives have to be performed keeping the bare quantities $\hat{a}_s$, $\hat{m}$ fixed. Additionally, we work in Feynman–gauge and therefore the gauge–parameter is not present in Eq. 2.102. In the following we will consider only one mass m. The composite operators (2.86–2.88) are renormalized introducing operator Z–factors

$$O^{\sf NS}_{q,r;\mu_1,\ldots,\mu_N} = Z^{\sf NS}(\mu^2) \hat{O}^{\sf NS}_{q,r;\mu_1,\ldots,\mu_N}, \tag{2.105}$$

$$O^{\sf S}_{i;\mu_1,\ldots,\mu_N} = Z^{\sf S}_{ij}(\mu^2) \hat{O}^{\sf S}_{j;\mu_1,\ldots,\mu_N}, \quad i = q, g, \tag{2.106}$$

where in the singlet case mixing occurs since these operators carry the same quantum numbers. The anomalous dimensions of the operators are defined by

$$\gamma_{qq}^{\sf NS} = \mu Z^{-1,{\sf NS}}(\mu^2) \frac{\partial}{\partial \mu} Z^{\sf NS}(\mu^2), \tag{2.107}$$

$$\gamma_{ij}^{\sf S} = \mu Z_{il}^{-1,{\sf S}}(\mu^2) \frac{\partial}{\partial \mu} Z_{lj}^{\sf S}(\mu^2)\,. \tag{2.108}$$

We begin by considering the partonic structure functions calculated with external fields l. Here we would like to point out that we calculate matrix elements of currents, operators, etc. and not vacuum expectation values of time-ordered products with the external fields included. The anomalous dimensions of the latter therefore do not contribute [105] and they are parts of the anomalous dimensions of the composite operators, respectively. The RGE reads

$$\mathcal{D}(\mu^2) \mathcal{F}^l_{(2,L)}(N, Q^2) = 0\,. \tag{2.109}$$

On the partonic level, Eq. 2.93 takes the form

$$\mathcal{F}^{l}_{(2,L)}(N,Q^2) = \sum_j C_{j,(2,L)}\left(N,\frac{Q^2}{\mu^2}\right)\langle l|O_j(\mu^2)|l\rangle. \tag{2.110}$$

From the operator renormalization constants of the O_i, Eqs. 2.105, 2.106 the following RGE is derived for the matrix elements [152]

$$\sum_j \left(\mathcal{D}(\mu^2)\,\delta_{ij} + \frac{1}{2}\gamma_{ij}^{\mathsf{S},NS}\right)\langle l|O_j(\mu^2)|l\rangle = 0, \tag{2.111}$$

where we write the S and NS case in one equation for brevity and we remind the reader that in the latter case, $i, j, l = q$ only. Combining Eqs. 2.109–2.111 one can determine the RGE for the Wilson coefficients. It reads

$$\sum_i \left(\mathcal{D}(\mu^2)\,\delta_{ij} - \frac{1}{2}\gamma_{ij}^{\mathsf{S},NS}\right) C_{i,(2,L)}\left(N,\frac{Q^2}{\mu^2}\right) = 0\,. \tag{2.112}$$

The structure functions, which are observables, obey the same RGE as on the partonic level

$$\mathcal{D}(\mu^2)F_{(2,L)}(N,Q^2) = \mu^2\frac{d}{d\mu^2}F_{(2,L)}(N,Q^2) = 0\,. \tag{2.113}$$

Using the factorization of the structure functions into Wilson coefficients and parton densities, Eqs. 2.94, 2.97 together with the RGE derived for the Wilson coefficients in Eq. 2.112, one obtains from the above formula the QCD evolution equations for the parton densities, cf. e.g. [89, 91, 152, 159–161]

$$\frac{d}{d\ln\mu^2} f_i^{\mathsf{S},NS}(n_f,N,\mu^2) = -\frac{1}{2}\sum_j \gamma_{ij}^{\mathsf{S},NS} f_j^{\mathsf{S},NS}(n_f,N,\mu^2)\,. \tag{2.114}$$

Eq. 2.114 describes the change of the parton densities w.r.t. the scale μ. In the more familiar matrix notation, these equations read

$$\frac{d}{d\ln\mu^2}\begin{pmatrix}\Sigma(n_f,N,\mu^2)\\ G(n_f,N,\mu^2)\end{pmatrix} = -\frac{1}{2}\begin{pmatrix}\gamma_{qq} & \gamma_{qg}\\ \gamma_{gq} & \gamma_{gg}\end{pmatrix}\begin{pmatrix}\Sigma(n_f,N,\mu^2)\\ G(n_f,N,\mu^2)\end{pmatrix}, \tag{2.115}$$

$$\frac{d}{d\ln\mu^2}\Delta_k(n_f,N,\mu^2) = -\frac{1}{2}\gamma_{qq}^{\mathsf{NS}}\Delta_k(n_f,N,\mu^2), \tag{2.116}$$

where we have used the definition for the parton densities in Eqs. 2.95, 2.96. The anomalous dimensions in the above equations can be calculated order by order in perturbation theory. At LO [68–70] and NLO [189–195] they have been known for a long time. The NNLO anomalous dimension were calculated first for fixed moments in Refs. [178–180] and the complete result for all moments has been obtained in Refs. [196, 197].[8] As described, the PDFs are non-perturbative quantities and have to be extracted at a certain scale from experimental data using the factorization relation (2.94). If the scale μ^2 is large enough to apply perturbation theory, the evolution equations can be used to calculate the PDFs at another perturbative scale, which provides a detailed QCD test comparing to precision data. There are similar evolution equations for the structure functions and Wilson coefficients, cf. e.g. [89, 91, 152, 159–161]. Different groups analyze the evolution of the parton distribution functions based on precision data from deep–inelastic scattering experiments and other hard scattering cross sections. Analyzes were performed by the Dortmund group [198–206] by Alekhin et. al. [207, 208] Blümlein et. al. [209, 210] the MSTW—[211] QTEQ—[212] and the NNPDF–collaborations [213]. The PDFs determined in this way can e.g. be used as input data for the pp collisions at the LHC, since they are universal quantities and only relate to the structure of the proton and not to the particular kind of scattering events considered. Apart from performing precision analyzes of the PDFs, one can also use the evolution equations to determine a_s more precisely [198, 199, 205–211].

The evolution Eqs. 2.114–2.116 are written for moments only. The representation in x-space is obtained by using (2.65–2.67) and is usually expressed in terms of the splitting functions $P_{ij}(x)$, [94]. At the level of twist–2 the latter are connected to the anomalous dimensions by the Mellin–transform

$$\gamma_{ij}(N) = -\mathbf{M}[P_{ij}](N)\ . \tag{2.117}$$

The behavior of parton distribution functions in the small x region attracted special interest due to possibly new dynamical contributions, such as Glauber-model based screening corrections [214–220] and the so-called BFKL contributions, a 'leading singularity' resummation in the anomalous dimensions for all orders in the strong coupling constant [221–225]. For both effects there is no evidence yet in the data both for $F_2(x, Q^2)$ and $F_L(x, Q^2)$, beyond the known perturbative contributions to $O(a_s^3)$. This does not exclude that at even smaller values of x contributions of this kind will be found. The BFKL contributions were investigated on the basis of a consistent renormalization group treatment, together with the fixed order contributions in Refs. [226–230]. One main characteristic, comparing with the fixed order case, is that several sub-leading series, which are unknown, are required to stabilize the results, see also [231]. This aspect also has to be studied within the framework of recent approaches [232, 233].

[8] Note that from our convention in Eqs. 2.107, 2.108 follows a relative factor 2 between the anomalous dimensions considered in this work compared to Refs. [196, 197].

References

1. L.W. Mo, C. Peck, 8-GeV/c *Spectrometer*, SLAC-TN-65-029 (1965)
2. R.E. Taylor, Nucleon form–factors above 6-GeV, in *Proceedings of the International Symposium on Electron and Photon Interactions at High Energies*, SLAC, September 5–9, 1967, (SLAC, Stanford CA, 1967), SLAC-PUB-0372, pp. 78
3. D.H. Coward et al., Phys. Rev. Lett. **20**, 292 (1968)
4. W.K.H. Panofsky, Low q^2 electrodynamics, elastic and inelastic electron (and muon) scattering, *Proceedings of 14th International Conference on High-Energy Physics*, Vienna, 1968, J. Prentki and J. Steinberger, eds., (CERN, Geneva, 1968), pp. 23
5. E.D. Bloom et al., Phys. Rev. Lett. **23**, 930 (1969)
6. M. Breidenbach et al., Phys. Rev. Lett. **23**, 935 (1969)
7. H.W. Kendall, Rev. Mod. Phys. **63**, 597 (1991)
8. R.E. Taylor, Rev. Mod. Phys. **63**, 573 (1991)
9. J.I. Friedman, Rev. Mod. Phys. **63**, 615 (1991)
10. W. Albrecht et al., Phys. Lett. **B28**, 225 (1968)
11. W. Albrecht et al., Nucl. Phys. **B13**, 1 (1969)
12. W. Albrecht, et al., Separation of sigma-L and sigma-t in the region of deep inelastic electron - proton scattering, DESY-69-046 (1969)
13. J.D. Bjorken, Phys. Rev. **179**, 1547 (1969)
14. R.P. Feynman, The behavior of hadron collisions at extreme energies, in *Proceedings of 3rd International Conference on High Energy Collisions*, Stony Brook, 1969, C.N. Yang, J.A. Cole, M. Good, R. Hwa, and J. Lee-Franzini, eds., (Gordon and Breach, New York, 1970), pp. 237
15. R.P. Feynman, Phys. Rev. Lett. **23**, 1415 (1969)
16. J.D. Bjorken, E.A. Paschos, Phys. Rev. **185**, 1975 (1969)
17. M. Gell-Mann, Phys. Lett. **8**, 214 (1964)
18. G. Zweig, An $SU(3)$ model for the strong interaction symmetry and its breaking, CERN-TH-401, 412 (1964)
19. J.S. Bell, R. Jackiw, Nuovo Cim. **A60**, 47 (1969)
20. S.L. Adler, Phys. Rev. **177**, 2426 (1969)
21. R.A. Bertlmann, *Anomalies in quantum field theory*, (Clarendon, Oxford, 1996) pp. 566.
22. S. Stein et al., Phys. Rev. **D12**, 1884 (1975)
23. W.B. Atwood et al., Phys. Lett. **B64**, 479 (1976)
24. A. Bodek et al., Phys. Rev. **D20**, 1471 (1979)
25. M.D. Mestayer et al., Phys. Rev. **D27**, 285 (1983)
26. EMC collaboration, O. Allkofer, et al., Nucl. Instr. Meth. **179**, 445 (1981)
27. EMC collaboration, J. J. Aubert et al., Nucl. Phys. **B259**. 189 (1985)
28. D. Bollini et al., Nucl. Instr. Meth. **204**, 333 (1983)
29. BCDMS collaboration, A.C. Benvenuti, et al., Nucl. Instr. Meth. **A226**, 330 (1984)
30. BCDMS collaboration, A.C. Benvenuti, et al., Phys. Lett. **B195**, 91 (1987)
31. BCDMS collaboration, A.C. Benvenuti, et al., Phys. Lett. **B223**, 485 (1989)
32. BCDMS collaboration, A.C. Benvenuti, et al., Phys. Lett. **B237**, 592 (1990)
33. NMC collaboration, P. Amaudruz, et al., Nucl. Phys. **B371**, 3 (1992)
34. NMC collaboration, P. Amaudruz, et al., Phys. Lett. **B295**, 159 (1992)
35. NMC collaboration, M. Arneodo, et al., Phys. Lett. **B364**, 107 (1995), hep-ph/9509406
36. R. Clifft, N. Doble, Proposed Design of a High-Energy, High Intensity Muon Beam for the SPS North Experimental Area, CERN/LAB. II/EA/74-2 (1974)
37. C. Chang et al., Phys. Rev. Lett. **35**, 901 (1975)
38. Y. Watanabe et al., Phys. Rev. Lett. **35**, 898 (1975)
39. D.J. Fox et al., Phys. Rev. Lett. **33**, 1504 (1974)
40. H.L. Anderson et al., Phys. Rev. **D20**, 2645 (1979)
41. E665 collaboration, M.R. Adams, et al., Nucl. Instrum. Meth **A291**, 533 (1990)

42. E665 collaboration, M.R. Adams, et al., Phys. Rev. **D54** , 3006 (1996)
43. T. Sloan, R. Voss, G. Smadja, Phys. Rept. **162**, 45 (1988)
44. CHARM collaboration, M. Jonker, et al., Phys. Lett. **B109**, 133 (1982)
45. CHARM collaboration, F. Bergsma, et al., Phys. Lett. **B123**, 269 (1983)
46. CHARM collaboration, F. Bergsma, et al., Phys. Lett. **B153**, 111 (1985)
47. J.P. Berge et al., Z. Phys. **D49**, 187 (1991)
48. M. Holder et al., Nucl. Instrum. Meth. **151**, 69 (1978)
49. CDHSW collaboration, W. Von Ruden, IEEE Trans. Nucl. Sci. **29**, 360 (1982)
50. Birmingham-CERN-Imperial College-München(MPI)-Oxford collaboration, G. T. Jones, et al., Z. Phys. **C62**, 575 (1994)
51. G. Harigel, *BEBC user's handbook.* (CERN, Geneva, 1977)
52. CCFR collaboration, M.H. Shaevitz, et al., Nucl. Phys. Proc. Suppl. **38**, 188 (1995)
53. W.K. Sakumoto et al., Nucl. Instrum. Meth. **A294**, 179 (1990)
54. B.J. King et al., Nucl. Instrum. Meth. A **302**, 254 (1991)
55. Aachen-Bonn-CERN-London-Oxford-Saclay collaboration, P.C. Bosetti, et al., Nucl. Phys. **B142**, 1 (1978)
56. de J.G.H. Groot et al., Z. Phys. C **1**, 143 (1979)
57. S.M. Heagy et al., Phys. Rev. D **23**, 1045 (1981)
58. Gargamelle SPS collaboration, J.G. Morfin, et al., Phys. Lett. **B104**, 235 (1981)
59. Aachen-Bonn-CERN-Democritos-London-Oxford-Saclay collaboration, P.C. Bosetti, et al., Nucl. Phys. **B203**, 362 (1982)
60. H. Abramowicz et al., Z. Phys. C **17**, 283 (1983)
61. D. MacFarlane et al., Z. Phys. C **26**, 1 (1984)
62. D. Allasia et al., Z. Phys. C **28**, 321 (1985)
63. M. Diemoz, F. Ferroni, E. Longo, Phys. Rept. **130**, 293 (1986)
64. F. Eisele, Rept. Prog. Phys. **49**, 233 (1986)
65. S.R. Mishra, F. Sciulli, Ann. Rev. Nucl. Part. Sci. **39**, 259 (1989)
66. Winter K., ed., Neutrino physics, Camb. Monogr. Part. Phys. Nucl. Phys. Cosmol. Vol. **1**, (Cambridge University Press, Cambridge, 1991), 670 p
67. N. Schmitz, *Neutrinophysik*, (Teubner, Stuttgart, 1997) pp. 478.
68. D.J. Gross, F. Wilczek, Phys. Rev. D **8**, 3633 (1973)
69. D.J. Gross, F. Wilczek, Phys. Rev. **D9**, 980 (1974)
70. H. Georgi, H.D. Politzer, Phys. Rev. **D9**, 416 (1974)
71. HERA - a proposal for a large electron proton colliding beam facility at DESY, (Hamburg, DESY, 1981), DESY HERA 81–10, 292 p
72. R.D. Peccei, ed., Proceedings, HERA Workshop, Hamburg, F.R. Germany, October 12–14, 1987. Vol. **1,2**, 937 p
73. W. Buchmüler, G. Ingelman, eds., Proceedings, Physics at HERA Workshop, Hamburg, F.R. Germany, October 29–30, 1991. Vol. **1–3**, 1566 p
74. J. Blümlein, T. Riemann, eds., Deep inelastic scattering. *Proceedings, Zeuthen Workshop on Elementary Particle Theory, Teupitz, Germany, April* 6–10, 1992, Nucl. Phys. Proc. Suppl. **B29A**, 295 p. (1992)
75. HERA - The new frontier for QCD. Proceedings, Workshop, Durham, UK, March 21–26, 1993, J. Phys. **G19** (1993), pp. 1427
76. J.F. Mathiot, J. Tran Thanh Van, eds., Prepared for 6th Rencontres de Blois: The Heart of the Matter: from Nuclear Interactions to Quark - Gluon Dynamics, Blois, France, 20–25 Jun 1994, (Ed. Frontieres, Gif–sur–Yvette, 1995), 556 p
77. J. Blümlein, W.D. Nowak, eds., Prospects of spin physics at HERA. Proc., Workshop, Zeuthen, Germany, August 28–31, 1995, DESY-95-200, (DESY, Hamburg, 1995), 387 p
78. G. Ingelman, A. De Roeck, R. Klanner, eds., Future physics at HERA. Proceedings, Workshop, Hamburg, Germany, September 25, 1995-May 31, 1996. Vol. **1,2,** DESY-96-235, 1231 p

79. J. Blümlein, A. De Roeck, T. Gehrmann, W. D. Nowak, eds., Deep inelastic scattering off polarized targets: Theory meets experiment. Physics with polarized protons at HERA. Proceedings, Workshops, SPIN'97, Zeuthen, Germany, September 1–5, 1997 and Hamburg, Germany, March-September 1997, DESY-97-200
80. J. Blümlein, W.D. Nowak, G. Schnell, eds., Transverse spin physics, Proceedings, Topical Workshop, Zeuthen, Germany, July 9–11, 2001, DESY-Zeuthen-01-01, Aug 2001. 374 p
81. H1 collaboration, I. Abt, et al., The H1 detector at HERA, DESY-93-103 (1993), 194 p
82. ZEUS collaboration, M. Derrick, et al., Phys. Lett. **B303**, 183 (1993)
83. HERMES collaboration, K. Ackerstaff, et al., Nucl. Instrum. Meth. **A417**, 230 (1998), hep-ex/9806008
84. E. Hartouni et al., HERA-B: An experiment to study CP violation in the B system using an internal target at the HERA proton ring. Design report, DESY-PRC-95-01 (1995), 491 p
85. K.G. Wilson, Phys. Rev. **179**, 1499 (1969)
86. W. Zimmermann, Lect. on Elementary Particle Physics and Quantum Field Theory, Brandeis Summer Inst., Vol. **1**, (MIT Press, Cambridge, 1970), pp. 395.
87. Y. Frishman, Annals Phys. **66**, 373 (1971)
88. R.A. Brandt, G. Preparata, Nucl. Phys. **B27**, 541 (1972)
89. E. Reya, Phys. Rept. **69**, 195 (1981)
90. T. Muta, Foundations of Quantum Chromodynamics, World Sci. Lect. Notes Phys. 57, (World Scientific, Singapore, 1998), 2nd edition
91. R.G. Roberts, *The Structure of the proton: Deep inelastic scattering*, (Cambridge University Press, Cambridge, 1990) pp. 182.
92. J.B. Kogut, L. Susskind, Phys. Rept. **8**, 75 (1973)
93. T.-M. Yan, Ann. Rev. Nucl. Part. Sci. **26**, 199 (1976)
94. G. Altarelli, G. Parisi, Nucl. Phys. **B126**, 298 (1977)
95. J. Blümlein, M. Klein, Nucl. Instrum. Meth. **A329**, 112 (1993)
96. J. Blümlein, Z. Phys. **C65**, 293 (1995), hep-ph/9403342
97. A. Arbuzov, D.Y. Bardin, J. Blüumlein, L. Kalinovskaya, T. Riemann, Comput. Phys. Commun. **94**, 128 (1996), hep-ph/9511434 and references therein
98. J.D. Bjorken, Phys. Rev. **D1**, 1376 (1970)
99. J. Blüumlein, M. Klein, T. Naumann, T. Riemann, Structure functions, quark distributions and $\Lambda_{\rm QCD}$ at HERA in *Proceedings of DESY Theory Workshop on Physics at HERA* (ed. R.D. Peccei), Hamburg, F.R. Germany, 12–14 October 1987, Vol **1**, 67 pp
100. A. Kwiatkowski, H. Spiesberger, H.J. Möhring, Comp. Phys. Commun. **69**, 155 (1992), and references therein
101. J. Engelen, P. Kooijman, Prog. Part. Nucl. Phys. **41**, 1 (1998)
102. H. Abramowicz, A. Caldwell, Rev. Mod. Phys. **71**, 1275 (1999), hep-ex/9903037
103. H1 collaboration, F.D. Aaron et al., Phys. Lett. **B665**, 139 (2008), hep-ex/0805.2809
104. J. Blüumlein, B. Geyer, D. Robaschik, Nucl. Phys. **B560**, 283 (1999), hep-ph/9903520
105. F.J. Yndurain, *The theory of quark and gluon interactions*, 4th edn. (Springer, Berlin, 2006) pp. 474.
106. R. Field, *Applications of perturbative QCD*, (Addison-Wesley, Redwood City, 1989) pp. 366.
107. LHPC collaboration, D. Dolgov, et al., Phys. Rev. **D66**, 034506 (2002), hep-lat/0201021
108. QCDSF collaboration, M.Göckeler, et al., PoS LAT **2007**, 147 (2007), hep-lat/0710.2489
109. ETM collaboration, R. Baron, et al., PoS LAT **2007**, 153 (2007), hep-lat/0710.1580
110. W. Bietenholz, et al., PoS LAT **2008**, 149 (2008), hep-lat/0808.3637
111. S.N. Syritsyn, et al., PoS LAT **2008**, 169 (2008), hep-lat/0903.3063
112. C. Itzykson, J. Zuber, *Quantum Field Theory*, (McGraw-Hill, New York, 1980) pp. 705.
113. J. Blümlein, N. Kochelev, Nucl. Phys. **B498**, 285 (1997), hep-ph/9612318
114. J. Blümlein, A.Tkabladze, Nucl. Phys. **B553**, 427 (1999), hep-ph/9812478
115. E. Rutherford, Phil. Mag. **21**, 669 (1911)
116. R.W. Mcallister, R. Hofstadter, Phys. Rev. **102**, 851 (1956)

117. D.N. Olson, H.F. Schopper, R.R. Wilson, Phys. Rev. Lett. **6**, 286 (1961)
118. R. Hofstadter, Electron scattering and nuclear and nucleon structure. A collection of reprints with an introduction, (New York, Benjamin, 1963), 690 p
119. R.P. Feynman, *Photon-hadron interactions*, (Benjamin Press, Reading, 1972) pp. 282.
120. C. Nash, Nucl. Phys. **B31**, 419 (1971)
121. P.V. Landshoff, J.C. Polkinghorne, Phys. Rept. **5**, 1 (1972)
122. J.D. Jackson, G.G. Ross, R.G. Roberts, Phys. Lett. **B226**, 159 (1989)
123. R.G. Roberts, G.G. Ross, Phys. Lett. **B373**, 235 (1996), hep-ph/9601235
124. J. Blümlein, N. Kochelev, Phys. Lett. **B381**, 296 (1996), hep-ph/9603397
125. J. Blümlein, V. Ravindran, and W.L. van Neerven, Phys. Rev. **D68**, 114004 (2003), hep-ph/0304292
126. F.E. Close, *An Introduction to Quarks and Partons*, (Academic Press, London, 1979) pp. 481.
127. S.B. Libby, G. Sterman, Phys. Rev. **D18**, 4737 (1978)
128. J.C. Collins, D.E. Soper, Ann. Rev. Nucl. Part. Sci. **37**, 383 (1987)
129. D. Amati, R. Petronzio, G. Veneziano, Nucl. Phys. **B140**, 54 (1978)
130. S.B. Libby, G. Sterman, Phys. Rev. **D18**, 3252 (1978)
131. A.H. Mueller, Phys. Rev. **D18**, 3705 (1978)
132. J.C. Collins, G. Sterman, Nucl. Phys. **B185**, 172 (1981)
133. G.T. Bodwin, Phys. Rev. **D31**, 2616 (1985); [Erratum-ibid.] **D34**, 3932 (1986)
134. J.C. Collins, D.E. Soper, G. Sterman, Nucl. Phys. **B261**, 104 (1985)
135. P. Mulders, Quantum Chromodynamics and Hard Scattering Processes, Lectures, Dutch Research School for Theoretical Physics, Dalfsen, January 1996, and Dutch Research School for Subatomic Physics, Beekbergen, February, 1996, (NIKHEF, Amsterdam, The Netherlands)
136. C.G. Callan, D.J. Gross, Phys. Rev. Lett. **22**, 156 (1969)
137. S.D. Drell, T.-M. Yan, Ann. Phys. **66**, 578 (1971)
138. J. Blümlein, Introduction into QCD, Lecture Notes (1997)
139. R. Jackiw, Canonical light-cone commutators and their applications in (49), pp 1
140. R.L. Jaffe, Deep Inelastic Scattering with Application to nuclear targets, Lectures presented at the Los Alamos School on Quark Nuclear Physics, Los Alamos, NM, 10–14 June 1985, M.B. Johnson and A. Picklesimer, eds., (Wiley, New York, 1986), pp. 82
141. Y. Frishman, Phys. Rept. **13**, 1 (1974)
142. B. Geyer, D. Robaschik, E. Wieczorek, Fortschr. Phys. **27**, 75 (1979)
143. A.H. Mueller, Phys. Rept. **73**, 237 (1981)
144. J. Blümlein, Comput. Phys. Commun. **133**, 76 (2000), hep-ph/0003100
145. J. Blümlein, S.-O. Moch, Phys. Lett. **B614**, 53 (2005), hep-ph/0503188
146. E. Carlson, Sur une classe de séries de Taylor, PhD Thesis, Uppsala, 1914
147. E. Titchmarsh, *Theory of Functions*. (Oxford University Press, Oxford, 1939) Chapt. 9.5
148. D.J. Gross, S.B. Treiman, Phys. Rev. **D4**, 1059 (1971)
149. W.A. Bardeen, A.J. Buras, D.W. Duke, T. Muta, Phys. Rev. **D18**, 3998 (1978)
150. L.D. Landau, Nucl. Phys. **13**, 181 (1959)
151. J.D. Bjorken, Experimental tests of quantum electrodynamics and spectral representations of Green's functions in perturbation theory, PhD Thesis, RX–1037 (1959)
152. A.J. Buras, Rev. Mod. Phys. **52**, 199 (1980)
153. A. Bassetto, M. Ciafaloni, G. Marchesini, Phys. Rept. **100**, 201 (1983)
154. E.C.G. Stückelberg, A. Petermann, Helv. Phys. Acta **24**, 317 (1951)
155. M. Gell-Mann, F.E. Low, Phys. Rev. **95**, 1300 (1954)
156. N.N. Bogolyubov, D.V. Shirkov, *Introduction to the theory of quantized fields*, (Interscience, New York, 1959) pp. 720.
157. K. Symanzik, Commun. Math. Phys. **18**, 227 (1970)
158. C.G. Callan, Phys. Rev. **D2**, 1541 (1970)
159. G. Altarelli, Ann. Rev. Nucl. Part. Sci. **39**, 357 (1989)
160. J.F. Owens, W.-K. Tung, Ann. Rev. Nucl. Part. Sci. **42**, 291 (1992)

161. J. Blümlein, On the Theoretical Status of Deep Inelastic Scattering, (1995), hep-ph/9512272
162. D.W. Duke, R.G. Roberts, Phys. Rept. **120**, 275 (1985)
163. S. Bethke, J.E. Pilcher, Ann. Rev. Nucl. Part. Sci. **42**, 251 (1992)
164. A. Zee, F. Wilczek, S.B. Treiman, Phys. Rev. **D10**, 2881 (1974)
165. W. Furmanski, R. Petronzio, Z. Phys. **C11**, 293 (1982), and references therein
166. D.W. Duke, J.D. Kimel, G.A. Sowell, Phys. Rev. **D25**, 71 (1982)
167. A. Devoto, D.W. Duke, J.D. Kimel, G.A. Sowell, Phys. Rev. **D30**, 541 (1984)
168. D.I. Kazakov, A.V. Kotikov, Nucl. Phys. **B307**, 721 (1988)
169. D.I. Kazakov, A.V. Kotikov, G. Parente, O.A. Sampayo, J. Sanchez Guillen, Phys. Rev. Lett. **65**, 1535 (1990)
170. J. Sanchez Guillen, J. Miramontes, M. Miramontes, G. Parente, O.A. Sampayo, Nucl. Phys. **B353**, 337 (1991)
171. van W.L. Neerven, E.B. Zijlstra, Phys. Lett. **B272**, 127 (1991)
172. van W.L. Neerven, E.B. Zijlstra, Phys. Lett. **B273**, 476 (1991)
173. van W.L. Neerven, E.B. Zijlstra, Nucl. Phys. **B383**, 525 (1992)
174. D.I. Kazakov, A.V. Kotikov, Phys. Lett. **B291**, 171 (1992)
175. S.A. Larin, J.A.M. Vermaseren, Z. Phys. **C57**, 93 (1993)
176. S. Moch, J.A.M. Vermaseren, Nucl. Phys. **B573**, 853 (2000), hep-ph/9912355
177. S.A. Larin, van T. Ritbergen, J.A.M. Vermaseren, Nucl. Phys. **B427**, 41 (1994)
178. S.A. Larin, P. Nogueira, T. van Ritbergen, J.A.M. Vermaseren, Nucl. Phys. **B492**, 338 (1997), hep-ph/9605317
179. A. Retey, J.A.M. Vermaseren, Nucl. Phys. **B604**, 281 (2001), hep-ph/0007294
180. J. Blümlein, J.A.M. Vermaseren, Phys. Lett. **B606**, 130 (2005), hep-ph/0411111
181. J.A.M. Vermaseren, A. Vogt, S. Moch, Nucl. Phys. **B724**, 3 (2005), hep-ph/0504242
182. S. Moch, J.A.M. Vermaseren, A. Vogt, Nucl. Phys. **B813**, 220 (2009), hep-ph/0812.4168
183. F. Bloch, A. Nordsieck, Phys. Rev. **52**, 54 (1937)
184. D.R. Yennie, S.C. Frautschi, H. Suura, Ann. Phys. **13**, 379 (1961)
185. T. Kinoshita, J. Math. Phys. **3**, 650 (1962)
186. T.D. Lee, M. Nauenberg, Phys. Rev. **133**, B1549 (1964)
187. A. Peterman, Phys. Rept. **53**, 157 (1979)
188. J.C. Collins, *Renormalization*, (Cambridge University Press, Cambridge, 1984) pp. 380.
189. E.G. Floratos, D.A. Ross, C.T. Sachrajda, Nucl. Phys. **B129**, 66 (1977); [Erratum-ibid.] **B139**, 545 (1978)
190. E.G. Floratos, D.A. Ross, C.T. Sachrajda, Nucl. Phys. **B152**, 493 (1979)
191. A. Gonzalez-Arroyo, C. Lopez, F.J. Yndurain, Nucl. Phys. **B153**, 161 (1979)
192. A. Gonzalez-Arroyo, C. Lopez, Nucl. Phys. **B166**, 429 (1980)
193. G. Curci, W. Furmanski, R. Petronzio, Nucl. Phys. **B175**, 27 (1980)
194. W. Furmanski, R. Petronzio, Phys. Lett. **B97**, 437 (1980)
195. R. Hamberg, van W.L. Neerven, Nucl. Phys. **B379**, 143 (1992)
196. S. Moch, J.A.M. Vermaseren, A. Vogt, Nucl. Phys. **B688**, 101 (2004), hep-ph/0403192
197. A. Vogt, S. Moch, J.A.M. Vermaseren, Nucl. Phys. **B691**, 129 (2004), hep-ph/0404111
198. M. Glück, E. Reya, C. Schuck, Nucl. Phys. **B754**, 178 (2006), hep-ph/0604116
199. M. Glück, E. Hoffmann, E. Reya, Z. Phys. **C13**, 119 (1982)
200. M. Glück, R.M. Godbole, E. Reya, Z. Phys. **C41**, 667 (1989)
201. M. Glück, E. Reya, A. Vogt, Z. Phys. **C48**, 471 (1990)
202. M. Glück, E. Reya, A. Vogt, Z. Phys. **C53**, 127 (1992)
203. M. Glück, E. Reya, A. Vogt, Z. Phys. **C67**, 433 (1995)
204. M. Glück, E. Reya, A. Vogt, Eur. Phys. J. **C5**, 461 (1998), hep-ph/9806404
205. M. Glück, P. Jimenez-Delgado, E. Reya, Eur. Phys. J. **C53**, 355 (2008), hep-ph/0709.0614
206. P. Jimenez-Delgado, E. Reya, Phys. Rev. **D79**, 074023 (2009), hep-ph/0810.4274
207. S. Alekhin, K. Melnikov, F. Petriello, Phys. Rev. **D74**, 054033 (2006), hep-ph/0606237
208. S. Alekhin, JETP Lett. **82**, 628 (2005), hep-ph/0508248

209. J. Blümlein, H. Böttcher, A. Guffanti, Nucl. Phys. Proc. Suppl. **135**, 152 (2004), hep-ph/0407089
210. J. Blümlein, H. Böttcher, A. Guffanti, Nucl. Phys. **B774**, 182 (2007), hep-ph/0607200
211. A.D. Martin, W.J. Stirling, R.S. Thorne, G. Watt, Eur. Phys. J. **C63**, 189 (2009), hep-ph/0901.0002
212. CTEQ collaboration, H.L. Lai et al., Eur. Phys. J. **C12**, 375 (2000), hep-ph/9903282
213. NNPDF collaboration, R.D. Ball et al., Nucl. Phys. **B809**, 1(2009), hep-ph/0808.1231
214. L.V. Gribov, E.M. Levin, M.G. Ryskin, Nucl. Phys. **B188**, 555 (1981)
215. A.H. Mueller, J.-W. Qiu, Nucl. Phys. **B268**, 427 (1986)
216. J.C. Collins, J. Kwiecinski, Nucl. Phys. **B335**, 89 (1990)
217. J. Bartels, G.A. Schuler, J. Blümlein, Z. Phys. **C50**, 91 (1991)
218. M. Altmann, M. Glück, E. Reya, Phys. Lett. **B285**, 359 (1992)
219. V. Del Duca, An introduction to the perturbative QCD pomeron and to jet physics at large rapidities, (1995), hep-ph/9503226
220. L.N. Lipatov, Phys. Rept. **286**, 131 (1997), hep-ph/9610276
221. V.S. Fadin, E.A. Kuraev, L.N. Lipatov, Phys. Lett. **B60**, 50 (1975)
222. I.I. Balitsky, L.N. Lipatov, Sov. J. Nucl. Phys. **28**, 822 (1978)
223. R. Kirschner, L.N. Lipatov, Nucl. Phys. **B213**, 122 (1983)
224. J. Bartels, B.I. Ermolaev, M.G. Ryskin, Z. Phys. **C72**, 627 (1996), hep-ph/9603204
225. V.S. Fadin, L.N. Lipatov, Phys. Lett. **B429**, 127 (1998), hep-ph/9802290
226. J. Blümlein, A. Vogt, Phys. Lett. **B370**, 149 (1996), hep-ph/9510410
227. J. Blümlein, A. Vogt, Acta Phys. Polon. **B27**, 1309 (1996), hep-ph/9603450
228. J. Blümlein, A. Vogt, Phys. Lett. **B386**, 350 (1996), hep-ph/9606254
229. J. Blümlein, A. Vogt, Phys. Rev. **D58**, 014020 (1998), hep-ph/9712546
230. J. Blümlein, V. Ravindran, W.L. van Neerven, A. Vogt, (1998), hep-ph/9806368
231. J. Blümlein, W.L. van Neerven, Phys. Lett. **B450**, 412 (1999), hep-ph/9811519
232. G. Altarelli, R.D. Ball, S. Forte, Nucl. Phys. **B799**, 199 (2008), hep-ph/0802.0032
233. M. Ciafaloni, D. Colferai, G.P. Salam, A.M. Stasto, JHEP **08**, 046 (2007), hep-ph/0707.1453

Chapter 3
Heavy Quark Production in DIS

In the Standard Model, the charm, bottom and top quark are treated as heavy quarks, all of which have a mass larger than the QCD-scale $\Lambda_{\text{QCD}}(n_f = 4) \approx 240\,\text{MeV}$, [1–7]. The up, down and strange quark are usually treated as massless. Because of confinement, the quarks can only be observed via the asymptotic states baryons and mesons, in which they are contained. In the following, we concentrate on the inclusive production of one species of a heavy quark, denoted by $Q(\overline{Q})$, with mass m. In the case of HERA kinematics, $Q = c$. This is justified to a certain extent by the observation that bottom quark contributions to DIS structure functions are much smaller, cf. [8].[1] Since the ratio $m_c^2/m_b^2 \approx 1/10$ is not small, there are regions in which both masses are potentially important. The description of these effects is beyond the scope of the formalism outlined below and of comparable order as the m_c^2/Q^2 corrections. Top-quark production in $l^{\pm}N$ scattering is usually treated as a semi-inclusive process, cf. [10–12].

Charmed mesons are more abundantly produced at HERA than baryons. D-mesons are bound states of charm and lighter quarks, e.g. $D_u = \overline{u}c,\ \ D_d = \overline{d}c$ etc. Furthermore also $c\overline{c}$ resonances contribute, such as J/ψ, by the observation of which charm was discovered [13–15]. The charm contributions to the structure functions are determined in experiment by tagging charm quarks in the final state, e.g. through the D-meson decay channel $D^* \to D^0\pi_s \to K\pi\pi_s$. In the case of DIS, the measured visible cross section is then extrapolated to the full inclusive phase space using theoretical models if structure functions are considered [8, 16–21].

Within the approach of this thesis, the main objective for studying heavy quark production in DIS is to provide a framework allowing for more precise measurements of a_s and of the parton densities and for a better description of the structure functions $F_2^{c\overline{c}}$, $F_2^{b\overline{b}}$. The current world data for the nucleon structure functions $F_2^{p,d}(x, Q^2)$ reached the precision of a few per cent over a wide kinematic region. Both the measurements of the heavy flavor contributions to the deep-inelastic structure functions, cf. [8, 19, 20, 22], and numerical studies [23–25], based on the leading, [26–30],

[1] Likewise, for even higher scales the b-quark could be considered as the heavy quark with u, d, s, c being effectively massless, cf. e.g. [9].

S. Klein, *Charm Production in Deep Inelastic Scattering*, Springer Theses,
DOI: 10.1007/978-3-642-23286-2_3,

and next-to-leading order heavy flavor Wilson coefficients [31–33], show that the scaling violations of the light and the heavy contributions to (2.97) exhibit a different behavior over a wide range of Q^2. This is both due to the logarithmic contributions $\ln^k(Q^2/m^2)$ and power corrections $\propto (m^2/Q^2)^k, k \geq 1$. Moreover, in the region of smaller values of x the heavy flavor contributions amount to 20–40%. Therefore, the precision measurement of the QCD parameter $\Lambda_{\rm QCD}$, [1–7, 30, 34–36], and of the parton distribution functions in deeply inelastic scattering requires the analysis at the level of the $O(a_s^3)$ corrections to control the theory-errors at the level of the experimental accuracy and below [2, 4, 34–36].

The precise value of $\Lambda_{\rm QCD}$, a fundamental parameter of the Standard Model, is of central importance for the quantitative understanding of all strongly interacting processes. Moreover, the possible unification of the gauge forces [37, 38], depends crucially on its value. In recent non-singlet analyzes [1, 2, 4], errors for $a_s(M_Z^2)$ of $O(1.5\%)$ were obtained, partially extending the analysis effectively to N^{3}LO. In the flavor singlet case the so far unknown 3-loop heavy flavor Wilson coefficients do yet prevent a consistent 3-loop analysis,[39–41]. Due to the large statistics in the lower x region, one may hope to eventually improve the accuracy of $a_s(M_Z^2)$ beyond the above value.

Of similar importance is the detailed knowledge of the PDFs for all hadron-induced processes, notably for the interpretation of all scattering cross sections measured at the TEVATRON and the LHC. For example, the process of Higgs-boson production at the LHC, cf. e.g. [42, 43], depends on the gluon density and its accuracy is widely determined by this distribution.

In Sect. 3.1, we describe the general framework of electroproduction of heavy quarks in DIS within the fixed-flavor-number-scheme (FFNS), treating only the light quarks and the gluon as constituents of the nucleon. In the following Sect. 3.2 , we outline the method, which we use to extract all but the power suppressed contributions $\propto (m^2/Q^2)^k, k \geq 1$ of the heavy flavor Wilson coefficients [44]. The latter are equivalent to the Wilson coefficients introduced in Sect. 2.3.3, including heavy quarks. Finally, in Sect. 3.3 we comment on the possibility to define heavy quark parton densities within a variable-flavor-number-scheme (VFNS) [45].

3.1 Electroproduction of Heavy Quarks

We study electroproduction of heavy quarks in unpolarized DIS via single photon exchange, cf. [26–30, 46], at sufficiently large virtualities Q^2, $Q^2 \geq 5\text{GeV}$. [2] Here, one can distinguish two possible production mechanisms for heavy quarks: extrinsic production and intrinsic heavy quark excitation. In the latter case, one introduces a heavy quark state in the nucleon wave function, i.e. the heavy quark is treated at

[2] One may however, also consider photoproduction of heavy quarks in ep collisions where $Q^2 \approx 0$, which is a widely hadronic process, cf. [47, 48], and especially important for the production of heavy quark resonances, as e.g. the J/ψ.

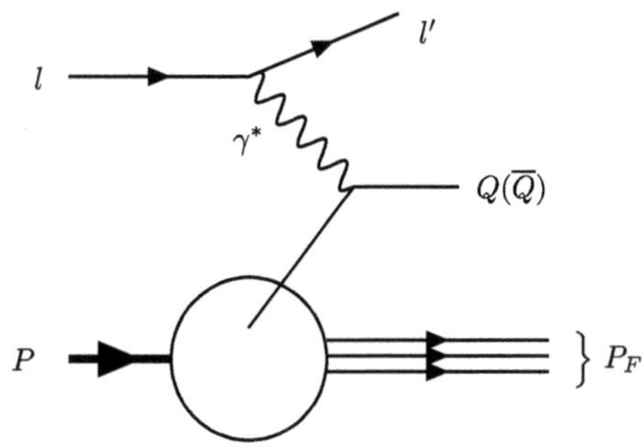

Fig. 3.1 LO intrinsic heavy quark production

the same level as the light quarks in the factorization of the structure functions, cf. Eqs. 2.97–2.100. The LO contribution is then given by the flavor excitation process shown in Fig. 3.1

$$\gamma^* + Q(\overline{Q}) \rightarrow Q(\overline{Q}). \tag{3.1}$$

Several experimental and theoretical studies suggest that the intrinsic contribution to the heavy flavor cross section is of the order of 1% or smaller [16, 49–52], and we will not consider it any further.

In extrinsic heavy flavor production, the heavy quarks are produced as final states in virtual gauge boson scattering off massless partons. This description is also referred to as the fixed flavor number scheme. At higher orders, one has to make the distinction between whether one considers the complete inclusive structure functions or only those heavy quark contributions, which can be determined in experiments by tagging the final state heavy quarks. In the former case, virtual corrections containing heavy quark loops have to be included into the theoretical calculation as well, cf. also Sect. 5.1.

We consider only twist-2 parton densities in the Bjorken limit. Therefore no transverse momentum effects in the initial parton distributions will be allowed, since these contributions are related, in the kinematic sense, to higher twist operators. From the conditions for the validity of the parton model, Eqs. 2.51, 2.56, it follows that in the region of not too small nor too large values of the Bjorken variable x, the partonic description holds for massless partons. Evidently, iff $Q^2(1-x)^2/m^2 \not\gg 1$ no partonic description for a potential heavy quark distribution can be obtained. The question under which circumstances one may introduce a heavy flavor parton density will be further discussed in Sect. 3.3. In a general kinematic region the parton densities in Eq. 2.97 are enforced to be massless and the heavy quark mass effects are contained in the inclusive Wilson coefficients. These are calculable perturbatively and denoted by

$$C_{i,(2,L)}^{\mathsf{S,PS,NS}}\left(\tau, n_f+1, \frac{Q^2}{\mu^2}, \frac{m^2}{\mu^2}\right). \tag{3.2}$$

The argument $n_f + 1$ denotes the presence of n_f light and one heavy flavor. τ is the partonic scaling variable defined in Eq. 2.38 and we will present some of the following equations in x-space rather than in Mellin space.

One may identify the massless flavor contributions in Eq. 3.2 and separate the Wilson coefficients into a purely light part $C_{i,(2,L)}$, cf. Eq. 2.97, and a heavy part

$$\begin{aligned} C_{i,(2,L)}^{\mathsf{S,PS,NS}}\left(\tau, n_f+1, \frac{Q^2}{\mu^2}, \frac{m^2}{\mu^2}\right) &= C_{i,(2,L)}^{\mathsf{S,PS,NS}}\left(\tau, n_f, \frac{Q^2}{\mu^2}\right) \\ &+ H_{i,(2,L)}^{\mathsf{S,PS}}\left(\tau, n_f+1, \frac{Q^2}{\mu^2}, \frac{m^2}{\mu^2}\right) \\ &+ L_{i,(2,L)}^{\mathsf{S,PS,NS}}\left(\tau, n_f+1, \frac{Q^2}{\mu^2}, \frac{m^2}{\mu^2}\right). \end{aligned} \quad (3.3)$$

Here, we denote the heavy flavor Wilson coefficients by $L_{i,j}$ and $H_{i,j}$, respectively, depending on whether the photon couples to a light (L) or heavy (H) quark line. From this it follows that the light flavor Wilson coefficients $C_{i,j}$ depend on n_f light flavors only, whereas $H_{i,j}$ and $L_{i,j}$ may contain light flavors in addition to the heavy quark, indicated by the argument $n_f + 1$. The perturbative series of the heavy flavor Wilson coefficients read

$$H_{g,(2,L)}^{\mathsf{S}}\left(\tau, n_f+1, \frac{Q^2}{\mu^2}, \frac{m^2}{\mu^2}\right) = \sum_{i=1}^{\infty} a_s^i H_{g,(2,L)}^{(i),\mathsf{S}}\left(\tau, n_f+1, \frac{Q^2}{\mu^2}, \frac{m^2}{\mu^2}\right), \quad (3.4)$$

$$H_{q,(2,L)}^{\mathsf{PS}}\left(\tau, n_f+1, \frac{Q^2}{\mu^2}, \frac{m^2}{\mu^2}\right) = \sum_{i=2}^{\infty} a_s^i H_{q,(2,L)}^{(i),\mathsf{PS}}\left(\tau, n_f+1, \frac{Q^2}{\mu^2}, \frac{m^2}{\mu^2}\right), \quad (3.5)$$

$$L_{g,(2,L)}^{\mathsf{S}}\left(\tau, n_f+1, \frac{Q^2}{\mu^2}, \frac{m^2}{\mu^2}\right) = \sum_{i=2}^{\infty} a_s^i L_{g,(2,L)}^{(i),\mathsf{S}}\left(\tau, n_f+1, \frac{Q^2}{\mu^2}, \frac{m^2}{\mu^2}\right), \quad (3.6)$$

$$L_{q,(2,L)}^{\mathsf{S}}\left(\tau, n_f+1, \frac{Q^2}{\mu^2}, \frac{m^2}{\mu^2}\right) = \sum_{i=2}^{\infty} a_s^i L_{q,(2,L)}^{(i),\mathsf{S}}\left(\tau, n_f+1, \frac{Q^2}{\mu^2}, \frac{m^2}{\mu^2}\right). \quad (3.7)$$

Note that we have not yet specified a scheme for treating a_s, but one has to use the same scheme when combining the above terms with the light flavor Wilson coefficients. At LO, only the term $H_{g,(2,L)}$ contributes via the photon–gluon fusion process shown in Fig. 3.2 ,

$$\gamma^* + g \to Q + \overline{Q}. \quad (3.8)$$

Fig. 3.2 LO extrinsic heavy quark production

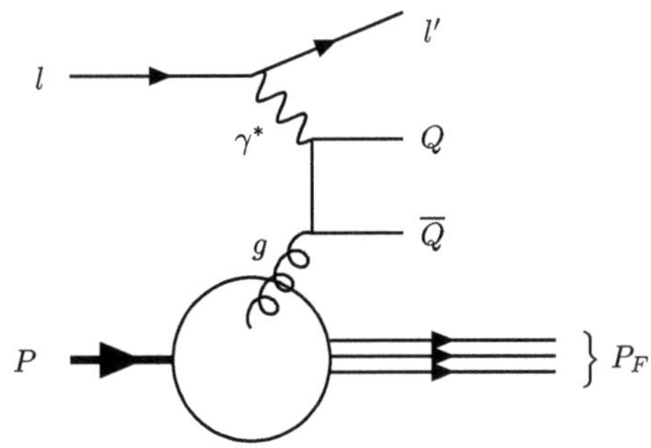

The LO Wilson coefficients corresponding to this process are given by [26–30, 46], [3]

$$H^{(1)}_{g,2}\left(\tau, \frac{m^2}{Q^2}\right) = 8T_F\left\{v\left[-\frac{1}{2}+4\tau(1-\tau)+2\frac{m^2}{Q^2}\tau(\tau-1)\right]\right.$$
$$+\left[-\frac{1}{2}+\tau-\tau^2+2\frac{m^2}{Q^2}\tau(3\tau-1)+4\frac{m^4}{Q^4}\tau^2\right]$$
$$\left.\times\ln\left(\frac{1-v}{1+v}\right)\right\}, \tag{3.9}$$

$$H^{(1)}_{g,L}\left(\tau, \frac{m^2}{Q^2}\right) = 16T_F\left[\tau(1-\tau)v+2\frac{m^2}{Q^2}\tau^2\ln\left(\frac{1-v}{1+v}\right)\right]. \tag{3.10}$$

The cms velocity v of the produced heavy quark pair is given by

$$v = \sqrt{1-\frac{4m^2\tau}{Q^2(1-\tau)}}. \tag{3.11}$$

The LO heavy flavor contributions to the structure functions are then

$$F^{Q\overline{Q}}_{(2,L)}(x, Q^2, m^2) = e_Q^2 a_s \int_{ax}^{1} \frac{dz}{z} H^{(1)}_{g,(2,L)}\left(\frac{x}{z}, \frac{m^2}{Q^2}\right)$$
$$\times G(n_f, z, Q^2), \quad a = 1+4m^2/Q^2, \tag{3.12}$$

where the integration boundaries follow from the kinematics of the process. Here e_Q denotes the electric charge of the heavy quark.

At $O(a_s^2)$, the terms $H^{\mathsf{PS}}_{q,(2,L)}$ and $L^{\mathsf{S}}_{q,(2,L)}$ contribute as well. They result from the process

[3] Eqs. 3.16, 3.17 in Ref. [53] contain misprints

the OMEs. Instead of absorbing these terms into the parton densities as well, they are used to reconstruct the asymptotic behavior of the heavy flavor Wilson coefficients. Here,

$$A_{ij}^{\mathsf{S,NS}}\Big(N, n_f+1, \frac{m^2}{\mu^2}\Big) = \langle j|O_i^{\mathsf{S,NS}}|j\rangle = \delta_{ij} + \sum_{i=1}^{\infty} a_s^i A_{ij}^{(i),\mathsf{S,NS}} \tag{3.17}$$

are the operator matrix elements of the local twist-2 operators being defined in Eqs. 2.86–2.88 between on-shell partonic states $|j\rangle$, $j = q, g$. As usual, the $\mathtt{S}$ contribution can be split into a $\mathtt{NS}$ and $\mathtt{PS}$ part via

$$A_{qq}^{\mathsf{S}} = A_{qq}^{\mathsf{NS}} + A_{qq}^{\mathsf{PS}}. \tag{3.18}$$

Due to the on-shell condition, all contributions but the $O(a_s^0)$ terms vanish [5] if no heavy quark is present in the virtual loops. This is due to the fact that integrals without scale vanish in dimensional regularization, cf. Sect. 4.1. Hence only those terms with a mass remain and these are referred to as massive OMEs. The calculation of these massive OMEs is the main objective of this thesis. In case of the gluon operator, (2.88), the contributing terms are denoted by $A_{gq,Q}$ and $A_{gg,Q}$, where the perturbative series of the former starts at $O(a_s^2)$ and the one of the latter at $O(a_s^0)$.[6] For the quark operator, one distinguishes whether the operator couples to a heavy or light quark. In the $\mathtt{NS}$-case, the operator by definition couples to the light quark. Thus there is only one term, $A_{qq,Q}^{\mathsf{NS}}$, which contributes at $O(a_s^0)$. In the $\mathtt{S}$ and $\mathtt{PS}$-case, two OMEs can be distinguished, $\{A_{qq,Q}^{\mathsf{PS}}, A_{qg,Q}^{\mathsf{S}}\}$ and $\{A_{Qq}^{\mathsf{PS}}, A_{Qg}^{\mathsf{S}}\}$, where in the former case the operator couples to a light quark and in the latter case to a heavy quark. The terms $A_{qi,Q}$ emerge for the first time at $O(a_s^3)$, A_{Qq}^{PS} at $O(a_s^2)$ and A_{Qg}^{S} at $O(a_s)$.

In this work we refer only to the even moments, cf. Sect. 2.3. In the non-singlet case we will obtain, however, besides the NS$^+$ contributions for the even moments also the NS$^-$ terms, which correspond to the odd moments.

Equation 3.16 can now be split into its parts by considering the different n_f-terms. We adopt the following notation for a function $f(n_f)$

$$\tilde{f}(n_f) \equiv \frac{f(n_f)}{n_f}. \tag{3.19}$$

This is necessary in order to separate the different types of contributions in Eq. 3.15, weighted by the electric charges of the light and heavy flavors, respectively. Since we concentrate on only the heavy flavor part, we define as well for later use

$$\hat{f}(n_f) \equiv f(n_f+1) - f(n_f), \tag{3.20}$$

[5] In Ref. [60] use was made of this fact to calculate the massless Wilson coefficients without having to calculate the massless OMEs.

[6] The $O(a_s^0)$ term of A_{gg} does not contain a heavy quark, but still remains in Eq. 3.16 because no loops have to be calculated.

with $\hat{\tilde{f}}(n_f) \equiv [\widehat{\tilde{f}(n_f)}]$. The following Eqs. 3.21–3.25 are the same as Eqs. 2.31–2.35 in Ref. [45]. We present these terms here again, however, since Ref. [45] contains a few inconsistencies regarding the $\tilde{f}$-description. Contrary to the latter reference, the argument corresponding to the number of flavors stands for all flavors, light or heavy. The separation for the NS-term is given by

$$C_{q,(2,L)}^{\mathsf{NS}}\Bigl(N,n_f,\frac{Q^2}{\mu^2}\Bigr)+L_{q,(2,L)}^{\mathsf{NS}}\Bigl(N,n_f+1,\frac{Q^2}{\mu^2},\frac{m^2}{\mu^2}\Bigr)$$
$$=A_{qq,Q}^{\mathsf{NS}}\Bigl(N,n_f+1,\frac{m^2}{\mu^2}\Bigr)C_{q,(2,L)}^{\mathsf{NS}}\Bigl(N,n_f+1,\frac{Q^2}{\mu^2}\Bigr). \tag{3.21}$$

Here and in the following, we omit the index "asymp" to denote the asymptotic heavy flavor Wilson coefficients, since no confusion is to be expected. For the remaining terms, we suppress for brevity the arguments N, Q^2/μ^2 and m^2/μ^2, all of which can be inferred from Eqs. 3.3, 3.16. Additionally, we will suppress from now on the index S and label only the NS and PS terms explicitly. The contributions to $L_{i,j}$ read

$$C_{q,(2,L)}^{\mathsf{PS}}(n_f)+L_{q,(2,L)}^{\mathsf{PS}}(n_f+1)$$
$$=\Bigl[A_{qq,Q}^{\mathsf{NS}}(n_f+1)+A_{qq,Q}^{\mathsf{PS}}(n_f+1)+A_{Qq}^{\mathsf{PS}}(n_f+1)\Bigr]n_f\tilde{C}_{q,(2,L)}^{\mathsf{PS}}(n_f+1)$$
$$+A_{qq,Q}^{\mathsf{PS}}(n_f+1)C_{q,(2,L)}^{\mathsf{NS}}(n_f+1)+A_{gq,Q}(n_f+1)n_f\tilde{C}_{g,(2,L)}(n_f+1), \tag{3.22}$$

$$C_{g,(2,L)}(n_f)+L_{g,(2,L)}(n_f+1)$$
$$=A_{gg,Q}(n_f+1)n_f\tilde{C}_{g,(2,L)}(n_f+1)+A_{qg,Q}(n_f+1)C_{q,(2,L)}^{\mathsf{NS}}(n_f+1)$$
$$+\bigl[A_{qg,Q}(n_f+1)+A_{Qg}(n_f+1)\bigr]n_f\tilde{C}_{q,(2,L)}^{\mathsf{PS}}(n_f+1). \tag{3.23}$$

The terms $H_{i,j}$ are given by

$$H_{q,(2,L)}^{\mathsf{PS}}(n_f+1)=A_{Qq}^{\mathsf{PS}}(n_f+1)\Bigl[C_{q,(2,L)}^{\mathsf{NS}}(n_f+1)+\tilde{C}_{q,(2,L)}^{\mathsf{PS}}(n_f+1)\Bigr]$$
$$+\Bigl[A_{qq,Q}^{\mathsf{NS}}(n_f+1)+A_{qq,Q}^{\mathsf{PS}}(n_f+1)\Bigr]\tilde{C}_{q,(2,L)}^{\mathsf{PS}}(n_f+1)$$
$$+A_{gq,Q}(n_f+1)\tilde{C}_{g,(2,L)}(n_f+1), \tag{3.24}$$

$$H_{g,(2,L)}(n_f+1)=A_{gg,Q}(n_f+1)\tilde{C}_{g,(2,L)}(n_f+1)$$
$$+A_{qg,Q}(n_f+1)\tilde{C}_{q,(2,L)}^{\mathsf{PS}}(n_f+1)$$
$$+A_{Qg}(n_f+1)\Bigl[C_{q,(2,L)}^{\mathsf{NS}}(n_f+1)+\tilde{C}_{q,(2,L)}^{\mathsf{PS}}(n_f+1)\Bigr]. \tag{3.25}$$

Expanding the above equations up to $O(a_s^3)$, we obtain, using Eqs. 3.19, 3.20, the heavy flavor Wilson coefficients in the asymptotic limit :

$$
\begin{aligned}
L_{q,(2,L)}^{\mathsf{NS}}(n_f+1) = a_s^2\Big[&A_{qq,Q}^{(2),\mathsf{NS}}(n_f+1)\delta_2+\hat{C}_{q,(2,L)}^{(2),\mathsf{NS}}(n_f)\Big]\\
+a_s^3\Big[&A_{qq,Q}^{(3),\mathsf{NS}}(n_f+1)\delta_2+A_{qq,Q}^{(2),\mathsf{NS}}(n_f+1)\\
&\times C_{q,(2,L)}^{(1),\mathsf{NS}}(n_f+1)+\hat{C}_{q,(2,L)}^{(3),\mathsf{NS}}(n_f)\Big],
\end{aligned}
\tag{3.26}
$$

$$
\begin{aligned}
L_{q,(2,L)}^{\mathsf{PS}}(n_f+1) = a_s^3\Big[&A_{qq,Q}^{(3),\mathsf{PS}}(n_f+1)\delta_2+A_{gq,Q}^{(2)}(n_f)\\
&\times n_f\widetilde{C}_{g,(2,L)}^{(1)}(n_f+1)+n_f\hat{\widetilde{C}}_{q,(2,L)}^{(3),\mathsf{PS}}(n_f)\Big],
\end{aligned}
\tag{3.27}
$$

$$
\begin{aligned}
L_{g,(2,L)}^{\mathsf{S}}(n_f+1) = a_s^2&A_{gg,Q}^{(1)}(n_f+1)n_f\widetilde{C}_{g,(2,L)}^{(1)}(n_f+1)\\
+a_s^3\Big[&A_{qg,Q}^{(3)}(n_f+1)\delta_2+A_{gg,Q}^{(1)}(n_f+1)\\
&\times n_f\widetilde{C}_{g,(2,L)}^{(2)}(n_f+1)+A_{gg,Q}^{(2)}(n_f+1)\\
&\times n_f\widetilde{C}_{g,(2,L)}^{(1)}(n_f+1)+A_{Qg}^{(1)}(n_f+1)\\
&\times n_f\widetilde{C}_{q,(2,L)}^{(2),\mathsf{PS}}(n_f+1)+n_f\hat{\widetilde{C}}_{g,(2,L)}^{(3)}(n_f)\Big],
\end{aligned}
\tag{3.28}
$$

$$
\begin{aligned}
H_{q,(2,L)}^{\mathsf{PS}}(n_f+1) = a_s^2\Big[&A_{Qq}^{(2),\mathsf{PS}}(n_f+1)\delta_2+\widetilde{C}_{q,(2,L)}^{(2),\mathsf{PS}}(n_f+1)\Big]\\
+a_s^3\Big[&A_{Qq}^{(3),\mathsf{PS}}(n_f+1)\delta_2+\widetilde{C}_{q,(2,L)}^{(3),\mathsf{PS}}(n_f+1)\\
&+A_{gq,Q}^{(2)}(n_f+1)\widetilde{C}_{g,(2,L)}^{(1)}(n_f+1)\\
&+A_{Qq}^{(2),\mathsf{PS}}(n_f+1)C_{q,(2,L)}^{(1),\mathsf{NS}}(n_f+1)\Big],
\end{aligned}
\tag{3.29}
$$

$$
\begin{aligned}
H_{g,(2,L)}^{\mathsf{S}}(n_f+1) = a_s\Big[&A_{Qg}^{(1)}(n_f+1)\delta_2+\widetilde{C}_{g,(2,L)}^{(1)}(n_f+1)\Big]\\
+a_s^2\Big[&A_{Qg}^{(2)}(n_f+1)\delta_2+A_{Qg}^{(1)}(n_f+1)\\
&\times C_{q,(2,L)}^{(1),\mathsf{NS}}(n_f+1)+A_{gg,Q}^{(1)}(n_f+1)\\
&\times \widetilde{C}_{g,(2,L)}^{(1)}(n_f+1)+\widetilde{C}_{g,(2,L)}^{(2)}(n_f+1)\Big]\\
+a_s^3\Big[&A_{Qg}^{(3)}(n_f+1)\delta_2+A_{Qg}^{(2)}(n_f+1)C_{q,(2,L)}^{(1),\mathsf{NS}}(n_f+1)\\
&+A_{gg,Q}^{(2)}(n_f+1)\widetilde{C}_{g,(2,L)}^{(1)}(n_f+1)
\end{aligned}
$$

$$+ A_{Qg}^{(1)}(n_f+1)$$
$$\times \left\{C_{q,(2,L)}^{(2),\mathsf{NS}}(n_f+1) + \widetilde{C}_{q,(2,L)}^{(2),\mathsf{PS}}(n_f+1)\right\}$$
$$+ A_{gg,Q}^{(1)}(n_f+1)\widetilde{C}_{g,(2,L)}^{(2)}(n_f+1) + \widetilde{C}_{g,(2,L)}^{(3)}(n_f+1)\Big]. \qquad (3.30)$$

Note that δ_2 has been defined in Eq. 2.101. The above equations include radiative corrections due to heavy quark loops to the Wilson coefficients. Therefore, in order to compare e.g. with the calculation in Refs. [31–33], these terms still have to be subtracted. Since the light flavor Wilson coefficients were calculated in the $\overline{\mathsf{MS}}$-scheme, the same scheme has to be used for the massive OMEs. It should also be thoroughly used for renormalization to derive consistent results in QCD analyzes of deep-inelastic scattering data and to be able to compare to other analyzes directly. This means that one has to take special attendance of which scheme for the definition of a_s was used. In Sect. 4.4 we will describe a scheme for a_s, to which one is naturally led in the course of renormalization. We refer to this scheme as MOM-scheme and present the transformation formula to the $\overline{\mathsf{MS}}$ as well. How this affects the asymptotic heavy flavor Wilson coefficients is described in Sect. 5.1, where we compare Eqs. 3.26–3.30 to those presented in Ref. [44].

3.3 Heavy Quark Parton Densities

The FFNS forms a general starting point to describe and to calculate the heavy flavor contributions to the DIS structure functions. Approaching higher values of Q^2, one may think of the heavy quark becoming effectively light and thus acquiring an own parton density. Different variable flavor scheme treatments were considered in the past, cf. e.g. [61, 62]. Here we follow [45] to obtain a description in complete accordance with the renormalization group in the $\overline{\mathsf{MS}}$-scheme. In the kinematic region in which the factorization relation (3.16) holds, one may redefine the results obtained in the FFNS, which allows for a partonic description at the level of (n_f+1) flavors.

In the strict sense, only massless particles can be interpreted as partons in hard scattering processes since the lifetime of these quantum-fluctuations off the hadronic background $\tau_{\mathsf{life}} \propto 1/(k_\perp^2 + m_Q^2)$ has to be large against the interaction time $\tau_{\mathsf{int}} \propto 1/Q^2$ in the infinite momentum frame, [63], cf. also Sect. 2.2.1. In the massive case, τ_{life} is necessarily finite and there exists a larger scale Q_0^2 below which any partonic description fails. From this it follows, that the heavy quark effects are genuinely described by the process dependent Wilson coefficients. Since parton-densities are process independent quantities, only process independent pieces out of the Wilson coefficients can be used to define them for heavy quarks at all. Clearly this is impossible in the region close to threshold but requires $Q^2/m_Q^2 = r \gg 1$, with $r \gtrsim 10$ in case of $F_2(x, Q^2)$. For $F_L(x, Q^2)$ the corresponding ratio even turns out to be $r \gtrsim 800$ [25, 44, 59]. Heavy flavor parton distributions can thus be constructed only for scales $\mu^2 \gg m_Q^2$. This is done under the further assumption that for the other heavy flavors the masses m_{Q_i} form a hierarchy $m_{Q_1}^2 \ll m_{Q_2}^2 \ll$ etc. Their use in observables is restricted to a region, in which the power corrections can be safely neglected. This range may strongly depend on the observable considered as the examples of F_2 and F_L show. Also in case of the structure functions associated to transverse virtual gauge boson polarizations, like $F_2(x, Q^2)$, the factorization (3.16) only occurs far above threshold, $Q^2_{\rm thr} \sim 4m_Q^2 x/(1-x)$, and at even larger scales for $F_L(x, Q^2)$.

In order to maintain the process independence of the parton distributions, we define them for $(n_f + 1)$ flavors from the light flavor parton distribution functions for n_f flavors together with the massive operator matrix elements. The following set of parton densities is obtained in Mellin-space, [45] :

$$\begin{aligned}
&f_k(n_f+1, N, \mu^2, m^2) + f_{\overline{k}}(n_f+1, N, \mu^2, m^2) \\
&= A_{qq,Q}^{\rm NS}\left(N, n_f+1, \frac{\mu^2}{m^2}\right) \cdot \left[f_k(n_f, N, \mu^2) + f_{\overline{k}}(n_f, N, \mu^2)\right] \\
&\quad + \frac{1}{n_f} A_{qq,Q}^{\rm PS}\left(N, n_f+1, \frac{\mu^2}{m^2}\right) \cdot \Sigma(n_f, N, \mu^2) \\
&\quad + \frac{1}{n_f} A_{qg,Q}\left(N, n_f+1, \frac{\mu^2}{m^2}\right) \cdot G(n_f, N, \mu^2),
\end{aligned} \tag{3.31}$$

$$\begin{aligned}
&f_Q(n_f+1, N, \mu^2, m^2) + f_{\overline{Q}}(n_f+1, N, \mu^2, m^2) \\
&= A_{Qq}^{\rm PS}\left(N, n_f+1, \frac{\mu^2}{m^2}\right) \cdot \Sigma(n_f, N, \mu^2) \\
&\quad + A_{Qg}\left(N, n_f+1, \frac{\mu^2}{m^2}\right) \cdot G(n_f, N, \mu^2).
\end{aligned} \tag{3.32}$$

The flavor singlet, non-singlet and gluon densities for $(n_f + 1)$ flavors are given by

$$\begin{aligned}
\Sigma(n_f+1, N, \mu^2, m^2) = &\left[A_{qq,Q}^{\rm NS}\left(N, n_f+1, \frac{\mu^2}{m^2}\right) + A_{qq,Q}^{\rm PS}\left(N, n_f+1, \frac{\mu^2}{m^2}\right)\right. \\
&\left. + A_{Qq}^{\rm PS}\left(N, n_f+1, \frac{\mu^2}{m^2}\right)\right] \cdot \Sigma(n_f, N, \mu^2) \\
&+ \left[A_{qg,Q}\left(N, n_f+1, \frac{\mu^2}{m^2}\right) + A_{Qg}\left(N, n_f+1, \frac{\mu^2}{m^2}\right)\right] \\
&\times G(n_f, N, \mu^2),
\end{aligned} \tag{3.33}$$

$$\begin{aligned}\Delta_k(n_f+1, N, \mu^2, m^2) = & f_k(n_f+1, N, \mu^2, m^2) \\ & + f_{\overline{k}}(n_f+1, N, \mu^2, m^2) \\ & - \frac{1}{n_f+1}\Sigma(n_f+1, N, \mu^2, m^2), \end{aligned} \tag{3.34}$$

$$\begin{aligned}G(n_f+1, N, \mu^2, m^2) = & A_{gq,Q}\left(N, n_f+1, \frac{\mu^2}{m^2}\right) \cdot \Sigma(n_f, N, \mu^2) \\ & + A_{gg,Q}\left(N, n_f+1, \frac{\mu^2}{m^2}\right) \cdot G(n_f, N, \mu^2). \end{aligned} \tag{3.35}$$

Note, that the new parton densities depend on the renormalized heavy quark mass $m^2 = m^2(a_s^2(\mu^2))$. As will be outlined in Chaps. 4, 5, the corresponding relations for the operator matrix elements depend on the mass-renormalization scheme. This has to be taken into account in QCD-analyzes, in particular, m^2 cannot be chosen constant. The quarkonic and gluonic operators obtained in the light-cone expansion can be normalized arbitrarily. It is, however, convenient to chose the relative factor such, that the non-perturbative nucleon-state expectation values, $\Sigma(n_f, N, \mu^2)$ and $G(n_f, N, \mu^2)$, obey

$$\Sigma(n_f, N=2, \mu^2) + G(n_f, N=2, \mu^2) = 1 \tag{3.36}$$

due to 4-momentum conservation. As a consequence, the OMEs fulfill the relations, [45],

$$A_{qq,Q}^{\mathsf{NS}}(N=2) + A_{qq,Q}^{\mathsf{PS}}(N=2) + A_{Qq}^{\mathsf{PS}}(N=2) + A_{gq,Q}(N=2) = 1, \tag{3.37}$$

$$A_{qg,Q}(N=2) + A_{Qg}(N=2) + A_{gg,Q}(N=2) = 1. \tag{3.38}$$

The above scenario can be easily followed up to 2-loop order. Also here diagrams contribute which carry two different heavy quark flavors. At this level, the additional heavy degree of freedom may be absorbed into the coupling constant and thus decoupled temporarily. Beginning with 3-loop order the situation becomes more involved since there are graphs in which two different heavy quark flavors occur in nested topologies, i.e., the corresponding diagrams depend on the ratio $\rho = m_c^2/m_b^2$ yielding power corrections in ρ. There is no strong hierarchy between these two masses. The above picture, leading to heavy flavor parton distributions whenever $Q^2 \gg m^2$ will not hold anymore, since, in case of the two-flavor graphs, one cannot decide immediately whether they belong to the c- or the b-quark distribution. Hence, the partonic description can only be maintained within a certain approximation by assuming $\rho \ll 1$.

Conversely, one may extend the kinematic regime for deep-inelastic scattering to define the distribution functions (3.31)–(3.35) upon knowing the power corrections which occur in the heavy flavor Wilson coefficients $\mathbf{H}_{i,j} = H_{i,j}, L_{i,j}$. This is the case for 2-loop order. We separate

Chapter 4
Renormalization of Composite Operator Matrix Elements

Before renormalizing the massive OMEs, they have to be calculated applying a suitable regularization scheme, for which we apply dimensional regularization in $D=4+\varepsilon$ dimensions, see Sect. 4.1. The unrenormalized massive OMEs are then denoted by a double–hat and are expanded into a perturbative series in the bare coupling constant $\hat{a}_s$ [1] via

$$\begin{aligned}\hat{\hat{A}}_{ij}\Big(\frac{\hat{m}^2}{\mu^2},\varepsilon,N\Big)&=\delta_{ij}+\sum_{l=1}^{\infty}\hat{a}_s^l\hat{\hat{A}}_{ij}^{(l)}\Big(\frac{\hat{m}^2}{\mu^2},\varepsilon,N\Big)\\&=\delta_{ij}+\sum_{l=1}^{\infty}\hat{a}_s^l\Big(\frac{\hat{m}^2}{\mu^2}\Big)^{l\varepsilon/2}\hat{\hat{A}}_{ij}^{(l)}\Big(\hat{m}^2=\mu^2,\varepsilon,N\Big)\end{aligned}\tag{4.1}$$

The OMEs in Eq. 4.1 depend on ε, the Mellin–Parameter N, the bare mass $\hat{m}$ and the renormalization scale $\mu=\mu_R$. Also the factorization scale μ_F will be identified with μ in the following. Note that in the last line of (4.1), the dependence on the ratio of the mass and the renormalization scale was made explicit for each order in $\hat{a}_s$. The possible values of the indices ij have been described in Sect. 3.2, below Eq. (3.17).

The factorization between the massive OMEs and the massless Wilson coefficients (3.16) requires the external legs of the operator matrix elements to be on–shell,

$$p^2=0,\tag{4.2}$$

where p denotes the external momentum. Unlike in the massless case, where the scale of the OMEs is set by an off–shell momentum $-p^2<0$, in our framework the internal heavy quark mass yields the scale. In the former case, one observes a mixing of the physical OMEs with non-gauge invariant (NGI) operators, cf. [1–3],

[1] We would like to remind the reader of the definition of the hat–symbol for a function f, Eq. 3.20, which is not to be confused with the hat–symbol denoting unrenormalized quantities

S. Klein, *Charm Production in Deep Inelastic Scattering*, Springer Theses,
DOI: 10.1007/978-3-642-23286-2_4, © Springer-Verlag Berlin Heidelberg 2012

and contributions originating in the violation of the equations of motion (EOM). Terms of this kind do not contribute in the present case, as will be discussed in Sect. 4.2.

Renormalizing the OMEs then consists of four steps. First, mass and charge renormalization have to performed. The former is done in the on–mass–shell–scheme and described in Sect. 4.3. For the latter, we present the final result in the $\overline{\mathsf{MS}}$–scheme, but in an intermediate step, we adopt an on–shell subtraction scheme (MOM–scheme) for the gluon propagator, cf. Sect. 4.4. This is necessary to maintain condition (4.2), i.e., to keep the external massless partons on–shell. Note, that there are other, differing MOM–schemes used in the literature, cf. e.g. [4].

After mass and coupling constant renormalization, we denote the OMEs with a single hat, $\hat{A}_{ij}$. The remaining singularities are then connected to the composite operators and the particle kinematics of the corresponding Feynman–diagrams. One can distinguish between ultraviolet (UV) and collinear (C) divergences. In Sect. 4.5, we describe how the former are renormalized via the operator Z–factors. The UV–finite OMEs are denoted by a bar, $\bar{A}_{ij}$. Finally, the C–divergences are removed via mass factorization, cf. Sect. 4.6. The renormalized OMEs are then denoted by A_{ij}. Section 4.7 contains the general structure of the massive OMEs up to $O(a_s^3)$ in terms of renormalization constants and lower order contributions.

4.1 Regularization Scheme

When evaluating momentum integrals of Feynman diagrams in $D=4$ dimensions, one encounters singularities, which have to be regularized. A convenient method is to apply D-dimensional regularization, [5–8]. The dimensionality of space–time is analytically continued to values $D \neq 4$, for which the corresponding integrals converge. After performing a Wick rotation, integrals in Euclidean space of the form

$$\int \frac{d^D k}{(2\pi)^D} \frac{(k^2)^r}{(k^2+R^2)^m} = \frac{1}{(4\pi)^{D/2}} \frac{\Gamma(r+D/2)\Gamma(m-r-D/2)}{\Gamma(D/2)\Gamma(m)} (R^2)^{r+D/2-m} \tag{4.3}$$

are obtained. Note that within dimensional regularization, this integral vanishes if $R=0$, i.e., if it does not contain a scale, [9]. The properties of the Γ–function in the complex plane are well known, see Appendix C. Therefore one can analytically continue the right-hand side of Eq. 4.3 from integer values of D to arbitrary complex values. In order to recover the physical space–time dimension, we set $D=4+\varepsilon$. The singularities can now be isolated by expanding the Γ–functions into Laurent-series around $\varepsilon=0$. Note that this method regularizes both UV- and C-singularities and one could in principle distinguish their origins by a label, $\varepsilon_{UV}, \varepsilon_C$, but we treat all singularities by a common parameter ε in the following. Additionally, all other quantities have to be considered in D dimensions. This applies for the metric tensor $g_{\mu\nu}$ and the Clifford-Algebra of γ–matrices, see Appendix A. Also the bare coupling constant $\hat{g}_s$, which is dimensionless in $D = 4$, has to be continued to D dimensions. Due to this it acquires the dimension of mass,

$$\hat{g}_{s,D} = \mu^{-\varepsilon/2} \hat{g}_s, \tag{4.4}$$

which is described by a scale μ corresponding to the renormalization scale in Eq. 4.1. From now on, Eq. 4.4 is understood to have been applied and we set

$$\frac{\hat{g}_s^2}{(4\pi)^2} = \hat{a}_s. \tag{4.5}$$

Dimensional regularization has the advantage, unlike the Pauli–Villars regularization, [10], that it obeys all physical requirements such as Lorentz-invariance, gauge invariance and unitarity, [5, 11]. Hence it is suitable to be applied in perturbative calculations in quantum field theory including Yang–Mills fields.

Using dimensional regularization, the poles of the unrenormalized results appear as terms $1/\varepsilon^i$, where in the calculations in this thesis i can run from 1 to the number of loops. In order to remove these singularities, one has to perform renormalization and mass factorization. To do this, a suitable scheme has to be chosen. The most commonly used schemes in perturbation theory are the MS-scheme, [12], and the $\overline{\text{MS}}$-scheme, [13], to which we will refer in the following.

In the MS-scheme only the pole terms in ε are subtracted. More generally, the $\overline{\text{MS}}$-scheme makes use of the observation that $1/\varepsilon$–poles always appear in combination with the spherical factor

$$S_\varepsilon \equiv \exp\left[\frac{\varepsilon}{2}(\gamma_E - \ln(4\pi))\right], \tag{4.6}$$

which may be bracketed out for each loop order. Here γ_E denotes the Euler-Mascheroni constant

$$\gamma_E \equiv \lim_{N\to\infty}\left(\sum_{k=1}^{N}\frac{1}{k} - \ln(N)\right) \approx 0.577215664901\ldots. \tag{4.7}$$

By subtracting the poles in the form $S_\varepsilon/\varepsilon$ in the $\overline{\text{MS}}$-scheme, no terms containing $\ln^k(4\pi)$, γ_E^k will appear in the renormalized result, simplifying the expression. This is due to the fact that for a k–loop calculation, one will always obtain the overall term

$$\frac{\Gamma(1 - k\frac{\varepsilon}{2})}{(4\pi)^{\frac{k\varepsilon}{2}}} = S_\varepsilon^k \exp\left(\sum_{i=2}^{\infty}\frac{\zeta_i}{i}\left(\frac{k\varepsilon}{2}\right)^i\right), \tag{4.8}$$

with ζ_i being Riemann's ζ–values, cf. Appendix C. In the following, we will always assume that the $\overline{\text{MS}}$-scheme is applied and set $S_\varepsilon \equiv 1$.

4.2 Projectors

We consider the expectation values of the local operators (2.86)–(2.88) between partonic states j

$$G_{ij,Q} = \langle j|O_i|j\rangle_Q. \tag{4.9}$$

Here, $i,\ j = q,\ g$ and the subscript Q denotes the presence of one heavy quark. In case of massless QCD, one has to take the external parton j of momentum p off–shell, $p^2 < 0$, which implies that the OMEs derived from Eq. 4.9 are not gauge invariant. As has been outlined in Ref. [14], they acquire unphysical parts which are due to the breakdown of the EOM and the mixing with additional NGI operators. The EOM terms may be dealt with by applying a suitable projection operator to eliminate them, [14]. The NGI terms are more difficult to deal with, since they affect the renormalization constants and one has to consider additional ghost—and alien—OMEs, see [1–3, 14, 15] for details.

In the case of massive OMEs, these difficulties do not occur. The external particles are massless and taken to be on–shell. Hence the EOM are not violated. Additionally, the OMEs remain gauge invariant quantities, since the external states are physical and therefore no mixing with NGI–operators occurs, [1, 9, 14, 15].

The computation of the Green's functions will reveal trace terms which do not contribute since the local operators are traceless and symmetric under the Lorentz group. It is convenient to project these terms out from the beginning by contracting with an external source term

$$J_N \equiv \Delta_{\mu_1} \ldots \Delta_{\mu_N}. \tag{4.10}$$

Here Δ_μ is a light-like vector, $\Delta^2 = 0$. In this way, the Feynman–rules for composite operators can be derived, cf. Appendix B. In addition, one has to amputate the external field. Note that we nonetheless choose to renormalize the mass and the coupling multiplicative and include self–energy insertions containing massive lines on external legs into our calculation. The Green's functions in momentum space corresponding to the OMEs with external gluons are then given by

$$\epsilon^\mu(p) G^{ab}_{Q,\mu\nu} \epsilon^\nu(p) = \epsilon^\mu(p) J_N \langle A^a_\mu(p)|O_{Q;\mu_1\ldots\mu_N}|A^b_\nu(p)\rangle \epsilon^\nu(p), \tag{4.11}$$

$$\epsilon^\mu(p) G^{ab}_{q,Q,\mu\nu} \epsilon^\nu(p) = \epsilon^\mu(p) J_N \langle A^a_\mu(p)|O_{q;\mu_1\ldots\mu_N}|A^b_\nu(p)\rangle_Q \epsilon^\nu(p), \tag{4.12}$$

$$\epsilon^\mu(p) G^{ab}_{g,Q,\mu\nu} \epsilon^\nu(p) = \epsilon^\mu(p) J_N \langle A^a_\mu(p)|O_{g;\mu_1\ldots\mu_N}|A^b_\nu(p)\rangle_Q \epsilon^\nu(p). \tag{4.13}$$

In Eqs. 4.11–4.13, A^a_μ denote the external gluon fields with color index a, Lorentz index μ and momentum p. The polarization vector of the external gluon is given by $\epsilon^\mu(p)$. Note that in Eq. 4.11, the operator couples to the heavy quark.

In Eqs. 4.12, 4.13 it couples to a light quark or gluon, respectively, with the heavy quark still being present in virtual loops.

In the flavor non–singlet case, there is only one term which reads

$$\overline{u}(p,s)G^{ij,\mathsf{NS}}_{q,Q}\lambda_r u(p,s)=J_N\langle\overline{\Psi}_i(p)|O^{\mathsf{NS}}_{q,r;\mu_1\ldots\mu_N}|\Psi^j(p)\rangle_Q, \tag{4.14}$$

with $u(p,s)$, $\overline{u}(p,s)$ being the bi–spinors of the external massless quark and anti–quark, respectively. The remaining Green's functions with an outer quark are given by

$$\overline{u}(p,s)G^{ij}_{Q}u(p,s)=J_N\langle\overline{\Psi}_i(p)|O_{Q,\mu_1\ldots\mu_N}|\Psi^j(p)\rangle, \tag{4.15}$$

$$\overline{u}(p,s)G^{ij}_{q,Q}u(p,s)=J_N\langle\overline{\Psi}_i(p)|O_{q,\mu_1\ldots\mu_N}|\Psi^j(p)\rangle_Q, \tag{4.16}$$

$$\overline{u}(p,s)G^{ij}_{g,Q}u(p,s)=J_N\langle\overline{\Psi}_i(p)|O_{g,\mu_1\ldots\mu_N}|\Psi^j(p)\rangle_Q, \tag{4.17}$$

Note that in the quarkonic case the fields $\overline{\Psi}$, Ψ with color indices *i,j* stand for the external light quarks only. Further, we remind that the S—contributions are split up according to Eq. (3.18), which is of relevance for Eq. (4.16).

The above tensors have the general form, cf. [14, 16],

$$\hat{G}^{ab}_{Q,\mu\nu}=\hat{\hat{A}}_{Qg}\left(\frac{\hat{m}^2}{\mu^2},\varepsilon,N\right)\delta^{ab}(\Delta\cdot p)^N\left[-g_{\mu\nu}+\frac{p_\mu\Delta_\nu+\Delta_\mu p_\nu}{\Delta\cdot p}\right], \tag{4.18}$$

$$\hat{G}^{ab}_{l,Q,\mu\nu}=\hat{\hat{A}}_{lg,Q}\left(\frac{\hat{m}^2}{\mu^2},\varepsilon,N\right)\delta^{ab}(\Delta\cdot p)^N\left[-g_{\mu\nu}+\frac{p_\mu\Delta_\nu+\Delta_\mu p_\nu}{\Delta\cdot p}\right],\quad l=g,q, \tag{4.19}$$

$$\hat{G}^{ij}_{Q}=\hat{\hat{A}}^{\mathsf{PS}}_{Qq}\left(\frac{\hat{m}^2}{\mu^2},\varepsilon,N\right)\delta^{ij}(\Delta\cdot p)^{N-1}\not{\Delta}, \tag{4.20}$$

$$\hat{G}^{ij,r}_{l,Q}=\hat{\hat{A}}^{r}_{lq,Q}\left(\frac{\hat{m}^2}{\mu^2},\varepsilon,N\right)\delta^{ij}(\Delta\cdot p)^{N-1}\not{\Delta},\quad l=g,q,\quad r=\mathsf{S},\ \mathsf{NS},\ \mathsf{PS} \tag{4.21}$$

Here, we have denoted the Green's function with a hat to signify that the above equations are written on the unrenormalized level. In order to simplify the evaluation, it is useful to define projection operators which, applied to the Green's function, yield the corresponding OME. For outer gluons, one defines

$$P^{(1)}_g\hat{G}^{ab}_{l,(Q),\mu\nu}\equiv-\frac{\delta_{ab}}{N_c^2-1}\frac{g^{\mu\nu}}{D-2}(\Delta\cdot p)^{-N}\hat{G}^{ab}_{l,(Q),\mu\nu}, \tag{4.22}$$

We briefly summarize the main steps in the massless case for n_f flavors in the $\overline{\mathsf{MS}}$–scheme. The bare coupling constant $\hat{a}_s$ is expressed by the renormalized coupling $a_s^{\overline{\mathsf{MS}}}$ via

$$\begin{aligned}\hat{a}_s &= Z_g^{\overline{\mathsf{MS}}^2}(\varepsilon, n_f) a_s^{\overline{\mathsf{MS}}}(\mu^2) \\ &= a_s^{\overline{\mathsf{MS}}}(\mu^2)\left[1 + \delta a_{s,1}^{\overline{\mathsf{MS}}}(n_f) a_s^{\overline{\mathsf{MS}}}(\mu^2) + \delta a_{s,2}^{\overline{\mathsf{MS}}}(n_f) a_s^{\overline{\mathsf{MS}}^2}(\mu^2)\right] + O\left(a_s^{\overline{\mathsf{MS}}^3}\right).\end{aligned} \quad (4.33)$$

The coefficients in Eq. (4.33) are, [23–26] and [27, 28],

$$\delta a_{s,1}^{\overline{\mathsf{MS}}}(n_f) = \frac{2}{\varepsilon}\beta_0(n_f), \quad (4.34)$$

$$\delta a_{s,2}^{\overline{\mathsf{MS}}}(n_f) = \frac{4}{\varepsilon^2}\beta_0^2(n_f) + \frac{1}{\varepsilon}\beta_1(n_f), \quad (4.35)$$

with

$$\beta_0(n_f) = \frac{11}{3}C_A - \frac{4}{3}T_F n_f, \quad (4.36)$$

$$\beta_1(n_f) = \frac{34}{3}C_A^2 - 4\left(\frac{5}{3}C_A + C_F\right)T_F n_f. \quad (4.37)$$

From the above equations, one can determine the β–function, Eq. 2.103, which describes the running of the strong coupling constant and leads to asymptotic freedom in case of QCD, [23, 24]. It can be calculated using the fact that the bare strong coupling constant does not depend on the renormalization scale μ. Using Eq. 4.4, one obtains

$$0 = \frac{d\hat{a}_{s,D}}{d\ln\mu^2} = \frac{d}{d\ln\mu^2}\hat{a}_s\mu^{-\varepsilon} = \frac{d}{d\ln\mu^2}a_s(\mu^2)Z_g^2(\varepsilon, n_f, \mu^2)\mu^{-\varepsilon}, \quad (4.38)$$

$$\Longrightarrow \beta = \frac{\varepsilon}{2}a_s(\mu^2) - 2a_s(\mu^2)\frac{d}{d\ln\mu^2}\ln Z_g(\varepsilon, n_f, \mu^2). \quad (4.39)$$

Note that in Eq. 4.39 we have not specified a scheme yet and kept a possible μ–dependence for Z_g, which is not present in case of the $\overline{\mathsf{MS}}$–scheme. From (4.39), one can calculate the expansion coefficients of the β–function. Combining it with the result for $Z_g^{\overline{\mathsf{MS}}}$ in Eqs. 4.34, 4.35, one obtains in the $\overline{MS}$-scheme for n_f light flavors, cf. [23–28],

$$\beta^{\overline{\mathsf{MS}}}(n_f) = -\beta_0(n_f)a_s^{\overline{\mathsf{MS}}^2} - \beta_1(n_f)a_s^{\overline{\mathsf{MS}}^3} + O\left(a_s^{\overline{\mathsf{MS}}^4}\right). \quad (4.40)$$

Additionally, it follows

$$\frac{da_s(\mu^2)}{d\ln(\mu^2)} = \frac{1}{2}\varepsilon a_s(\mu^2) - \sum_{k=0}^{\infty} \beta_k a_s^{k+2}(\mu^2). \tag{4.41}$$

The factorization relation (3.16) strictly requires that the external massless particles are on–shell. Massive loop corrections to the gluon propagator violate this condition, which has to be enforced subtracting the corresponding corrections. These can be uniquely absorbed into the strong coupling constant applying the background field method, [29–31], to maintain the Slavnov-Taylor identities of QCD. We thus determine the coupling constant renormalization in the $\overline{\mathsf{MS}}$-scheme as far as the light flavors and the gluon are concerned. In addition, we make the choice that the heavy quark decouples in the running coupling constant $a_s(\mu^2)$ for $\mu^2 < m^2$ and thus from the renormalized OMEs. This implies the requirement that $\Pi_H(0, m^2) = 0$, where $\Pi_H(p^2, m^2)$ is the contribution to the gluon self-energy due to the heavy quark loops, [16]. Since this condition introduces higher order terms in ε into Z_g, we left the $\overline{\mathsf{MS}}$–scheme. This new scheme is a MOM–scheme. After mass renormalization in the on–shell–scheme via Eq. 4.27, we obtain for the heavy quark contributions to the gluon self–energy in the background field formalism

$$\begin{aligned}
&\hat{\Pi}^{\mu\nu}_{H,ab,BF}(p^2, m^2, \mu^2, \varepsilon, \hat{a}_s) \\
&\quad = i(-p^2 g^{\mu\nu} + p^\mu p^\nu)\delta_{ab}\hat{\Pi}_{H,BF}(p^2, m^2, \mu^2, \varepsilon, \hat{a}_s), \\
&\hat{\Pi}_{H,BF}(0, m^2, \mu^2, \varepsilon, \hat{a}_s) \\
&\quad = \hat{a}_s \frac{2\beta_{0,Q}}{\varepsilon}\left(\frac{m^2}{\mu^2}\right)^{\varepsilon/2} \exp\Bigl(\sum_{i=2}^{\infty} \frac{\zeta_i}{i}\left(\frac{\varepsilon}{2}\right)^i\Bigr) \\
&\qquad + \hat{a}_s^2\left(\frac{m^2}{\mu^2}\right)^{\varepsilon}\Biggl[\frac{1}{\varepsilon}\Bigl(-\frac{20}{3}T_F C_A - 4T_F C_F\Bigr) - \frac{32}{9}T_F C_A + 15 T_F C_F \\
&\qquad\qquad + \varepsilon\Bigl(-\frac{86}{27}T_F C_A - \frac{31}{4}T_F C_F - \frac{5}{3}\zeta_2 T_F C_A - \zeta_2 T_F C_F\Bigr)\Biggr],
\end{aligned} \tag{4.42}$$

with

$$\beta_{0,Q} = \hat{\beta}_0(n_f) = -\frac{4}{3}T_F. \tag{4.43}$$

Note that Eq. 4.42 holds only up to order $O(\varepsilon)$, although we have partially included higher orders in ε in order to keep the expressions shorter. We have used the Feynman–rules of the background field formalism as given in Ref. [32]. In the following, we define

$$f(\varepsilon) \equiv \left(\frac{m^2}{\mu^2}\right)^{\varepsilon/2} \exp\Bigl(\sum_{i=2}^{\infty} \frac{\zeta_i}{i}\left(\frac{\varepsilon}{2}\right)^i\Bigr). \tag{4.44}$$

The renormalization constant of the background field Z_A is related to Z_g via

$$Z_A = Z_g^{-2}. \tag{4.45}$$

The light flavor contributions to Z_A, $Z_{A,l}$, can thus be determined by combining Eqs. 4.34, 4.35, 4.45. The heavy flavor part follows from the condition

$$\Pi_{H,BF}(0, m^2) + Z_{A,H} \equiv 0, \tag{4.46}$$

which ensures that the on–shell gluon remains strictly massless. Thus we newly define the renormalization constant of the strong coupling with n_f light and one heavy flavor as

$$Z_g^{\mathrm{MOM}}(\varepsilon, n_f + 1, \mu^2, m^2) \equiv \frac{1}{(Z_{A,l} + Z_{A,H})^{1/2}} \tag{4.47}$$

and obtain

$$\begin{aligned} Z_g^{\mathrm{MOM}^2}(\varepsilon, n_f + 1, \mu^2, m^2) = 1 &+ a_s^{\mathrm{MOM}}(\mu^2)\Big[\frac{2}{\varepsilon}(\beta_0(n_f) + \beta_{0,Q} f(\varepsilon))\Big] \\ &+ a_s^{\mathrm{MOM}^2}(\mu^2)\Big[\frac{\beta_1(n_f)}{\varepsilon} + \frac{4}{\varepsilon^2}(\beta_0(n_f) + \beta_{0,Q} f(\varepsilon))^2 \\ &+ \frac{1}{\varepsilon}\left(\frac{m^2}{\mu^2}\right)^{\varepsilon}\left(\beta_{1,Q} + \varepsilon\beta_{1,Q}^{(1)} + \varepsilon^2\beta_{1,Q}^{(2)}\right)\Big] + O\left(a_s^{\mathrm{MOM}^3}\right), \end{aligned} \tag{4.48}$$

with

$$\beta_{1,Q} = \hat{\beta}_1(n_f) = -4\left(\frac{5}{3}C_A + C_F\right)T_F, \tag{4.49}$$

$$\beta_{1,Q}^{(1)} = -\frac{32}{9}T_F C_A + 15 T_F C_F, \tag{4.50}$$

$$\beta_{1,Q}^{(2)} = -\frac{86}{27}T_F C_A - \frac{31}{4}T_F C_F - \zeta_2\left(\frac{5}{3}T_F C_A + T_F C_F\right). \tag{4.51}$$

The coefficients corresponding to Eq. 4.33 then read in the MOM–scheme

$$\delta a_{s,1}^{\mathrm{MOM}} = \Big[\frac{2\beta_0(n_f)}{\varepsilon} + \frac{2\beta_{0,Q}}{\varepsilon} f(\varepsilon)\Big], \tag{4.52}$$

$$\begin{aligned} \delta a_{s,2}^{\mathrm{MOM}} = \Big[&\frac{\beta_1(n_f)}{\varepsilon} + \left(\frac{2\beta_0(n_f)}{\varepsilon} + \frac{2\beta_{0,Q}}{\varepsilon} f(\varepsilon)\right)^2 \\ &+ \frac{1}{\varepsilon}\left(\frac{m^2}{\mu^2}\right)^{\varepsilon}\left(\beta_{1,Q} + \varepsilon\beta_{1,Q}^{(1)} + \varepsilon^2\beta_{1,Q}^{(2)}\right)\Big] + O(\varepsilon^2). \end{aligned} \tag{4.53}$$

Since the $\overline{\mathrm{MS}}$–scheme is commonly used, we transform our results back from the MOM–description into the $\overline{\mathrm{MS}}$–scheme, in order to be able to compare to other analyzes. This is achieved by observing that the bare coupling does not change under this transformation and one obtains the condition

$$Z_g^{\overline{\mathsf{MS}}^2}(\varepsilon, n_f+1) a_s^{\overline{\mathsf{MS}}}(\mu^2) = Z_g^{\mathrm{MOM}^2}(\varepsilon, n_f+1, \mu^2, m^2) a_s^{\mathrm{MOM}}(\mu^2). \qquad (4.54)$$

The following relations hold :

$$a_s^{\mathrm{MOM}} = a_s^{\overline{\mathsf{MS}}} - \beta_{0,Q} \ln\left(\frac{m^2}{\mu^2}\right) a_s^{\overline{\mathsf{MS}}^2} + \left[\beta_{0,Q}^2 \ln^2\left(\frac{m^2}{\mu^2}\right) - \beta_{1,Q} \ln\left(\frac{m^2}{\mu^2}\right) - \beta_{1,Q}^{(1)}\right] a_s^{\overline{\mathsf{MS}}^3} + O\left(a_s^{\overline{\mathsf{MS}}^4}\right), \qquad (4.55)$$

or,

$$a_s^{\overline{\mathsf{MS}}} = a_s^{\mathrm{MOM}} + a_s^{\mathrm{MOM}^2}\left(\delta a_{s,1}^{\mathrm{MOM}} - \delta a_{s,1}^{\overline{\mathsf{MS}}}(n_f+1)\right) + a_s^{\mathrm{MOM}^3} \times \left(\delta a_{s,2}^{\mathrm{MOM}} - \delta a_{s,2}^{\overline{\mathsf{MS}}}(n_f+1) - 2\delta a_{s,1}^{\overline{\mathsf{MS}}}(n_f+1)\left[\delta a_{s,1}^{\mathrm{MOM}} - \delta a_{s,1}^{\overline{\mathsf{MS}}}(n_f+1)\right]\right) + O\left(a_s^{\mathrm{MOM}^4}\right), \qquad (4.56)$$

vice versa. Equation 4.56 is valid to all orders in ε. Here, $a_s^{\overline{\mathsf{MS}}} = a_s^{\overline{\mathsf{MS}}}(n_f+1)$. Applying the on–shell–scheme for mass renormalization and the described MOM–scheme for the renormalization of the coupling, one obtains as general formula for mass and coupling constant renormalization up to $O\left(a_s^{\mathrm{MOM}^3}\right)$

$$\begin{aligned}\hat{A}_{ij} = {} & \delta_{ij} + a_s^{\mathrm{MOM}} \hat{\hat{A}}_{ij}^{(1)} + a_s^{\mathrm{MOM}^2}\left[\hat{\hat{A}}_{ij}^{(2)} + \delta m_1 \left(\frac{m^2}{\mu^2}\right)^{\varepsilon/2} m \frac{d}{dm} \hat{\hat{A}}_{ij}^{(1)} + \delta a_{s,1}^{\mathrm{MOM}} \hat{\hat{A}}_{ij}^{(1)}\right] \\ & + a_s^{\mathrm{MOM}^3}\left[\hat{\hat{A}}_{ij}^{(3)} + \delta a_{s,2}^{\mathrm{MOM}} \hat{\hat{A}}_{ij}^{(1)} + 2\delta a_{s,1}^{\mathrm{MOM}}\left(\hat{\hat{A}}_{ij}^{(2)} + \delta m_1 \left(\frac{m^2}{\mu^2}\right)^{\varepsilon/2} m \frac{d}{dm} \hat{\hat{A}}_{ij}^{(1)}\right)\right. \\ & + \delta m_1 \left(\frac{m^2}{\mu^2}\right)^{\varepsilon/2} m \frac{d}{dm} \hat{\hat{A}}_{ij}^{(2)} + \delta m_2 \left(\frac{m^2}{\mu^2}\right)^{\varepsilon} m \frac{d}{dm} \hat{\hat{A}}_{ij}^{(1)} \\ & \left. + \frac{\delta m_1^2}{2}\left(\frac{m^2}{\mu^2}\right)^{\varepsilon} m^2 \frac{d^2}{dm^2} \hat{\hat{A}}_{ij}^{(1)}\right],\end{aligned} \qquad (4.57)$$

where we have suppressed the dependence on m, ε and N in the arguments.[3]

4.5 Operator Renormalization

The renormalization of the UV singularities of the composite operators is being performed introducing the corresponding Z_{ij}-factors, which have been defined in

[3] Here we corrected a typographical error in [17], Eq. 4.48.

Eqs. 2.157, 2.106. We consider first only n_f massless flavors, cf. [14], and do then include subsequently one heavy quark. In the former case, renormalization proceeds in the $\overline{\mathsf{MS}}$–scheme via

$$A_{qq}^{\mathsf{NS}}\Big(\frac{-p^2}{\mu^2}, a_s^{\overline{\mathsf{MS}}}, n_f, N\Big) = Z_{qq}^{-1,\mathsf{NS}}(a_s^{\overline{\mathsf{MS}}}, n_f, \varepsilon, N)\hat{A}_{qq}^{\mathsf{NS}}\Big(\frac{-p^2}{\mu^2}, a_s^{\overline{\mathsf{MS}}}, n_f, \varepsilon, N\Big), \tag{4.58}$$

$$A_{ij}\Big(\frac{-p^2}{\mu^2}, a_s^{\overline{\mathsf{MS}}}, n_f, N\Big) = Z_{il}^{-1}(a_s^{\overline{\mathsf{MS}}}, n_f, \varepsilon, N)\hat{A}_{lj}\left(\frac{-p^2}{\mu^2}, a_s^{\overline{\mathsf{MS}}}, n_f, \varepsilon, N\right),\ i, j = q, g, \tag{4.59}$$

with p a space–like momentum. As is well known, operator mixing occurs in the singlet case, Eq. 4.59. As mentioned before, we neglected all terms being associated to EOM and NGI parts, since they do not contribute in the renormalization of the massive on–shell operator matrix elements. The NS and PS contributions are separated via

$$Z_{qq}^{-1} = Z_{qq}^{-1,\mathsf{PS}} + Z_{qq}^{-1,\mathsf{NS}}, \tag{4.60}$$

$$A_{qq} = A_{qq}^{\mathsf{PS}} + A_{qq}^{\mathsf{NS}} \tag{4.61}$$

The anomalous dimensions γ_{ij} of the operators are defined in Eqs. 2.107, 2.108 and can be expanded in a perturbative series as follows

$$\gamma_{ij}^{\mathsf{S,PS,NS}}(a_s^{\overline{\mathsf{MS}}}, n_f, N) = \sum_{l=1}^{\infty} a_s^{\overline{\mathsf{MS}}^l}\gamma_{ij}^{(l),\mathsf{S,PS,NS}}(n_f, N). \tag{4.62}$$

Here, the PS contribution starts at $O(a_s^2)$. In the following, we do not write the dependence on the Mellin–variable N for the OMEs, the operator Z–factors and the anomalous dimensions explicitly. Further, we will suppress the dependence on ε for unrenormalized quantities and Z–factors. From Eqs. 2.107, 2.108, one can determine the relation between the anomalous dimensions and the Z–factors order by order in perturbation theory. In the general case, one finds up to $O\left(a_s^{\overline{\mathsf{MS}}^3}\right)$

$$\begin{aligned}Z_{ij}(a_s^{\overline{\mathsf{MS}}}, n_f) = \delta_{ij} &+ a_s^{\overline{\mathsf{MS}}}\frac{\gamma_{ij}^{(0)}}{\varepsilon} + a_s^{\overline{\mathsf{MS}}^2}\left\{\frac{1}{\varepsilon^2}\Big(\frac{1}{2}\gamma_{il}^{(0)}\gamma_{lj}^{(0)} + \beta_0\gamma_{ij}^{(0)}\Big) + \frac{1}{2\varepsilon}\gamma_{ij}^{(1)}\right\}\\ &+ a_s^{\overline{\mathsf{MS}}^3}\left\{\frac{1}{\varepsilon^3}\Big(\frac{1}{6}\gamma_{il}^{(0)}\gamma_{lk}^{(0)}\gamma_{kj}^{(0)} + \beta_0\gamma_{il}^{(0)}\gamma_{lj}^{(0)} + \frac{4}{3}\beta_0^2\gamma_{ij}^{(0)}\Big)\right.\\ &\quad + \frac{1}{\varepsilon^2}\Big(\frac{1}{6}\gamma_{il}^{(1)}\gamma_{lj}^{(0)} + \frac{1}{3}\gamma_{il}^{(0)}\gamma_{lj}^{(1)} + \frac{2}{3}\beta_0\gamma_{ij}^{(1)}\\ &\quad \left. + \frac{2}{3}\beta_1\gamma_{ij}^{(0)}\Big) + \frac{\gamma_{ij}^{(2)}}{3\varepsilon}\right\}.\end{aligned} \tag{4.63}$$

The NS and PS Z–factors are given by[4]

[4] In Eq. 4.65 we corrected typographical errors contained in Eq. 4.34, [17].

$$
\begin{aligned}
Z_{qq}^{\mathsf{NS}}(a_s^{\overline{\mathsf{MS}}}, n_f) = 1 &+ a_s^{\overline{\mathsf{MS}}}\frac{\gamma_{qq}^{(0),\mathsf{NS}}}{\varepsilon} + a_s^{\overline{\mathsf{MS}}^2}\left\{\frac{1}{\varepsilon^2}\left(\frac{1}{2}\gamma_{qq}^{(0),\mathsf{NS}^2} + \beta_0\gamma_{qq}^{(0),\mathsf{NS}}\right) + \frac{1}{2\varepsilon}\gamma_{qq}^{(1),\mathsf{NS}}\right\} \\
&+ a_s^{\overline{\mathsf{MS}}^3}\left\{\frac{1}{\varepsilon^3}\left(\frac{1}{6}\gamma_{qq}^{(0),\mathsf{NS}^3} + \beta_0\gamma_{qq}^{(0),\mathsf{NS}^2} + \frac{4}{3}\beta_0^2\gamma_{qq}^{(0),\mathsf{NS}}\right)\right. \\
&\quad + \frac{1}{\varepsilon^2}\left(\frac{1}{2}\gamma_{qq}^{(0),\mathsf{NS}}\gamma_{qq}^{(1),\mathsf{NS}} + \frac{2}{3}\beta_0\gamma_{qq}^{(1),\mathsf{NS}} + \frac{2}{3}\beta_1\gamma_{qq}^{(0),\mathsf{NS}}\right) \\
&\quad \left. + \frac{1}{3\varepsilon}\gamma_{qq}^{(2),\mathsf{NS}}\right\},
\end{aligned}
\tag{4.64}
$$

$$
\begin{aligned}
Z_{qq}^{\mathsf{PS}}(a_s^{\overline{\mathsf{MS}}}, n_f) = a_s^{\overline{\mathsf{MS}}^2}&\left\{\frac{1}{2\varepsilon^2}\gamma_{qg}^{(0)}\gamma_{gq}^{(0)} + \frac{1}{2\varepsilon}\gamma_{qq}^{(1),\mathsf{PS}}\right\} \\
+ a_s^{\overline{\mathsf{MS}}^3}&\left\{\frac{1}{\varepsilon^3}\left(\frac{1}{3}\gamma_{qq}^{(0)}\gamma_{qg}^{(0)}\gamma_{gq}^{(0)} + \frac{1}{6}\gamma_{qg}^{(0)}\gamma_{gg}^{(0)}\gamma_{gq}^{(0)} + \beta_0\gamma_{qg}^{(0)}\gamma_{gq}^{(0)}\right)\right. \\
&+ \frac{1}{\varepsilon^2}\left(\frac{1}{3}\gamma_{qg}^{(0)}\gamma_{gq}^{(1)} + \frac{1}{6}\gamma_{qg}^{(1)}\gamma_{gq}^{(0)} + \frac{1}{2}\gamma_{qq}^{(0)}\gamma_{qq}^{(1),\mathsf{PS}}\right. \\
&\left.\left. + \frac{2}{3}\beta_0\gamma_{qq}^{(1),\mathsf{PS}}\right) + \frac{\gamma_{qq}^{(2),\mathsf{PS}}}{3\varepsilon}\right\}.
\end{aligned}
\tag{4.65}
$$

All quantities in Eqs. 4.63–4.65 refer to n_f light flavors and renormalize the massless off–shell OMEs given in Eqs. 4.58, 4.59.

In the next step, we consider an additional heavy quark with mass m. We keep the external momentum artificially off–shell for the moment, in order to deal with the UV–singularities only. For the additional massive quark, one has to account for the renormalization of the coupling constant we defined in Eqs. 4.52, 4.53. The Z–factors including one massive quark are then obtained by taking Eqs. 4.63–4.65 at $(n_f + 1)$ flavors and performing the scheme transformation given in (4.56). The emergence of $\delta a_{s,k}^{\mathrm{MOM}}$ in Z_{ij} is due to the finite mass effects and cancels singularities which emerge for real radiation and virtual processes at $p^2 \to 0$. Thus one obtains up to $O(a_s^{\mathrm{MOM}^3})$

$$
\begin{aligned}
Z_{ij}^{-1}(a_s^{\mathrm{MOM}}, n_f + 1, \mu^2) = \delta_{ij} &- a_s^{\mathrm{MOM}}\frac{\gamma_{ij}^{(0)}}{\varepsilon} \\
+ a_s^{\mathrm{MOM}^2}&\left[\frac{1}{\varepsilon}\left(-\frac{1}{2}\gamma_{ij}^{(1)} - \delta a_{s,1}^{\mathrm{MOM}}\gamma_{ij}^{(0)}\right) + \frac{1}{\varepsilon^2}\left(\frac{1}{2}\gamma_{il}^{(0)}\gamma_{lj}^{(0)} + \beta_0\gamma_{ij}^{(0)}\right)\right] \\
+ a_s^{\mathrm{MOM}^3}&\left[\frac{1}{\varepsilon}\left(-\frac{1}{3}\gamma_{ij}^{(2)} - \delta a_{s,1}^{\mathrm{MOM}}\gamma_{ij}^{(1)} - \delta a_{s,2}^{\mathrm{MOM}}\gamma_{ij}^{(0)}\right)\right. \\
&+ \frac{1}{\varepsilon^2}\left(\frac{4}{3}\beta_0\gamma_{ij}^{(1)} + 2\delta a_{s,1}^{\mathrm{MOM}}\beta_0\gamma_{ij}^{(0)} + \frac{1}{3}\beta_1\gamma_{ij}^{(0)}\right.
\end{aligned}
$$

$$+\,\delta a_{s,1}^{\text{MOM}}\gamma_{il}^{(0)}\gamma_{lj}^{(0)}+\frac{1}{3}\gamma_{il}^{(1)}\gamma_{lj}^{(0)}+\frac{1}{6}\gamma_{il}^{(0)}\gamma_{lj}^{(1)}\Big)$$
$$+\frac{1}{\varepsilon^3}\Big(-\frac{4}{3}\beta_0^2\gamma_{ij}^{(0)}-\beta_0\gamma_{il}^{(0)}\gamma_{lj}^{(0)}-\frac{1}{6}\gamma_{il}^{(0)}\gamma_{lk}^{(0)}\gamma_{kj}^{(0)}\Big)\Big], \tag{4.66}$$

and

$$Z_{qq}^{-1,\mathsf{NS}}(a_s^{\text{MOM}},n_f+1,\mu^2)=1-a_s^{\text{MOM}}\frac{\gamma_{qq}^{(0),\mathsf{NS}}}{\varepsilon}$$
$$+\,a_s^{\text{MOM}^2}\Big[\frac{1}{\varepsilon}\Big(-\frac{1}{2}\gamma_{qq}^{(1),\mathsf{NS}}-\delta a_{s,1}^{\text{MOM}}\gamma_{qq}^{(0),\mathsf{NS}}\Big)$$
$$+\frac{1}{\varepsilon^2}\Big(\beta_0\gamma_{qq}^{(0),\mathsf{NS}}+\frac{1}{2}\gamma_{qq}^{(0),\mathsf{NS}^2}\Big)\Big]$$
$$+\,a_s^{\text{MOM}^3}\Big[\frac{1}{\varepsilon}\Big(-\frac{1}{3}\gamma_{qq}^{(2),\mathsf{NS}}-\delta a_{s,1}^{\text{MOM}}\gamma_{qq}^{(1),\mathsf{NS}}-\delta a_{s,2}^{\text{MOM}}\gamma_{qq}^{(0),\mathsf{NS}}\Big)$$
$$+\frac{1}{\varepsilon^2}\Big(\frac{4}{3}\beta_0\gamma_{qq}^{(1),\mathsf{NS}}+2\delta a_{s,1}^{\text{MOM}}\beta_0\gamma_{qq}^{(0),\mathsf{NS}}$$
$$+\frac{1}{3}\beta_1\gamma_{qq}^{(0),\mathsf{NS}}+\frac{1}{2}\gamma_{qq}^{(0),\mathsf{NS}}\gamma_{qq}^{(1),\mathsf{NS}}+\delta a_{s,1}^{\text{MOM}}\gamma_{qq}^{(0),\mathsf{NS}^2}\Big)$$
$$+\frac{1}{\varepsilon^3}\Big(-\frac{4}{3}\beta_0^2\gamma_{qq}^{(0),\mathsf{NS}}-\beta_0\gamma_{qq}^{(0),\mathsf{NS}^2}-\frac{1}{6}\gamma_{qq}^{(0),\mathsf{NS}^3}\Big)\Big], \tag{4.67}$$

$$Z_{qq}^{-1,\mathsf{PS}}(a_s^{\text{MOM}},n_f+1,\mu^2)=a_s^{\text{MOM}^2}\Big[\frac{1}{\varepsilon}\Big(-\frac{1}{2}\gamma_{qq}^{(1),\mathsf{PS}}\Big)+\frac{1}{\varepsilon^2}\Big(\frac{1}{2}\gamma_{qg}^{(0)}\gamma_{gq}^{(0)}\Big)\Big]$$
$$+\,a_s^{\text{MOM}^3}\Big[\frac{1}{\varepsilon}\Big(-\frac{1}{3}\gamma_{qq}^{(2),\mathsf{PS}}-\delta a_{s,1}^{\text{MOM}}\gamma_{qq}^{(1),\mathsf{PS}}\Big)$$
$$+\frac{1}{\varepsilon^2}\Big(\frac{1}{6}\gamma_{qg}^{(0)}\gamma_{gq}^{(1)}+\frac{1}{3}\gamma_{gq}^{(0)}\gamma_{qg}^{(1)}+\frac{1}{2}\gamma_{qq}^{(0)}\gamma_{qq}^{(1),\mathsf{PS}}$$
$$+\frac{4}{3}\beta_0\gamma_{qq}^{(1),\mathsf{PS}}+\delta a_{s,1}^{\text{MOM}}\gamma_{qg}^{(0)}\gamma_{gq}^{(0)}\Big)$$
$$+\frac{1}{\varepsilon^3}\Big(-\frac{1}{3}\gamma_{qg}^{(0)}\gamma_{gq}^{(0)}\gamma_{qq}^{(0)}-\frac{1}{6}\gamma_{gq}^{(0)}\gamma_{qg}^{(0)}\gamma_{gg}^{(0)}-\beta_0\gamma_{qg}^{(0)}\gamma_{gq}^{(0)}\Big)\Big]. \tag{4.68}$$

The above equations are given for n_f+1 flavors. One re-derives the expressions for n_f light flavors by setting $(n_f+1)=\,:n_f$ and $\delta a_s^{\text{MOM}}=\delta a_s^{\overline{\mathsf{MS}}}$. As a next step, we split the OMEs into a part involving only light flavors and the heavy flavor part

$$\hat{A}_{ij}(p^2,m^2,\mu^2,a_s^{\text{MOM}},n_f+1)=\hat{A}_{ij}\Big(\frac{-p^2}{\mu^2},a_s^{\overline{\mathsf{MS}}},n_f\Big)$$
$$+\,\hat{A}_{ij}^{Q}(p^2,m^2,\mu^2,a_s^{\text{MOM}},n_f+1). \tag{4.69}$$

In (4.69, 4.70), the light flavor part depends on $a_s^{\overline{\mathrm{MS}}}$, since the prescription adopted for coupling constant renormalization only applies to the massive part. $\hat{A}_{ij}^{Q}$ denotes any massive OME we consider. The correct UV–renormalization prescription for the massive contribution is obtained by subtracting from Eq. 4.69 the terms applying to the light part only :

$$\begin{aligned}
\bar{A}_{ij}^{Q}(p^2, m^2, \mu^2, a_s^{\mathrm{MOM}}, n_f+1) = {} & Z_{il}^{-1}(a_s^{\mathrm{MOM}}, n_f+1, \mu^2) \\
& \times \hat{A}_{ij}^{Q}(p^2, m^2, \mu^2, a_s^{\mathrm{MOM}}, n_f+1) \\
& + Z_{il}^{-1}(a_s^{\mathrm{MOM}}, n_f+1, \mu^2)\hat{A}_{ij}\left(\frac{-p^2}{\mu^2}, a_s^{\overline{\mathrm{MS}}}, n_f\right) \\
& - Z_{il}^{-1}(a_s^{\overline{\mathrm{MS}}}, n_f, \mu^2)\hat{A}_{ij}\left(\frac{-p^2}{\mu^2}, a_s^{\overline{\mathrm{MS}}}, n_f\right),
\end{aligned} \tag{4.70}$$

where

$$Z_{ij}^{-1} = \delta_{ij} + \sum_{k=1}^{\infty} a_s^k Z_{ij}^{-1,(k)}. \tag{4.71}$$

In the limit $p^2 = 0$, integrals without a scale vanish within dimensional regularization. Hence for the light flavor OMEs only the term δ_{ij} remains and one obtains the UV–finite massive OMEs after expanding in a_s

$$\begin{aligned}
& \bar{A}_{ij}^{Q}\left(\frac{m^2}{\mu^2}, a_s^{\mathrm{MOM}}, n_f+1\right) \\
& = a_s^{\mathrm{MOM}}\left(\hat{A}_{ij}^{(1),Q}\left(\frac{m^2}{\mu^2}\right) + Z_{ij}^{-1,(1)}(n_f+1, \mu^2) - Z_{ij}^{-1,(1)}(n_f)\right) \\
& \quad + a_s^{\mathrm{MOM}2}\left(\hat{A}_{ij}^{(2),Q}\left(\frac{m^2}{\mu^2}\right) + Z_{ij}^{-1,(2)}(n_f+1, \mu^2) - Z_{ij}^{-1,(2)}(n_f)\right. \\
& \qquad \left. + Z_{ik}^{-1,(1)}(n_f+1, \mu^2)\hat{A}_{kj}^{(1),Q}\left(\frac{m^2}{\mu^2}\right)\right) \\
& \quad + a_s^{\mathrm{MOM}3}\left(\hat{A}_{ij}^{(3),Q}\left(\frac{m^2}{\mu^2}\right) + Z_{ij}^{-1,(3)}(n_f+1, \mu^2) - Z_{ij}^{-1,(3)}(n_f)\right. \\
& \qquad + Z_{ik}^{-1,(1)}(n_f+1, \mu^2)\hat{A}_{kj}^{(2),Q}\left(\frac{m^2}{\mu^2}\right) \\
& \qquad \left. + Z_{ik}^{-1,(2)}(n_f+1, \mu^2)\hat{A}_{kj}^{(1),Q}\left(\frac{m^2}{\mu^2}\right)\right).
\end{aligned} \tag{4.72}$$

The Z–factors at n_f+1 flavors refer to Eqs. 4.66–4.68, whereas those at n_f flavors correspond to the massless case.

4.6 Mass Factorization

Finally, we have to remove the collinear singularities contained in $\bar{A}_{ij}$, which emerge in the limit $p^2 = 0$. They are absorbed into the parton distribution functions and are not present in case of the off–shell massless OMEs. As a generic renormalization formula, generalizing Eqs. 4.58, 4.59, one finds

$$A_{ij} = Z^{-1}_{il} \hat{A}_{lk} \Gamma^{-1}_{kj}. \tag{4.73}$$

The renormalized operator matrix elements are obtained by

$$A_{ij}^Q\left(\frac{m^2}{\mu^2}, a_s^{\rm MOM}, n_f+1\right) = \bar{A}_{il}^Q\left(\frac{m^2}{\mu^2}, a_s^{\rm MOM}, n_f+1\right)\Gamma^{-1}_{lj}. \tag{4.74}$$

If all quarks were massless, the identity, [16],

$$\Gamma_{ij} = Z^{-1}_{ij}. \tag{4.75}$$

would hold. However, due to the presence of a heavy quark Q, the transition functions $\Gamma(n_f)$ refer only to massless sub-graphs. Hence the Γ–factors contribute up to $O(a_s^2)$ only and do not involve the special scheme adopted for the renormalization of the coupling. Due to Eq. 4.75, they can be read off from Eqs. 4.63–4.65. The renormalized operator matrix elements are then given by:

$$\begin{aligned}
&A_{ij}^Q\left(\frac{m^2}{\mu^2}, a_s^{\rm MOM}, n_f+1\right)\\
&= a_s^{\rm MOM}\left(\hat{A}_{ij}^{(1),Q}\left(\frac{m^2}{\mu^2}\right) + Z_{ij}^{-1,(1)}(n_f+1) - Z_{ij}^{-1,(1)}(n_f)\right)\\
&+ a_s^{\rm MOM2}\Biggl(\hat{A}_{ij}^{(2),Q}\left(\frac{m^2}{\mu^2}\right) + Z_{ij}^{-1,(2)}(n_f+1)\\
&\qquad - Z_{ij}^{-1,(2)}(n_f) + Z_{ik}^{-1,(1)}(n_f+1)\hat{A}_{kj}^{(1),Q}\left(\frac{m^2}{\mu^2}\right)\\
&\qquad + \left[\hat{A}_{il}^{(1),Q}\left(\frac{m^2}{\mu^2}\right) + Z_{il}^{-1,(1)}(n_f+1) - Z_{il}^{-1,(1)}(n_f)\right]\Gamma_{lj}^{-1,(1)}(n_f)\Biggr)\\
&+ a_s^{\rm MOM3}\Biggl(\hat{A}_{ij}^{(3),Q}\left(\frac{m^2}{\mu^2}\right) + Z_{ij}^{-1,(3)}(n_f+1)\\
&\qquad - Z_{ij}^{-1,(3)}(n_f) + Z_{ik}^{-1,(1)}(n_f+1)\hat{A}_{kj}^{(2),Q}\left(\frac{m^2}{\mu^2}\right)\\
&\qquad + Z_{ik}^{-1,(2)}(n_f+1)\hat{A}_{kj}^{(1),Q}\left(\frac{m^2}{\mu^2}\right)\\
&\qquad + \left[\hat{A}_{il}^{(1),Q}\left(\frac{m^2}{\mu^2}\right) + Z_{il}^{-1,(1)}(n_f+1) - Z_{il}^{-1,(1)}(n_f)\right]\Gamma_{lj}^{-1,(2)}(n_f)
\end{aligned}$$

$$+\left[\hat{A}_{il}^{(2),Q}\left(\frac{m^2}{\mu^2}\right)+Z_{il}^{-1,(2)}(n_f+1)-Z_{il}^{-1,(2)}(n_f)\right.$$

$$\left.+Z_{ik}^{-1,(1)}(n_f+1)\hat{A}_{kl}^{(1),Q}\left(\frac{m^2}{\mu^2}\right)\right]\Gamma_{lj}^{-1,(1)}(n_f)\Big)+O\left(a_s^{\mathrm{MOM}^4}\right) \quad (4.76)$$

From (4.76) it is obvious that the renormalization of A_{ij}^Q to $O(a_s^3)$ requires the 1–loop terms up to $O(\varepsilon^2)$ and the 2–loop terms up to $O(\varepsilon)$, cf. [16, 17, 33–35]. These terms are calculated in Sect. 4.6. Finally, we transform the coupling constant back into the $\overline{\mathsf{MS}}$–scheme by using Eq. 4.55. We do not give the explicit formula here, but present the individual renormalized OMEs after this transformation in the next Section as perturbative series in $a_s^{\overline{\mathsf{MS}}}$,

$$A_{ij}^Q\left(\frac{m^2}{\mu^2},a_s^{\overline{\mathsf{MS}}},n_f+1\right)=\delta_{ij}+a_s^{\overline{\mathsf{MS}}}A_{ij}^{Q,(1)}\left(\frac{m^2}{\mu^2},n_f+1\right)$$
$$+a_s^{\overline{\mathsf{MS}}^2}A_{ij}^{Q,(2)}\left(\frac{m^2}{\mu^2},n_f+1\right)$$
$$+a_s^{\overline{\mathsf{MS}}^3}A_{ij}^{Q,(3)}\left(\frac{m^2}{\mu^2},n_f+1\right)+O\left(a_s^{\overline{\mathsf{MS}}^4}\right). \quad (4.77)$$

As stated in Sect. 4.3, one has to use the same scheme when combining the massive OMEs with the massless Wilson coefficients in the factorization formula (3.16). The effects of the transformation between the MOM–and $\overline{\mathsf{MS}}$–scheme are discussed in Sect. 4.5. The subscript Q was introduced in this Section to make the distinction between the massless and massive OMEs explicit and will be dropped from now on, since no confusion is expected. Comparing Eqs. 4.76 and 4.77, one notices that the term δ_{ij} is not present in the former because it was subtracted together with the light flavor contributions. However, as one infers from Eq. 3.16 and the discussion below, this term is necessary when calculating the massive Wilson coefficients in the asymptotic limit and we therefore have re–introduced it into Eq. 4.77.

4.7 General Structure of the Massive Operator Matrix Elements

In the following, we present the general structure of the unrenormalized and renormalized massive operator matrix elements for the specific partonic channels. The former are expressed as a Laurent–series in ε via

$$\hat{\hat{A}}_{ij}^{(l)}\left(\frac{\hat{m}^2}{\mu^2},\varepsilon,N\right)=\left(\frac{\hat{m}^2}{\mu^2}\right)^{l\varepsilon/2}\sum_{k=0}^{\infty}\frac{a_{ij}^{(l,k)}}{\varepsilon^{l-k}}. \quad (4.78)$$

Additionally, we set

$$a_{ij}^{(l,l)}\equiv a_{ij}^{(l)},\quad a_{ij}^{(l,l+1)}\equiv\overline{a}_{ij}^{(l)}, etc.. \quad (4.79)$$

The pole terms can all be expressed by known renormalization constants and lower order contributions to the massive OMEs, which provides us with a strong check on our calculation. In particular, the complete NLO anomalous dimensions, as well as their T_F–terms at NNLO, contribute at $O(a_s^3)$. The moments of the $O(\varepsilon^0)$–terms of the unrenormalized OMEs at the 3–loop level, $a_{ij}^{(3)}$, are a new result of this thesis and will be calculated in Section 7, cf. [36]. The $O(\varepsilon)$ terms at the 2–loop level, $\overline{a}_{ij}^{(2)}$, contribute to the non–logarithmic part of the renormalized 3–loop OMEs and are calculated for general values of N in Section 6, cf. [17, 35]. The pole terms and the $O(\varepsilon^0)$ terms, $a_{ij}^{(2)}$, at 2–loop have been calculated for the first time in Refs. [16, 34]. The terms involving the quark operator, (2.86, 2.87), were confirmed in [33] and the terms involving the gluon operator (2.88) by the present work, cf. [35]. In order to keep up with the notation used in [16, 34], we define the 2–loop terms $a_{ij}^{(2)}$, $\overline{a}_{ij}^{(2)}$ after performing mass renormalization in the on–shell–scheme. This we do not apply for the 3–loop terms. We choose to calculate one–particle reducible diagrams and therefore have to include external self–energies containing massive quarks into our calculation. Before presenting the operator matrix elements up to three loops, we first summarize the necessary self–energy contributions in the next Section. The remaining (Sects. 4.7.2–4.7.6), contain the general structure of the unrenormalized and renormalized massive OMEs up to 3–loops. In these Sections, we always proceed as follows: From Eqs. 4.57, 4.76, one predicts the pole terms of the respective unrenormalized OMEs by demanding that these terms have to cancel through renormalization. The unrenormalized expressions are then renormalized in the MOM–scheme. Finally, Eq. 4.55 is applied and the renormalized massive OMEs are presented in the $\overline{\text{MS}}$–scheme.

4.7.1 Self–Energy Contributions

The gluon and quark self-energy contributions due to heavy quark lines are given by

$$\hat{\Pi}_{\mu\nu}^{ab}(p^2,\hat{m}^2,\mu^2,\hat{a}_s)=i\delta^{ab}\Big[-g_{\mu\nu}p^2+p_\mu p_\nu\Big]\hat{\Pi}(p^2,\hat{m}^2,\mu^2,\hat{a}_s) \tag{4.80}$$

$$\hat{\Pi}(p^2,\hat{m}^2,\mu^2,\hat{a}_s)=\sum_{k=1}^{\infty}\hat{a}_s^k\hat{\Pi}^{(k)}(p^2,\hat{m}^2,\mu^2). \tag{4.81}$$

$$\hat{\Sigma}_{ij}(p^2,\hat{m}^2,\mu^2,\hat{a}_s)=i\delta_{ij}\not{p}\hat{\Sigma}(p^2,\hat{m}^2,\mu^2,\hat{a}_s), \tag{4.82}$$

$$\hat{\Sigma}(p^2,\hat{m}^2,\mu^2,\hat{a}_s)=\sum_{k=2}^{\infty}\hat{a}_s^k\hat{\Sigma}^{(k)}(p^2,\hat{m}^2,\mu^2). \tag{4.83}$$

Note, that the quark self–energy contributions start at 2–loop order. These self–energies are easily calculated using MATAD, [37], cf. Section 7. The expansion coefficients for $p^2 = 0$ of Eqs. 4.82, 4.83 are needed for the calculation of the gluonic and quarkonic OMEs, respectively. The contributions to the gluon vacuum polarization for general gauge parameter ξ are

$$\hat{\Pi}^{(1)}\Bigl(0,\frac{\hat{m}^2}{\mu^2}\Bigr) = T_F\Bigl(\frac{\hat{m}^2}{\mu^2}\Bigr)^{\varepsilon/2}\Bigl(-\frac{8}{3\varepsilon}\exp\Bigl(\sum_{i=2}^{\infty}\frac{\zeta_i}{i}\Bigl(\frac{\varepsilon}{2}\Bigr)^i\Bigr)\Bigr), \tag{4.84}$$

$$\begin{aligned}\hat{\Pi}^{(2)}\Bigl(0,\frac{\hat{m}^2}{\mu^2}\Bigr) = T_F\Bigl(\frac{\hat{m}^2}{\mu^2}\Bigr)^{\varepsilon}\Bigl(-\frac{4}{\varepsilon^2}C_A + \frac{1}{\varepsilon}\Bigl\{-12C_F + 5C_A\Bigr\} + C_A\Bigl(\frac{13}{12} - \zeta_2\Bigr)\\ -\frac{13}{3}C_F + \varepsilon\Bigl\{C_A\Bigl(\frac{169}{144} + \frac{5}{4}\zeta_2 - \frac{\zeta_3}{3}\Bigr)\\ + C_F\Bigl(-\frac{35}{12} - 3\zeta_2\Bigr)\Bigr\}\Bigr) + O(\varepsilon^2),\end{aligned} \tag{4.85}$$

$$\begin{aligned}&\hat{\Pi}^{(3)}\Bigl(0,\frac{\hat{m}^2}{\mu^2}\Bigr)\\ &= T_F\Bigl(\frac{\hat{m}^2}{\mu^2}\Bigr)^{3\varepsilon/2}\Bigl(\frac{1}{\varepsilon^3}\Bigl\{-\frac{32}{9}T_F C_A\Bigl(2n_f + 1\Bigr) + C_A^2\Bigl(\frac{164}{9} + \frac{4}{3}\xi\Bigr)\Bigr\}\\ &\quad + \frac{1}{\varepsilon^2}\Bigl\{\frac{80}{27}\Bigl(C_A - 6C_F\Bigr)n_f T_F + \frac{8}{27}\Bigl(35C_A - 48C_F\Bigr)T_F\\ &\quad + \frac{C_A^2}{27}\Bigl(-781 + 63\xi\Bigr) + \frac{712}{9}C_A C_F\Bigr\}\\ &\quad + \frac{1}{\varepsilon}\Bigl\{\frac{4}{27}\Bigl(C_A(-101 - 18\zeta_2) - 62C_F\Bigr)n_f T_F\\ &\quad + \frac{2}{27}\Bigl(C_A(-37 - 18\zeta_2) - 80C_F\Bigr)T_F\\ &\quad + C_A^2\Bigl(-12\zeta_3 + \frac{41}{6}\zeta_2 + \frac{3181}{108} + \frac{\zeta_2}{2}\xi\frac{137}{36}\xi\Bigr)\\ &\quad + +C_A C_F\Bigl(16\zeta_3 - \frac{1570}{27}\Bigr) + \frac{272}{3}C_F^2\Bigr\}\\ &\quad + n_f T_F\Bigl\{C_A\Bigl(\frac{56}{9}\zeta_3 + \frac{10}{9}\zeta_2 - \frac{3203}{243}\Bigr)\\ &\quad + C_F\Bigl(-\frac{20}{3}\zeta_2 - \frac{1942}{81}\Bigr)\Bigr\}\end{aligned}$$

$$
\begin{aligned}
&+T_F\Biggl\{C_A\Bigl(-\frac{295}{18}\zeta_3+\frac{35}{9}\zeta_2+\frac{6361}{486}\Bigr)\\
&+C_F\Bigl(-7\zeta_3-\frac{16}{3}\zeta_2-\frac{218}{81}\Bigr)\Biggr\}\\
&+C_A^2\Biggl\{4B_4-27\zeta_4+\frac{1969}{72}\zeta_3-\frac{781}{72}\zeta_2\\
&+\frac{42799}{3888}-\frac{7}{6}\zeta_3\xi+\frac{7}{8}\zeta_2\xi+\frac{3577}{432}\xi\Biggr\}\\
&+C_AC_F\Biggl\{-8B_4+36\zeta_4-\frac{1957}{12}\zeta_3+\frac{89}{3}\zeta_2+\frac{10633}{81}\Biggr\}\\
&+C_F^2\Biggl\{\frac{95}{3}\zeta_3+\frac{274}{9}\Biggr\}\Biggr)+O(\varepsilon),
\end{aligned}
\tag{4.86}
$$

and for the quark self–energy,

$$
\hat{\Sigma}^{(2)}\left(0,\frac{\hat{m}^2}{\mu^2}\right)=T_FC_F\left(\frac{\hat{m}^2}{\mu^2}\right)^{\varepsilon}\left\{\frac{2}{\varepsilon}+\frac{5}{6}+\left[\frac{89}{72}+\frac{\zeta_2}{2}\right]\varepsilon\right\}+O(\varepsilon^2),
\tag{4.87}
$$

$$
\begin{aligned}
&\hat{\Sigma}^{(3)}\left(0,\frac{\hat{m}^2}{\mu^2}\right)\\
&=T_FC_F\left(\frac{\hat{m}^2}{\mu^2}\right)^{3\varepsilon/2}\Biggl(\frac{8}{3\varepsilon^3}C_A\{1-\xi\}+\frac{1}{\varepsilon^2}\Biggl\{\frac{32}{9}T_F(n_f+2)-C_A\Bigl(\frac{40}{9}+4\xi\Bigr)-\frac{8}{3}C_F\Biggr\}\\
&+\frac{1}{\varepsilon}\Biggl\{\frac{40}{27}T_F(n_f+2)+C_A\Bigl\{\zeta_2+\frac{454}{27}-\zeta_2\xi-\frac{70}{9}\xi\Bigr\}-26C_F\Biggr\}\\
&+n_fT_F\Biggl\{\frac{4}{3}\zeta_2+\frac{674}{81}\Biggr\}+T_F\Biggl\{\frac{8}{3}\zeta_2+\frac{604}{81}\Biggr\}\\
&+C_A\Biggl\{\frac{17}{3}\zeta_3-\frac{5}{3}\zeta_2+\frac{1879}{162}+\frac{7}{3}\zeta_3\xi-\frac{3}{2}\zeta_2\xi-\frac{407}{27}\xi\Biggr\}\\
&+C_F\Biggl\{-8\zeta_3-\zeta_2-\frac{335}{18}\Biggr\}\Biggr)+O(\varepsilon),
\end{aligned}
\tag{4.88}
$$

see also [4, 38, 39]. In Eq. 4.86 the constant

$$
B_4=-4\zeta_2\ln^2(2)+\frac{2}{3}\ln^4(2)-\frac{13}{2}\zeta_4+16Li_4\left(\frac{1}{2}\right)\approx-1.762800087\ldots
\tag{4.89}
$$

appears due to genuine massive effects, cf. [40–44].

4.7.2 $A_{qq,Q}^{\mathsf{NS}}$

The lowest non–trivial NS–contribution is of $O(a_s^2)$,

$$A_{qq,Q}^{\mathsf{NS}} = 1 + a_s^2 A_{qq,Q}^{(2),\mathsf{NS}} + a_s^3 A_{qq,Q}^{(3),\mathsf{NS}} + O(a_s^4). \tag{4.90}$$

The expansion coefficients are obtained in the MOM–scheme from the bare quantities, using Eqs. 4.57, 4.76. After mass—and coupling constant renormalization, the OMEs are given by

$$A_{qq,Q}^{(2),NS,\mathrm{MOM}} = \hat{A}_{qq,Q}^{(2),\mathsf{NS},\mathrm{MOM}} + Z_{qq}^{-1,(2),\mathsf{NS}}(n_f+1) - Z_{qq}^{-1,(2),\mathsf{NS}}(n_f), \tag{4.91}$$

$$\begin{aligned} A_{qq,Q}^{(3),NS,\mathrm{MOM}} &= \hat{A}_{qq,Q}^{(3),\mathsf{NS},\mathrm{MOM}} + Z_{qq}^{-1,(3),\mathsf{NS}}(n_f+1) - Z_{qq}^{-1,(3),\mathsf{NS}}(n_f) \\ &+ Z_{qq}^{-1,(1),\mathsf{NS}}(n_f+1)\hat{A}_{qq,Q}^{(2),\mathsf{NS},\mathrm{MOM}} \\ &+ \Big[\hat{A}_{qq,Q}^{(2),\mathsf{NS},\mathrm{MOM}} + Z_{qq}^{-1,(2),\mathsf{NS}}(n_f+1) - Z_{qq}^{-1,(2),\mathsf{NS}}(n_f)\Big]\Gamma_{qq}^{-1,(1)}(n_f). \end{aligned} \tag{4.92}$$

From (4.57, 4.76, 4.91, 4.92), one predicts the pole terms of the unrenormalized OME. At second and third order they read

$$\hat{\hat{A}}_{qq,Q}^{(2),\mathsf{NS}} = \left(\frac{\hat{m}^2}{\mu^2}\right)^{\varepsilon}\left(\frac{\beta_{0,Q}\gamma_{qq}^{(0)}}{\varepsilon^2} + \frac{\hat{\gamma}_{qq}^{(1),\mathsf{NS}}}{2\varepsilon} + a_{qq,Q}^{(2),\mathsf{NS}} + \overline{a}_{qq,Q}^{(2),\mathsf{NS}}\varepsilon\right), \tag{4.93}$$

$$\begin{aligned} \hat{\hat{A}}_{qq,Q}^{(3),\mathsf{NS}} = \left(\frac{\hat{m}^2}{\mu^2}\right)^{3\varepsilon/2}\Bigg\{ &-\frac{4\gamma_{qq}^{(0)}\beta_{0,Q}}{3\varepsilon^3}\Big(\beta_0 + 2\beta_{0,Q}\Big) \\ &+ \frac{1}{\varepsilon^2}\Bigg(\frac{2\gamma_{qq}^{(1),\mathsf{NS}}\beta_{0,Q}}{3} - \frac{4\hat{\gamma}_{qq}^{(1),\mathsf{NS}}}{3}\Big[\beta_0 + \beta_{0,Q}\Big] \\ &+ \frac{2\beta_{1,Q}\gamma_{qq}^{(0)}}{3} - 2\delta m_1^{(-1)}\beta_{0,Q}\gamma_{qq}^{(0)}\Bigg) \\ &+ \frac{1}{\varepsilon}\Bigg(\frac{\hat{\gamma}_{qq}^{(2),\mathsf{NS}}}{3} - 4a_{qq,Q}^{(2),\mathsf{NS}}\Big[\beta_0 + \beta_{0,Q}\Big] + \beta_{1,Q}^{(1)}\gamma_{qq}^{(0)} \\ &+ \frac{\gamma_{qq}^{(0)}\beta_0\beta_{0,Q}\zeta_2}{2} - 2\delta m_1^{(0)}\beta_{0,Q}\gamma_{qq}^{(0)} \\ &- \delta m_1^{(-1)}\hat{\gamma}_{qq}^{(1),\mathsf{NS}}\Bigg) + a_{qq,Q}^{(3),\mathsf{NS}}\Bigg\}. \end{aligned} \tag{4.94}$$

Note, that we have already used the general structure of the unrenormalized lower order OME in the evaluation of the $O(\hat{a}_s^3)$ term, as we will always do in the following. Using Eqs. 4.57, 4.91, 4.92, one can renormalize the above expressions. In addition, we finally transform back to the $\overline{\text{MS}}$–scheme using Eq. 4.55. Thus one obtains the renormalized expansion coefficients of Eq. 4.90

$$A_{qq,Q}^{(2),\text{NS},\overline{\text{MS}}} = \frac{\beta_{0,Q}\gamma_{qq}^{(0)}}{4}\ln^2\left(\frac{m^2}{\mu^2}\right) + \frac{\hat{\gamma}_{qq}^{(1),\text{NS}}}{2}\ln\left(\frac{m^2}{\mu^2}\right) + a_{qq,Q}^{(2),\text{NS}} - \frac{\beta_{0,Q}\gamma_{qq}^{(0)}}{4}\zeta_2, \quad (4.95)$$

$$\begin{aligned}
A_{qq,Q}^{(3),\text{NS},\overline{\text{MS}}} = &-\frac{\gamma_{qq}^{(0)}\beta_{0,Q}}{6}\left(\beta_0 + 2\beta_{0,Q}\right)\ln^3\left(\frac{m^2}{\mu^2}\right)\\
&+\frac{1}{4}\left\{2\gamma_{qq}^{(1),\text{NS}}\beta_{0,Q} - 2\hat{\gamma}_{qq}^{(1),\text{NS}}\left(\beta_0+\beta_{0,Q}\right) + \beta_{1,Q}\gamma_{qq}^{(0)}\right\}\ln^2\left(\frac{m^2}{\mu^2}\right)\\
&+\frac{1}{2}\left\{\hat{\gamma}_{qq}^{(2),\text{NS}} - \left(4a_{qq,Q}^{(2),\text{NS}} - \zeta_2\beta_{0,Q}\gamma_{qq}^{(0)}\right)(\beta_0+\beta_{0,Q}) + \gamma_{qq}^{(0)}\beta_{1,Q}^{(1)}\right\}\\
&\times\ln\left(\frac{m^2}{\mu^2}\right) + 4\overline{a}_{qq,Q}^{(2),\text{NS}}(\beta_0+\beta_{0,Q}) - \gamma_{qq}^{(0)}\beta_{1,Q}^{(2)} - \frac{\gamma_{qq}^{(0)}\beta_0\beta_{0,Q}\zeta_3}{6}\\
&-\frac{\gamma_{qq}^{(1),\text{NS}}\beta_{0,Q}\zeta_2}{4} + 2\delta m_1^{(1)}\beta_{0,Q}\gamma_{qq}^{(0)} + \delta m_1^{(0)}\hat{\gamma}_{qq}^{(1),\text{NS}}\\
&+2\delta m_1^{(-1)}a_{qq,Q}^{(2),\text{NS}} + a_{qq,Q}^{(3),\text{NS}}.
\end{aligned} \quad (4.96)$$

Note that in the NS–case, one is generically provided with even and odd moments due to a Ward–identity relating the results in the polarized and unpolarized case. The former refer to the anomalous dimensions $\gamma_{qq}^{\text{NS},+}$ and the latter to $\gamma_{qq}^{\text{NS},-}$, respectively, as given in Eqs. 3.5, 3.7 and Eqs. 3.6, 3.8 in Ref. [45]. The relations above also apply to other twist–2 non-singlet massive OMEs, as to transversity, for which the 2- and 3–loop heavy flavor corrections are given in Section 9, cf. also [46].

4.7.3 A_{Qq}^{PS} and $A_{qq,Q}^{\text{PS}}$

There are two different PS–contributions, cf. the discussion below Eq. 3.18,

$$A_{Qq}^{\text{PS}} = a_s^2 A_{Qq}^{(2),\text{PS}} + a_s^3 A_{Qq}^{(3),\text{PS}} + O(a_s^4), \quad (4.97)$$

$$A_{qq,Q}^{\text{PS}} = a_s^3 A_{qq,Q}^{(3),\text{PS}} + O(a_s^4). \quad (4.98)$$

Separating these contributions is not straightforward, since the generic renormalization formula for operator renormalization and mass factorization, Eq. 4.76, applies to the sum of these terms only. At $O(a_s^2)$, this problem does not occur and renormalization proceeds in the MOM–scheme via

$$A_{Qq}^{(2),PS,\mathrm{MOM}} = \hat{A}_{Qq}^{(2),\mathrm{PS},\mathrm{MOM}} + Z_{qq}^{-1,(2),\mathrm{PS}}(n_f+1) - Z_{qq}^{-1,(2),\mathrm{PS}}(n_f) + \Big[\hat{A}_{Qg}^{(1),\mathrm{MOM}} + Z_{qg}^{-1,(1)}(n_f+1) - Z_{qg}^{-1,(1)}(n_f)\Big]\Gamma_{gq}^{-1,(1)}(n_f). \tag{4.99}$$

The unrenormalized expression is given by

$$\hat{\hat{A}}_{Qq}^{(2),\mathrm{PS}} = \left(\frac{\hat{m}^2}{\mu^2}\right)^{\varepsilon}\left(-\frac{\hat{\gamma}_{qg}^{(0)}\gamma_{gq}^{(0)}}{2\varepsilon^2} + \frac{\hat{\gamma}_{qq}^{(1),\mathrm{PS}}}{2\varepsilon} + a_{Qq}^{(2),\mathrm{PS}} + \overline{a}_{Qq}^{(2),\mathrm{PS}}\varepsilon\right). \tag{4.100}$$

The renormalized result in the $\overline{\mathsf{MS}}$–scheme reads

$$A_{Qq}^{(2),PS,\overline{\mathsf{MS}}} = -\frac{\hat{\gamma}_{qg}^{(0)}\gamma_{gq}^{(0)}}{8}\ln^2\left(\frac{m^2}{\mu^2}\right) + \frac{\hat{\gamma}_{qq}^{(1),\mathrm{PS}}}{2}\ln\left(\frac{m^2}{\mu^2}\right) + a_{Qq}^{(2),\mathrm{PS}} + \frac{\hat{\gamma}_{qg}^{(0)}\gamma_{gq}^{(0)}}{8}\zeta_2. \tag{4.101}$$

The corresponding renormalization relation at third order is given by

$$\begin{aligned}
A_{Qq}^{(3),\mathrm{PS},\mathrm{MOM}} &+ A_{qq,Q}^{(3),PS,\mathrm{MOM}} = \hat{A}_{Qq}^{(3),\mathrm{PS},\mathrm{MOM}} + \hat{A}_{qq,Q}^{(3),\mathrm{PS},\mathrm{MOM}} \\
&+ Z_{qq}^{-1,(3),\mathrm{PS}}(n_f+1) - Z_{qq}^{-1,(3),\mathrm{PS}}(n_f) \\
&+ Z_{qq}^{-1,(1)}(n_f+1)\hat{A}_{Qq}^{(2),\mathrm{PS},\mathrm{MOM}} + Z_{qg}^{-1,(1)}(n_f+1)\hat{A}_{gq,Q}^{(2),\mathrm{MOM}} \\
&+ \Big[\hat{A}_{Qg}^{(1),\mathrm{MOM}} + Z_{qg}^{-1,(1)}(n_f+1) - Z_{qg}^{-1,(1)}(n_f)\Big]\Gamma_{gq}^{-1,(2)}(n_f) \\
&+ \Big[\hat{A}_{Qq}^{(2),\mathrm{PS},\mathrm{MOM}} + Z_{qq}^{-1,(2),\mathrm{PS}}(n_f+1) - Z_{qq}^{-1,(2),\mathrm{PS}}(n_f)\Big]\Gamma_{qq}^{-1,(1)}(n_f) \\
&+ \Big[\hat{A}_{Qg}^{(2),\mathrm{MOM}} + Z_{qg}^{-1,(2)}(n_f+1) - Z_{qg}^{-1,(2)}(n_f) + Z_{qq}^{-1,(1)}(n_f+1)\,A_{Qg}^{(1),\mathrm{MOM}} \\
&\quad + Z_{qg}^{-1,(1)}(n_f+1)A_{gg,Q}^{(1),\mathrm{MOM}}\Big]\Gamma_{gq}^{-1,(1)}(n_f).
\end{aligned} \tag{4.102}$$

Taking into account the structure of the UV– and collinear singularities of the contributing Feynman–diagrams, these two contributions can be separated. For the bare quantities we obtain

$$\begin{aligned}
\hat{\hat{A}}_{Qq}^{(3),\mathrm{PS}} = \left(\frac{\hat{m}^2}{\mu^2}\right)^{3\varepsilon/2}\Bigg[&\frac{\hat{\gamma}_{qg}^{(0)}\gamma_{gq}^{(0)}}{6\varepsilon^3}\left(\gamma_{gg}^{(0)} - \gamma_{qq}^{(0)} + 6\beta_0 + 16\beta_{0,Q}\right) \\
&+ \frac{1}{\varepsilon^2}\Bigg(-\frac{4\hat{\gamma}_{qq}^{(1),\mathrm{PS}}}{3}\Big[\beta_0 + \beta_{0,Q}\Big] - \frac{\gamma_{gq}^{(0)}\hat{\gamma}_{qg}^{(1)}}{3} \\
&+ \frac{\hat{\gamma}_{qg}^{(0)}}{6}\Big[2\hat{\gamma}_{gq}^{(1)} - \gamma_{gq}^{(1)}\Big] + \delta m_1^{(-1)}\hat{\gamma}_{qg}^{(0)}\gamma_{gq}^{(0)}\Bigg) \\
&+ \frac{1}{\varepsilon}\Bigg(\frac{\hat{\gamma}_{qq}^{(2),\mathrm{PS}}}{3} - n_f\frac{\hat{\tilde{\gamma}}_{qq}^{(2),\mathrm{PS}}}{3} + \hat{\gamma}_{qg}^{(0)}a_{gq,Q}^{(2)} - \gamma_{gq}^{(0)}a_{Qg}^{(2)}
\end{aligned}$$

$$- 4(\beta_0 + \beta_{0,Q}) a_{Qq}^{(2),\mathsf{PS}} - \frac{\hat{\gamma}_{qg}^{(0)} \gamma_{gq}^{(0)} \zeta_2}{16} \Big[\gamma_{gg}^{(0)} - \gamma_{qq}^{(0)} + 6\beta_0 \Big] + \delta m_1^{(0)} \hat{\gamma}_{qg}^{(0)} \gamma_{gq}^{(0)} - \delta m_1^{(-1)} \hat{\gamma}_{qq}^{(1),\mathsf{PS}} \Big) + a_{Qq}^{(3),\mathsf{PS}} \Big], \tag{4.103}$$

$$\hat{\hat{A}}_{qq,Q}^{(3),\mathsf{PS}} = n_f \Big(\frac{\hat{m}^2}{\mu^2}\Big)^{3\varepsilon/2} \Bigg[\frac{2\hat{\gamma}_{qg}^{(0)} \gamma_{gq}^{(0)} \beta_{0,Q}}{3\varepsilon^3} + \frac{1}{3\varepsilon^2} \Big(2\hat{\gamma}_{qq}^{(1),\mathsf{PS}} \beta_{0,Q} + \hat{\gamma}_{qg}^{(0)} \hat{\gamma}_{gq}^{(1)} \Big) + \frac{1}{\varepsilon} \Big(\frac{\hat{\tilde{\gamma}}_{qq}^{(2),\mathsf{PS}}}{3} + \hat{\gamma}_{qg}^{(0)} a_{gq,Q}^{(2)} - \frac{\hat{\gamma}_{qg}^{(0)} \gamma_{gq}^{(0)} \beta_{0,Q} \zeta_2}{4} \Big) + \frac{a_{qq,Q}^{(3),\mathsf{PS}}}{n_f} \Bigg]. \tag{4.104}$$

The renormalized terms in the $\overline{\mathsf{MS}}$–scheme are given by

$$\begin{aligned} A_{Qq}^{(3),\mathsf{PS},\overline{\mathsf{MS}}} &= \frac{\hat{\gamma}_{qg}^{(0)} \gamma_{gq}^{(0)}}{48} \Big\{ \gamma_{gg}^{(0)} - \gamma_{qq}^{(0)} + 6\beta_0 + 16\beta_{0,Q} \Big\} \ln^3\Big(\frac{m^2}{\mu^2}\Big) \\ &+ \frac{1}{8} \Big\{ -4\hat{\gamma}_{qq}^{(1),\mathsf{PS}} \Big(\beta_0 + \beta_{0,Q}\Big) + \hat{\gamma}_{qg}^{(0)} \Big(\hat{\gamma}_{gq}^{(1)} - \gamma_{gq}^{(1)}\Big) - \gamma_{gq}^{(0)} \hat{\gamma}_{qg}^{(1)} \Big\} \ln^2\Big(\frac{m^2}{\mu^2}\Big) \\ &+ \frac{1}{16} \Big\{ 8\hat{\gamma}_{qq}^{(2),\mathsf{PS}} - 8n_f \hat{\tilde{\gamma}}_{qq}^{(2),\mathsf{PS}} - 32 a_{Qq}^{(2),\mathsf{PS}} (\beta_0 + \beta_{0,Q}) \\ &\quad + 8\hat{\gamma}_{qg}^{(0)} a_{gq,Q}^{(2)} - 8\gamma_{gq}^{(0)} a_{Qg}^{(2)} - \hat{\gamma}_{qg}^{(0)} \gamma_{gq}^{(0)} \zeta_2 \\ &\quad \times \Big(\gamma_{gg}^{(0)} - \gamma_{qq}^{(0)} + 6\beta_0 + 8\beta_{0,Q}\Big) \Big\} \ln\Big(\frac{m^2}{\mu^2}\Big) \\ &+ 4(\beta_0 + \beta_{0,Q}) \overline{a}_{Qq}^{(2),\mathsf{PS}} + \gamma_{gq}^{(0)} \overline{a}_{Qg}^{(2)} \\ &- \hat{\gamma}_{qg}^{(0)} \overline{a}_{gq,Q}^{(2)} + \frac{\gamma_{gq}^{(0)} \hat{\gamma}_{qg}^{(0)} \zeta_3}{48} \Big(\gamma_{gg}^{(0)} - \gamma_{qq}^{(0)} + 6\beta_0\Big) + \frac{\hat{\gamma}_{qg}^{(0)} \gamma_{gq}^{(1)} \zeta_2}{16} \\ &- \delta m_1^{(1)} \hat{\gamma}_{qg}^{(0)} \gamma_{gq}^{(0)} + \delta m_1^{(0)} \hat{\gamma}_{qq}^{(1),\mathsf{PS}} + 2\delta m_1^{(-1)} a_{Qq}^{(2),\mathsf{PS}} + a_{Qq}^{(3),\mathsf{PS}}. \end{aligned} \tag{4.105}$$

$$\begin{aligned} A_{qq,Q}^{(3),\mathsf{PS},\overline{\mathsf{MS}}} = n_f \Bigg\{ & \frac{\gamma_{gq}^{(0)} \hat{\gamma}_{qg}^{(0)} \beta_{0,Q}}{12} \ln^3\Big(\frac{m^2}{\mu^2}\Big) + \frac{1}{8} \Big(4\hat{\gamma}_{qq}^{(1),\mathsf{PS}} \beta_{0,Q} + \hat{\gamma}_{qg}^{(0)} \hat{\gamma}_{gq}^{(1)} \Big) \ln^2\Big(\frac{m^2}{\mu^2}\Big) \\ & + \frac{1}{4} \Big(2\hat{\tilde{\gamma}}_{qq}^{(2),\mathsf{PS}} + \hat{\gamma}_{qg}^{(0)} \Big\{ 2a_{gq,Q}^{(2)} - \gamma_{gq}^{(0)} \beta_{0,Q} \zeta_2 \Big\} \Big) \ln\Big(\frac{m^2}{\mu^2}\Big) \\ & - \hat{\gamma}_{qg}^{(0)} \overline{a}_{gq,Q}^{(2)} + \frac{\gamma_{gq}^{(0)} \hat{\gamma}_{qg}^{(0)} \beta_{0,Q} \zeta_3}{12} - \frac{\hat{\gamma}_{qq}^{(1),\mathsf{PS}} \beta_{0,Q} \zeta_2}{4} \Bigg\} + a_{qq,Q}^{(3),\mathsf{PS}}. \end{aligned} \tag{4.106}$$

4.7.4 A_{Qg} and $A_{qg,Q}$

The OME A_{Qg} is the most complex expression. As in the PS–case, there are two different contributions

$$A_{Qg} = a_s A_{Qg}^{(1)} + a_s^2 A_{Qg}^{(2)} + a_s^3 A_{Qg}^{(3)} + O(a_s^4). \tag{4.107}$$

$$A_{qg,Q} = a_s^3 A_{qg,Q}^{(3)} + O(a_s^4). \tag{4.108}$$

In the MOM–scheme the 1– and 2–loop contributions obey the following relations

$$A_{Qg}^{(1),\mathrm{MOM}} = \hat{A}_{Qg}^{(1),\mathrm{MOM}} + Z_{qg}^{-1,(1)}(n_f+1) - Z_{qg}^{-1,(1)}(n_f), \tag{4.109}$$

$$\begin{aligned} A_{Qg}^{(2),\mathrm{MOM}} = \hat{A}_{Qg}^{(2),\mathrm{MOM}} &+ Z_{qg}^{-1,(2)}(n_f+1) - Z_{qg}^{-1,(2)}(n_f) \\ &+ Z_{qg}^{-1,(1)}(n_f+1)\hat{A}_{gg,Q}^{(1),\mathrm{MOM}} + Z_{qq}^{-1,(1)}(n_f+1)\hat{A}_{Qg}^{(1),\mathrm{MOM}} \\ &+ \Big[\hat{A}_{Qg}^{(1),\mathrm{MOM}} + Z_{qg}^{-1,(1)}(n_f+1) - Z_{qg}^{-1,(1)}(n_f)\Big]\Gamma_{gg}^{-1,(1)}(n_f). \end{aligned} \tag{4.110}$$

The unrenormalized terms are given by

$$\hat{\hat{A}}_{Qg}^{(1)} = \left(\frac{\hat{m}^2}{\mu^2}\right)^{\varepsilon/2} \frac{\hat{\gamma}_{qg}^{(0)}}{\varepsilon} \exp\left(\sum_{i=2}^{\infty} \frac{\zeta_i}{i}\left(\frac{\varepsilon}{2}\right)^i\right), \tag{4.111}$$

$$\begin{aligned} \hat{\hat{A}}_{Qg}^{(2)} = \left(\frac{\hat{m}^2}{\mu^2}\right)^{\varepsilon} \Bigg[&-\frac{\hat{\gamma}_{qg}^{(0)}}{2\varepsilon^2}\left(\gamma_{gg}^{(0)} - \gamma_{qq}^{(0)} + 2\beta_0 + 4\beta_{0,Q}\right) + \frac{\hat{\gamma}_{qg}^{(1)} - 2\delta m_1^{(-1)}\hat{\gamma}_{qg}^{(0)}}{2\varepsilon} + a_{Qg}^{(2)} \\ &- \delta m_1^{(0)}\hat{\gamma}_{qg}^{(0)} - \frac{\hat{\gamma}_{qg}^{(0)}\beta_{0,Q}\zeta_2}{2} + \varepsilon\left(\overline{a}_{Qg}^{(2)} - \delta m_1^{(1)}\hat{\gamma}_{qg}^{(0)} - \frac{\hat{\gamma}_{qg}^{(0)}\beta_{0,Q}\zeta_2}{12}\right)\Bigg]. \end{aligned} \tag{4.112}$$

Note that we have already made the one–particle reducible contributions to Eq. 4.112 explicit, which are given by the LO–term multiplied with the 1–loop gluon–self energy, cf. Eq. 4.84. Furthermore, Eq. 4.112 already contains terms in the $O(\varepsilon^0)$ and $O(\varepsilon)$ expressions which result from mass renormalization. At this stage of the renormalization procedure they should not be present, however, we have included them here in order to have the same notation as in Refs. [16, 34] at the 2–loop level. The renormalized terms then become in the $\overline{\mathrm{MS}}$–scheme

$$A_{Qg}^{(1),\overline{\mathrm{MS}}} = \frac{\hat{\gamma}_{qg}^{(0)}}{2}\ln\left(\frac{m^2}{\mu^2}\right), \tag{4.113}$$

$$A_{Qg}^{(2),\overline{\mathsf{MS}}} = -\frac{\hat{\gamma}_{qg}^{(0)}}{8}\left[\gamma_{gg}^{(0)} - \gamma_{qq}^{(0)} + 2\beta_0 + 4\beta_{0,Q}\right]\ln^2\left(\frac{m^2}{\mu^2}\right) + \frac{\hat{\gamma}_{qg}^{(1)}}{2}\ln\left(\frac{m^2}{\mu^2}\right)$$
$$+ a_{Qg}^{(2)} + \frac{\hat{\gamma}_{qg}^{(0)}\zeta_2}{8}\left(\gamma_{gg}^{(0)} - \gamma_{qq}^{(0)} + 2\beta_0\right). \tag{4.114}$$

The generic renormalization relation at the 3–loop level is given by

$$A_{Qg}^{(3),\mathrm{MOM}} + A_{qg,Q}^{(3),\mathrm{MOM}} = \hat{A}_{Qg}^{(3),\mathrm{MOM}} + \hat{A}_{qg,Q}^{(3),\mathrm{MOM}}$$
$$+ Z_{qg}^{-1,(3)}(n_f+1) - Z_{qg}^{-1,(3)}(n_f) + Z_{qg}^{-1,(2)}(n_f+1)\hat{A}_{gg,Q}^{(1),\mathrm{MOM}}$$
$$+ Z_{qg}^{-1,(1)}(n_f+1)\hat{A}_{gg,Q}^{(2),\mathrm{MOM}} + Z_{qq}^{-1,(2)}(n_f+1)\hat{A}_{Qg}^{(1),\mathrm{MOM}}$$
$$+ Z_{qq}^{-1,(1)}(n_f+1)\hat{A}_{Qg}^{(2),\mathrm{MOM}}$$
$$+ \left[\hat{A}_{Qg}^{(1),\mathrm{MOM}} + Z_{qg}^{-1,(1)}(n_f+1) - Z_{qg}^{-1,(1)}(n_f)\right]\Gamma_{gg}^{-1,(2)}(n_f)$$
$$+ \left[\hat{A}_{Qg}^{(2),\mathrm{MOM}} + Z_{qg}^{-1,(2)}(n_f+1) - Z_{qg}^{-1,(2)}(n_f)\right.$$
$$\left. + Z_{qq}^{-1,(1)}(n_f+1)A_{Qg}^{(1),\mathrm{MOM}} + Z_{qg}^{-1,(1)}(n_f+1)A_{gg,Q}^{(1),\mathrm{MOM}}\right]\Gamma_{gg}^{-1,(1)}(n_f)$$
$$+ \left[\hat{A}_{Qq}^{(2),\mathsf{PS},\mathrm{MOM}} + Z_{qq}^{-1,(2),\mathsf{PS}}(n_f+1) - Z_{qq}^{-1,(2),\mathsf{PS}}(n_f)\right]\Gamma_{qg}^{-1,(1)}(n_f)$$
$$+ \left[\hat{A}_{qq,Q}^{(2),\mathsf{NS},\mathrm{MOM}} + Z_{qq}^{-1,(2),\mathsf{NS}}(n_f+1) - Z_{qq}^{-1,(2),\mathsf{NS}}(n_f)\right]\Gamma_{qg}^{-1,(1)}(n_f). \tag{4.115}$$

Similar to the PS–case, the different contributions can be separated and one obtains the following unrenormalized results

$$\hat{\hat{A}}_{Qg}^{(3)} = \left(\frac{\hat{m}^2}{\mu^2}\right)^{3\varepsilon/2}\left[\frac{\hat{\gamma}_{qg}^{(0)}}{6\varepsilon^3}\left((n_f+1)\gamma_{gq}^{(0)}\hat{\gamma}_{qg}^{(0)} + \gamma_{qq}^{(0)}\left[\gamma_{qq}^{(0)} - 2\gamma_{gg}^{(0)} - 6\beta_0 - 8\beta_{0,Q}\right] + 8\beta_0^2\right.\right.$$
$$\left. + 28\beta_{0,Q}\beta_0 + 24\beta_{0,Q}^2 + \gamma_{gg}^{(0)}\left[\gamma_{gg}^{(0)} + 6\beta_0 + 14\beta_{0,Q}\right]\right)$$
$$+ \frac{1}{6\varepsilon^2}\left(\hat{\gamma}_{qg}^{(1)}\left[2\gamma_{qq}^{(0)} - 2\gamma_{gg}^{(0)} - 8\beta_0 - 10\beta_{0,Q}\right]\right.$$
$$+ \hat{\gamma}_{qg}^{(0)}\left[\hat{\gamma}_{qq}^{(1),\mathsf{PS}}\{1-2n_f\} + \gamma_{qq}^{(1),\mathsf{NS}} + \hat{\gamma}_{qq}^{(1),\mathsf{NS}}\right.$$
$$\left. + 2\hat{\gamma}_{gg}^{(1)} - \gamma_{gg}^{(1)} - 2\beta_1 - 2\beta_{1,Q}\right]$$
$$\left. + 6\delta m_1^{(-1)}\hat{\gamma}_{qg}^{(0)}\left[\gamma_{gg}^{(0)} - \gamma_{qq}^{(0)} + 3\beta_0 + 5\beta_{0,Q}\right]\right)$$
$$+ \frac{1}{\varepsilon}\left(\frac{\hat{\gamma}_{qg}^{(2)}}{3} - n_f\frac{\hat{\hat{\gamma}}_{qg}^{(2)}}{3} + \hat{\gamma}_{qg}^{(0)}\left[a_{gg,Q}^{(2)} - n_f a_{Qq}^{(2),\mathsf{PS}}\right]\right.$$
$$+ a_{Qg}^{(2)}\left[\gamma_{qq}^{(0)} - \gamma_{gg}^{(0)} - 4\beta_0 - 4\beta_{0,Q}\right]$$

$$
\begin{aligned}
&+\frac{\hat{\gamma}_{qg}^{(0)}\zeta_2}{16}\Big[\gamma_{gg}^{(0)}\Big\{2\gamma_{qq}^{(0)}-\gamma_{gg}^{(0)}-6\beta_0+2\beta_{0,Q}\Big\}\\
&-(n_f+1)\gamma_{gq}^{(0)}\hat{\gamma}_{qg}^{(0)}+\gamma_{qq}^{(0)}\Big\{-\gamma_{qq}^{(0)}+6\beta_0\Big\}\\
&-8\beta_0^2+4\beta_{0,Q}\beta_0+24\beta_{0,Q}^2\Big]\\
&+\frac{\delta m_1^{(-1)}}{2}\Big[-2\hat{\gamma}_{qg}^{(1)}+3\delta m_1^{(-1)}\hat{\gamma}_{qg}^{(0)}+2\delta m_1^{(0)}\hat{\gamma}_{qg}^{(0)}\Big]\\
&+\delta m_1^{(0)}\hat{\gamma}_{qg}^{(0)}\Big[\gamma_{gg}^{(0)}-\gamma_{qq}^{(0)}+2\beta_0+4\beta_{0,Q}\Big]\\
&-\delta m_2^{(-1)}\hat{\gamma}_{qg}^{(0)}\Big)+a_{Qg}^{(3)}\Big].
\end{aligned}
\quad (4.116)
$$

$$
\begin{aligned}
\hat{\hat{A}}_{qg,Q}^{(3)}=n_f\Big(\frac{\hat{m}^2}{\mu^2}\Big)^{3\varepsilon/2}&\Big[\frac{\hat{\gamma}_{qg}^{(0)}}{6\varepsilon^3}\Big(\gamma_{gq}^{(0)}\hat{\gamma}_{qg}^{(0)}+2\beta_{0,Q}\Big[\gamma_{gg}^{(0)}-\gamma_{qq}^{(0)}+2\beta_0\Big]\Big)\\
&+\frac{1}{\varepsilon^2}\Big(\frac{\hat{\gamma}_{qg}^{(0)}}{6}\Big[2\hat{\gamma}_{gg}^{(1)}+\hat{\gamma}_{qq}^{(1),\mathsf{PS}}-2\hat{\gamma}_{qq}^{(1),\mathsf{NS}}+4\beta_{1,Q}\Big]+\frac{\hat{\gamma}_{qg}^{(1)}\beta_{0,Q}}{3}\Big)\\
&+\frac{1}{\varepsilon}\Big(\frac{\hat{\hat{\gamma}}_{qg}^{(2)}}{3}+\hat{\gamma}_{qg}^{(0)}\Big[a_{gg,Q}^{(2)}-a_{qq,Q}^{(2),\mathsf{NS}}+\beta_{1,Q}^{(1)}\Big]\\
&-\frac{\hat{\gamma}_{qg}^{(0)}\zeta_2}{16}\Big[\gamma_{gq}^{(0)}\hat{\gamma}_{qg}^{(0)}+2\beta_{0,Q}\Big\{\gamma_{gg}^{(0)}-\gamma_{qq}^{(0)}\\
&+2\beta_0\Big\}\Big]\Big)+\frac{a_{qg,Q}^{(3)}}{n_f}\Big].
\end{aligned}
\quad (4.117)
$$

The renormalized expressions are

$$
\begin{aligned}
A_{Qg}^{(3),\overline{\mathsf{MS}}}=\frac{\hat{\gamma}_{qg}^{(0)}}{48}&\Big\{(n_f+1)\gamma_{gq}^{(0)}\hat{\gamma}_{qg}^{(0)}+\gamma_{gg}^{(0)}\Big(\gamma_{gg}^{(0)}-2\gamma_{qq}^{(0)}+6\beta_0+14\beta_{0,Q}\Big)\\
&+\gamma_{qq}^{(0)}\Big(\gamma_{qq}^{(0)}-6\beta_0-8\beta_{0,Q}\Big)+8\beta_0^2+28\beta_{0,Q}\beta_0+24\beta_{0,Q}^2\Big\}\ln^3\Big(\frac{m^2}{\mu^2}\Big)\\
+\frac{1}{8}&\Big\{\hat{\gamma}_{qg}^{(1)}\Big(\gamma_{qq}^{(0)}-\gamma_{gg}^{(0)}-4\beta_0-6\beta_{0,Q}\Big)+\hat{\gamma}_{qg}^{(0)}\Big(\hat{\gamma}_{gg}^{(1)}-\gamma_{gg}^{(1)}\\
&+(1-n_f)\hat{\gamma}_{qq}^{(1),\mathsf{PS}}+\gamma_{qq}^{(1),\mathsf{NS}}+\hat{\gamma}_{qq}^{(1),\mathsf{NS}}-2\beta_1-2\beta_{1,Q}\Big)\Big\}\ln^2\Big(\frac{m^2}{\mu^2}\Big)\\
+&\Big\{\frac{\hat{\gamma}_{qg}^{(2)}}{2}-n_f\frac{\hat{\hat{\gamma}}_{qg}^{(2)}}{2}+\frac{a_{Qg}^{(2)}}{2}\Big(\gamma_{qq}^{(0)}-\gamma_{gg}^{(0)}-4\beta_0-4\beta_{0,Q}\Big)\\
&+\frac{\hat{\gamma}_{qg}^{(0)}}{2}\Big(a_{gg,Q}^{(2)}-n_f a_{Qq}^{(2),\mathsf{PS}}\Big)+\frac{\hat{\gamma}_{qg}^{(0)}\zeta_2}{16}\Big(-(n_f+1)\gamma_{gq}^{(0)}\hat{\gamma}_{qg}^{(0)}\\
&+\gamma_{gg}^{(0)}\Big[2\gamma_{qq}^{(0)}-\gamma_{gg}^{(0)}-6\beta_0-6\beta_{0,Q}\Big]-4\beta_0[2\beta_0+3\beta_{0,Q}]
\end{aligned}
$$

$$
\begin{aligned}
&+\gamma_{qq}^{(0)}\Big[-\gamma_{qq}^{(0)}+6\beta_0+4\beta_{0,Q}\Big]\Big)\Big\}\ln\Big(\frac{m^2}{\mu^2}\Big)\\
&+\overline{a}_{Qg}^{(2)}\Big(\gamma_{gg}^{(0)}-\gamma_{qq}^{(0)}+4\beta_0+4\beta_{0,Q}\Big)+\hat{\gamma}_{qg}^{(0)}\Big(n_f\overline{a}_{Qq}^{(2),\mathsf{PS}}-\overline{a}_{gg,Q}^{(2)}\Big)\\
&+\frac{\hat{\gamma}_{qg}^{(0)}\zeta_3}{48}\Big((n_f+1)\gamma_{gq}^{(0)}\hat{\gamma}_{qg}^{(0)}+\gamma_{gg}^{(0)}\Big[\gamma_{gg}^{(0)}-2\gamma_{qq}^{(0)}+6\beta_0-2\beta_{0,Q}\Big]\\
&+\gamma_{qq}^{(0)}\Big[\gamma_{qq}^{(0)}-6\beta_0\Big]+8\beta_0^2-4\beta_0\beta_{0,Q}-24\beta_{0,Q}^2\Big)\\
&+\frac{\hat{\gamma}_{qg}^{(1)}\beta_{0,Q}\zeta_2}{8}+\frac{\hat{\gamma}_{qg}^{(0)}\zeta_2}{16}\Big(\gamma_{gg}^{(1)}-\hat{\gamma}_{qq}^{(1),\mathsf{NS}}-\gamma_{qq}^{(1),\mathsf{NS}}-\hat{\gamma}_{qq}^{(1),\mathsf{PS}}+2\beta_1+2\beta_{1,Q}\Big)\\
&+\frac{\delta m_1^{(-1)}}{8}\Big(16a_{Qg}^{(2)}+\hat{\gamma}_{qg}^{(0)}\Big[-24\delta m_1^{(0)}-8\delta m_1^{(1)}-\zeta_2\beta_0-9\zeta_2\beta_{0,Q}\Big]\Big)\\
&+\frac{\delta m_1^{(0)}}{2}\Big(2\hat{\gamma}_{qg}^{(1)}-\delta m_1^{(0)}\hat{\gamma}_{qg}^{(0)}\Big)+\delta m_1^{(1)}\hat{\gamma}_{qg}^{(0)}\Big(\gamma_{qq}^{(0)}-\gamma_{gg}^{(0)}-2\beta_0-4\beta_{0,Q}\Big)\\
&+\delta m_2^{(0)}\hat{\gamma}_{qg}^{(0)}+a_{Qg}^{(3)}.
\end{aligned}
\tag{4.118}
$$

$$
\begin{aligned}
A_{qg,Q}^{(3),\overline{\mathsf{MS}}}=n_f\Bigg[&\frac{\hat{\gamma}_{qg}^{(0)}}{48}\Big\{\gamma_{gq}^{(0)}\hat{\gamma}_{qg}^{(0)}+2\beta_{0,Q}\Big(\gamma_{gg}^{(0)}-\gamma_{qq}^{(0)}+2\beta_0\Big)\Big\}\ln^3\Big(\frac{m^2}{\mu^2}\Big)\\
&+\frac{1}{8}\Big\{2\hat{\gamma}_{qg}^{(1)}\beta_{0,Q}+\hat{\gamma}_{qg}^{(0)}\Big(\hat{\gamma}_{qq}^{(1),\mathsf{PS}}-\hat{\gamma}_{qq}^{(1),\mathsf{NS}}+\hat{\gamma}_{gg}^{(1)}+2\beta_{1,Q}\Big)\Big\}\ln^2\Big(\frac{m^2}{\mu^2}\Big)\\
&+\frac{1}{2}\Big\{\hat{\hat{\gamma}}_{qg}^{(2)}+\hat{\gamma}_{qg}^{(0)}\Big(a_{gg,Q}^{(2)}-a_{qq,Q}^{(2),\mathsf{NS}}+\beta_{1,Q}^{(1)}\Big)\\
&-\frac{\hat{\gamma}_{qg}^{(0)}}{8}\zeta_2\Big(\gamma_{gq}^{(0)}\hat{\gamma}_{qg}^{(0)}+2\beta_{0,Q}\Big[\gamma_{gg}^{(0)}-\gamma_{qq}^{(0)}+2\beta_0\Big]\Big)\Big\}\ln\Big(\frac{m^2}{\mu^2}\Big)\\
&+\hat{\gamma}_{qg}^{(0)}\Big(\overline{a}_{qq,Q}^{(2),\mathsf{NS}}-\overline{a}_{gg,Q}^{(2)}-\beta_{1,Q}^{(2)}\Big)\\
&+\frac{\hat{\gamma}_{qg}^{(0)}}{48}\zeta_3\Big(\gamma_{gq}^{(0)}\hat{\gamma}_{qg}^{(0)}+2\beta_{0,Q}\Big[\gamma_{gg}^{(0)}-\gamma_{qq}^{(0)}+2\beta_0\Big]\Big)\\
&-\frac{\zeta_2}{16}\Big(\hat{\gamma}_{qg}^{(0)}\hat{\gamma}_{qq}^{(1),\mathsf{PS}}+2\hat{\gamma}_{qg}^{(1)}\beta_{0,Q}\Big)+\frac{a_{qg,Q}^{(3)}}{n_f}\Bigg].
\end{aligned}
\tag{4.119}
$$

4.7.5 $A_{gq,Q}$

The gq–contributions start at $O(a_s^2)$,

$$A_{gq,Q}=a_s^2A_{gq,Q}^{(2)}+a_s^3A_{gq,Q}^{(3)}+O(a_s^4). \tag{4.120}$$

The renormalization formulas in the MOM–scheme read

$$A_{gq,Q}^{(2),\mathrm{MOM}} = \hat{A}_{gq,Q}^{(2),\mathrm{MOM}} + Z_{gq}^{-1,(2)}(n_f+1) - Z_{gq}^{-1,(2)}(n_f) + \left(\hat{A}_{gg,Q}^{(1),\mathrm{MOM}} + Z_{gg}^{-1,(1)}(n_f+1) - Z_{gg}^{-1,(1)}(n_f)\right)\Gamma_{gq}^{-1,(1)}, \tag{4.121}$$

$$\begin{aligned} A_{gq,Q}^{(3),\mathrm{MOM}} = {} & \hat{A}_{gq,Q}^{(3),\mathrm{MOM}} + Z_{gq}^{-1,(3)}(n_f+1) - Z_{gq}^{-1,(3)}(n_f) + Z_{gg}^{-1,(1)}(n_f+1)\hat{A}_{gq,Q}^{(2),\mathrm{MOM}} \\ & + Z_{gq}^{-1,(1)}(n_f+1)\hat{A}_{qq}^{(2),\mathrm{MOM}} \\ & + \left[\hat{A}_{gg,Q}^{(1),\mathrm{MOM}} + Z_{gg}^{-1,(1)}(n_f+1) - Z_{gg}^{-1,(1)}(n_f)\right]\Gamma_{gq}^{-1,(2)}(n_f) \\ & + \left[\hat{A}_{gq,Q}^{(2),\mathrm{MOM}} + Z_{gq}^{-1,(2)}(n_f+1) - Z_{gq}^{-1,(2)}(n_f)\right]\Gamma_{qq}^{-1,(1)}(n_f) \\ & + \Big[\hat{A}_{gg,Q}^{(2),\mathrm{MOM}} + Z_{gg}^{-1,(2)}(n_f+1) - Z_{gg}^{-1,(2)}(n_f) \\ & \quad + Z_{gg}^{-1,(1)}(n_f+1)\hat{A}_{gg,Q}^{(1),\mathrm{MOM}} + Z_{gq}^{-1,(1)}(n_f+1)\hat{A}_{Qg}^{(1),\mathrm{MOM}}\Big]\Gamma_{gq}^{-1,(1)}(n_f), \end{aligned} \tag{4.122}$$

while the unrenormalized expressions are

$$\hat{\hat{A}}_{gq,Q}^{(2)} = \left(\frac{\hat{m}^2}{\mu^2}\right)^{\varepsilon}\left[\frac{2\beta_{0,Q}}{\varepsilon^2}\gamma_{gq}^{(0)} + \frac{\hat{\gamma}_{gq}^{(1)}}{2\varepsilon} + a_{gq,Q}^{(2)} + \overline{a}_{gq,Q}^{(2)}\varepsilon\right], \tag{4.123}$$

$$\begin{aligned} \hat{\hat{A}}_{gq,Q}^{(3)} = \left(\frac{\hat{m}^2}{\mu^2}\right)^{3\varepsilon/2}\Bigg\{ & -\frac{\gamma_{gq}^{(0)}}{3\varepsilon^3}\left(\gamma_{gq}^{(0)}\hat{\gamma}_{qg}^{(0)} + \left[\gamma_{qq}^{(0)} - \gamma_{gg}^{(0)} + 10\beta_0 + 24\beta_{0,Q}\right]\beta_{0,Q}\right) \\ & + \frac{1}{\varepsilon^2}\Bigg(\gamma_{gq}^{(1)}\beta_{0,Q} + \frac{\hat{\gamma}_{gq}^{(1)}}{3}\left[\gamma_{gg}^{(0)} - \gamma_{qq}^{(0)} - 4\beta_0 - 6\beta_{0,Q}\right] \\ & + \frac{\gamma_{gq}^{(0)}}{3}\left[\hat{\gamma}_{qq}^{(1),\mathsf{NS}} + \hat{\gamma}_{qq}^{(1),\mathsf{PS}} - \hat{\gamma}_{gg}^{(1)} + 2\beta_{1,Q}\right] \\ & - 4\delta m_1^{(-1)}\beta_{0,Q}\gamma_{gq}^{(0)}\Bigg) \\ & + \frac{1}{\varepsilon}\Bigg(\frac{\hat{\gamma}_{gq}^{(2)}}{3} + a_{gq,Q}^{(2)}\left[\gamma_{gg}^{(0)} - \gamma_{qq}^{(0)} - 6\beta_{0,Q} - 4\beta_0\right] \\ & + \gamma_{gq}^{(0)}\left[a_{qq,Q}^{(2),\mathsf{NS}} + a_{Qq}^{(2),\mathsf{PS}} - a_{gg,Q}^{(2)}\right] + \gamma_{gq}^{(0)}\beta_{1,Q}^{(1)} \\ & + \frac{\gamma_{gq}^{(0)}\zeta_2}{8}\left[\gamma_{gq}^{(0)}\hat{\gamma}_{qg}^{(0)} + \beta_{0,Q}(\gamma_{qq}^{(0)} - \gamma_{gg}^{(0)} + 10\beta_0)\right] \\ & - \delta m_1^{(-1)}\hat{\gamma}_{gq}^{(1)} - 4\delta m_1^{(0)}\beta_{0,Q}\gamma_{gq}^{(0)}\Bigg) + a_{gq,Q}^{(3)}\Bigg\}. \end{aligned} \tag{4.124}$$

The contributions to the renormalized operator matrix element are given by

$$A_{gq,Q}^{(2),\overline{\mathsf{MS}}} = \frac{\beta_{0,Q}\gamma_{gq}^{(0)}}{2}\ln^2\left(\frac{m^2}{\mu^2}\right) + \frac{\hat{\gamma}_{gq}^{(1)}}{2}\ln\left(\frac{m^2}{\mu^2}\right) + a_{gq,Q}^{(2)} - \frac{\beta_{0,Q}\gamma_{gq}^{(0)}}{2}\zeta_2, \tag{4.125}$$

$$A_{gq,Q}^{(3),\overline{\mathrm{MS}}} = -\frac{\gamma_{gq}^{(0)}}{24}\Biggl\{\gamma_{gq}^{(0)}\hat{\gamma}_{qg}^{(0)} + \Bigl(\gamma_{qq}^{(0)} - \gamma_{gg}^{(0)} + 10\beta_0 + 24\beta_{0,Q}\Bigr)\beta_{0,Q}\Biggr\}\ln^3\Bigl(\frac{m^2}{\mu^2}\Bigr)$$
$$+\frac{1}{8}\Biggl\{6\gamma_{gq}^{(1)}\beta_{0,Q} + \hat{\gamma}_{gq}^{(1)}\Bigl(\gamma_{gg}^{(0)} - \gamma_{qq}^{(0)} - 4\beta_0 - 6\beta_{0,Q}\Bigr) + \gamma_{gq}^{(0)}\Bigl(\hat{\gamma}_{qq}^{(1),\mathrm{NS}}$$
$$+\hat{\gamma}_{qq}^{(1),\mathrm{PS}} - \hat{\gamma}_{gg}^{(1)} + 2\beta_{1,Q}\Bigr)\Biggr\}\ln^2\Bigl(\frac{m^2}{\mu^2}\Bigr)$$
$$+\frac{1}{8}\Biggl\{4\hat{\gamma}_{gq}^{(2)} + 4a_{gq,Q}^{(2)}\Bigl(\gamma_{gg}^{(0)} - \gamma_{qq}^{(0)} - 4\beta_0 - 6\beta_{0,Q}\Bigr)$$
$$+4\gamma_{gq}^{(0)}\Bigl(a_{qq,Q}^{(2),\mathrm{NS}} + a_{Qq}^{(2),\mathrm{PS}} - a_{gg,Q}^{(2)} + \beta_{1,Q}^{(1)}\Bigr)$$
$$+\gamma_{gq}^{(0)}\zeta_2\Bigl(\gamma_{gq}^{(0)}\hat{\gamma}_{qg}^{(0)} + \Bigl[\gamma_{qq}^{(0)} - \gamma_{gg}^{(0)} + 12\beta_{0,Q} + 10\beta_0\Bigr]\beta_{0,Q}\Bigr)\Biggr\}\ln\Bigl(\frac{m^2}{\mu^2}\Bigr)$$
$$+\overline{a}_{gq,Q}^{(2)}\Bigl(\gamma_{qq}^{(0)} - \gamma_{gg}^{(0)} + 4\beta_0 + 6\beta_{0,Q}\Bigr) + \gamma_{gq}^{(0)}\Bigl(\overline{a}_{gg,Q}^{(2)} - \overline{a}_{Qq}^{(2),\mathrm{PS}} - \overline{a}_{qq,Q}^{(2),\mathrm{NS}}\Bigr)$$
$$-\gamma_{gq}^{(0)}\beta_{1,Q}^{(2)} - \frac{\gamma_{gq}^{(0)}\zeta_3}{24}\Bigl(\gamma_{gq}^{(0)}\hat{\gamma}_{qg}^{(0)} + \Bigl[\gamma_{qq}^{(0)} - \gamma_{gg}^{(0)} + 10\beta_0\Bigr]\beta_{0,Q}\Bigr)$$
$$-\frac{3\gamma_{gq}^{(1)}\beta_{0,Q}\zeta_2}{8} + 2\delta m_1^{(-1)}a_{gq,Q}^{(2)} + \delta m_1^{(0)}\hat{\gamma}_{gq}^{(1)} + 4\delta m_1^{(1)}\beta_{0,Q}\gamma_{gq}^{(0)} + a_{gq,Q}^{(3)}. \tag{4.126}$$

4.7.6 $A_{gg,Q}$

The gg–contributions start at $O(a_s^0)$,

$$A_{gg,Q} = 1 + a_s A_{gg,Q}^{(1)} + a_s^2 A_{gg,Q}^{(2)} + a_s^3 A_{gg,Q}^{(3)} + O(a_s^4). \tag{4.127}$$

The corresponding renormalization formulas read in the MOM–scheme

$$A_{gg,Q}^{(1),\mathrm{MOM}} = \hat{A}_{gg,Q}^{(1),\mathrm{MOM}} + Z_{gg}^{-1,(1)}(n_f+1) - Z_{gg}^{-1,(1)}(n_f), \tag{4.128}$$

$$A_{gg,Q}^{(2),\mathrm{MOM}} = \hat{A}_{gg,Q}^{(2),\mathrm{MOM}} + Z_{gg}^{-1,(2)}(n_f+1) - Z_{gg}^{-1,(2)}(n_f)$$
$$+ Z_{gg}^{-1,(1)}(n_f+1)\hat{A}_{gg,Q}^{(1),\mathrm{MOM}} + Z_{gq}^{-1,(1)}(n_f+1)\hat{A}_{Qg}^{(1),\mathrm{MOM}}$$
$$+\Bigl[\hat{A}_{gg,Q}^{(1),\mathrm{MOM}} + Z_{gg}^{-1,(1)}(n_f+1) - Z_{gg}^{-1,(1)}(n_f)\Bigr]\Gamma_{gg}^{-1,(1)}(n_f), \tag{4.129}$$

$$A_{gg,Q}^{(3),\mathrm{MOM}} = \hat{A}_{gg,Q}^{(3),\mathrm{MOM}} + Z_{gg}^{-1,(3)}(n_f+1) - Z_{gg}^{-1,(3)}(n_f)$$
$$+ Z_{gg}^{-1,(2)}(n_f+1)\hat{A}_{gg,Q}^{(1),\mathrm{MOM}} + Z_{gg}^{-1,(1)}(n_f+1)\hat{A}_{gg,Q}^{(2),\mathrm{MOM}}$$
$$+ Z_{gq}^{-1,(2)}(n_f+1)\hat{A}_{Qg}^{(1),\mathrm{MOM}} + Z_{gq}^{-1,(1)}(n_f+1)\hat{A}_{Qg}^{(2),\mathrm{MOM}}$$

$$
\begin{aligned}
&+\Big[\hat{A}_{gg,Q}^{(1),\mathrm{MOM}} + Z_{gg}^{-1,(1)}(n_f+1) - Z_{gg}^{-1,(1)}(n_f)\Big]\Gamma_{gg}^{-1,(2)}(n_f) \\
&+\Big[\hat{A}_{gg,Q}^{(2),\mathrm{MOM}} + Z_{gg}^{-1,(2)}(n_f+1) - Z_{gg}^{-1,(2)}(n_f) \\
&+ Z_{gq}^{-1,(1)}(n_f+1)A_{Qg}^{(1),\mathrm{MOM}} + Z_{gg}^{-1,(1)}(n_f+1)A_{gg,Q}^{(1),\mathrm{MOM}}\Big]\Gamma_{gg}^{-1,(1)}(n_f) \\
&+\Big[\hat{A}_{gq,Q}^{(2),\mathrm{MOM}} + Z_{gq}^{-1,(2)}(n_f+1) - Z_{gq}^{-1,(2)}(n_f)\Big]\Gamma_{qg}^{-1,(1)}(n_f).
\end{aligned}
\tag{4.130}
$$

The general structure of the unrenormalized 1–loop result is then given by

$$
\hat{\hat{A}}_{gg,Q}^{(1)} = \left(\frac{\hat{m}^2}{\mu^2}\right)^{\varepsilon/2}\left(\frac{\hat{\gamma}_{gg}^{(0)}}{\varepsilon} + a_{gg,Q}^{(1)} + \varepsilon\overline{a}_{gg,Q}^{(1)} + \varepsilon^2\overline{\overline{a}}_{gg,Q}^{(1)}\right). \tag{4.131}
$$

One obtains

$$
\hat{\hat{A}}_{gg,Q}^{(1)} = \left(\frac{\hat{m}^2}{\mu^2}\right)^{\varepsilon/2}\left(-\frac{2\beta_{0,Q}}{\varepsilon}\right)\exp\Big(\sum_{i=2}^{\infty}\frac{\zeta_i}{i}\Big(\frac{\varepsilon}{2}\Big)^i\Big). \tag{4.132}
$$

Using Eq. 4.132, the 2–loop term is given by

$$
\begin{aligned}
\hat{\hat{A}}_{gg,Q}^{(2)} = \left(\frac{\hat{m}^2}{\mu^2}\right)^{\varepsilon}\Bigg[&\frac{1}{2\varepsilon^2}\Big\{\gamma_{gq}^{(0)}\hat{\gamma}_{qg}^{(0)} + 2\beta_{0,Q}\Big(\gamma_{gg}^{(0)} + 2\beta_0 + 4\beta_{0,Q}\Big)\Big\} \\
&+\frac{\hat{\gamma}_{gg}^{(1)} + 4\delta m_1^{(-1)}\beta_{0,Q}}{2\varepsilon} + a_{gg,Q}^{(2)} + 2\delta m_1^{(0)}\beta_{0,Q} + \beta_{0,Q}^2\zeta_2 \\
&+\varepsilon\Big[\overline{a}_{gg,Q}^{(2)} + 2\delta m_1^{(1)}\beta_{0,Q} + \frac{\beta_{0,Q}^2\zeta_3}{6}\Big]\Bigg].
\end{aligned}
\tag{4.133}
$$

Again, we have made explicit one–particle reducible contributions and terms stemming from mass renormalization in order to refer to the notation of Refs. [16, 34], cf. the discussion below (4.112). The 3–loop contribution becomes

$$
\begin{aligned}
\hat{\hat{A}}_{gg,Q}^{(3)} = \left(\frac{\hat{m}^2}{\mu^2}\right)^{3\varepsilon/2}\Bigg[&\frac{1}{\varepsilon^3}\Bigg(-\frac{\gamma_{gq}^{(0)}\hat{\gamma}_{qg}^{(0)}}{6}\Big[\gamma_{gg}^{(0)} - \gamma_{qq}^{(0)} + 6\beta_0 + 4n_f\beta_{0,Q} + 10\beta_{0,Q}\Big] \\
&-\frac{2\gamma_{gg}^{(0)}\beta_{0,Q}}{3}\Big[2\beta_0 + 7\beta_{0,Q}\Big] - \frac{4\beta_{0,Q}}{3}\Big[2\beta_0^2 + 7\beta_{0,Q}\beta_0 + 6\beta_{0,Q}^2\Big]\Bigg) \\
&+\frac{1}{\varepsilon^2}\Bigg(\frac{\hat{\gamma}_{qg}^{(0)}}{6}\Big[\gamma_{gq}^{(1)} - (2n_f-1)\hat{\gamma}_{gq}^{(1)}\Big] + \frac{\gamma_{gq}^{(0)}\hat{\gamma}_{qg}^{(1)}}{3} \\
&-\frac{\hat{\gamma}_{gg}^{(1)}}{3}\Big[4\beta_0 + 7\beta_{0,Q}\Big] + \frac{2\beta_{0,Q}}{3}\Big[\gamma_{gg}^{(1)} + \beta_1 + \beta_{1,Q}\Big] \\
&+\frac{2\gamma_{gg}^{(0)}\beta_{1,Q}}{3} + \delta m_1^{(-1)}\Big[-\hat{\gamma}_{qg}^{(0)}\gamma_{gq}^{(0)} - 2\beta_{0,Q}\gamma_{gg}^{(0)} - 10\beta_{0,Q}^2 - 6\beta_{0,Q}\beta_0\Big]\Bigg) \\
&+\frac{1}{\varepsilon}\Bigg(\frac{\hat{\gamma}_{gg}^{(2)}}{3} - 2(2\beta_0 + 3\beta_{0,Q})a_{gg,Q}^{(2)} - n_f\hat{\gamma}_{qg}^{(0)}a_{gq,Q}^{(2)} + \gamma_{gq}^{(0)}a_{Qg}^{(2)}
\end{aligned}
$$

$$+\beta_{1,Q}^{(1)}\gamma_{gg}^{(0)}+\frac{\gamma_{gq}^{(0)}\hat{\gamma}_{qg}^{(0)}\zeta_2}{16}\Big[\gamma_{gg}^{(0)}-\gamma_{qq}^{(0)}+2(2n_f+1)\beta_{0,Q}+6\beta_0\Big]$$
$$+\frac{\beta_{0,Q}\zeta_2}{4}\Big[\gamma_{gg}^{(0)}\{2\beta_0-\beta_{0,Q}\}+4\beta_0^2-2\beta_{0,Q}\beta_0-12\beta_{0,Q}^2\Big]$$
$$+\delta m_1^{(-1)}\Big[-3\delta m_1^{(-1)}\beta_{0,Q}-2\delta m_1^{(0)}\beta_{0,Q}-\hat{\gamma}_{gg}^{(1)}\Big]$$
$$+\delta m_1^{(0)}\Big[-\hat{\gamma}_{qg}^{(0)}\gamma_{gq}^{(0)}-2\gamma_{gg}^{(0)}\beta_{0,Q}-4\beta_{0,Q}\beta_0-8\beta_{0,Q}^2\Big]$$
$$+2\delta m_2^{(-1)}\beta_{0,Q}\Big)+a_{gg,Q}^{(3)}\Big]. \quad (4.134)$$

The renormalized results are

$$A_{gg,Q}^{(1),\overline{\mathsf{MS}}}=-\beta_{0,Q}\ln\left(\frac{m^2}{\mu^2}\right), \quad (4.135)$$

$$A_{gg,Q}^{(2),\overline{\mathsf{MS}}}=\frac{1}{8}\Big\{2\beta_{0,Q}\Big(\gamma_{gg}^{(0)}+2\beta_0\Big)+\gamma_{gq}^{(0)}\hat{\gamma}_{qg}^{(0)}+8\beta_{0,Q}^2\Big\}\ln^2\Big(\frac{m^2}{\mu^2}\Big)+\frac{\hat{\gamma}_{gg}^{(1)}}{2}\ln\Big(\frac{m^2}{\mu^2}\Big)$$
$$-\frac{\zeta_2}{8}\Big[2\beta_{0,Q}\Big(\gamma_{gg}^{(0)}+2\beta_0\Big)+\gamma_{gq}^{(0)}\hat{\gamma}_{qg}^{(0)}\Big]+a_{gg,Q}^{(2)}, \quad (4.136)$$

$$A_{gg,Q}^{(3),\overline{\mathsf{MS}}}=\frac{1}{48}\Big\{\gamma_{gq}^{(0)}\hat{\gamma}_{qg}^{(0)}\Big(\gamma_{qq}^{(0)}-\gamma_{gg}^{(0)}-6\beta_0-4n_f\beta_{0,Q}-10\beta_{0,Q}\Big)$$
$$-4\Big(\gamma_{gg}^{(0)}\Big[2\beta_0+7\beta_{0,Q}\Big]+4\beta_0^2+14\beta_{0,Q}\beta_0+12\beta_{0,Q}^2\Big)\beta_{0,Q}\Big\}\ln^3\Big(\frac{m^2}{\mu^2}\Big)$$
$$+\frac{1}{8}\Big\{\hat{\gamma}_{qg}^{(0)}\Big(\gamma_{gq}^{(1)}+(1-n_f)\hat{\gamma}_{gq}^{(1)}\Big)+\gamma_{gq}^{(0)}\hat{\gamma}_{qg}^{(1)}+4\gamma_{gg}^{(1)}\beta_{0,Q}$$
$$-4\hat{\gamma}_{gg}^{(1)}[\beta_0+2\beta_{0,Q}]+4[\beta_1+\beta_{1,Q}]\beta_{0,Q}+2\gamma_{gg}^{(0)}\beta_{1,Q}\Big\}\ln^2\Big(\frac{m^2}{\mu^2}\Big)$$
$$+\frac{1}{16}\Big\{8\hat{\gamma}_{gg}^{(2)}-8n_f a_{gq,Q}^{(2)}\hat{\gamma}_{qg}^{(0)}-16a_{gg,Q}^{(2)}(2\beta_0+3\beta_{0,Q})+8\gamma_{gq}^{(0)}a_{Qg}^{(2)}$$
$$+8\gamma_{gg}^{(0)}\beta_{1,Q}^{(1)}+\gamma_{gq}^{(0)}\hat{\gamma}_{qg}^{(0)}\zeta_2\Big(\gamma_{gg}^{(0)}-\gamma_{qq}^{(0)}+6\beta_0+4n_f\beta_{0,Q}+6\beta_{0,Q}\Big)$$
$$+4\beta_{0,Q}\zeta_2\Big(\gamma_{gg}^{(0)}+2\beta_0\Big)\Big(2\beta_0+3\beta_{0,Q}\Big)\Big\}\ln\Big(\frac{m^2}{\mu^2}\Big)$$
$$+2(2\beta_0+3\beta_{0,Q})\overline{a}_{gg,Q}^{(2)}+n_f\hat{\gamma}_{qg}^{(0)}\overline{a}_{gq,Q}^{(2)}-\gamma_{gq}^{(0)}\overline{a}_{Qg}^{(2)}-\beta_{1,Q}^{(2)}\gamma_{gg}^{(0)}$$
$$+\frac{\gamma_{gq}^{(0)}\hat{\gamma}_{qg}^{(0)}\zeta_3}{48}\Big(\gamma_{qq}^{(0)}-\gamma_{gg}^{(0)}-2[2n_f+1]\beta_{0,Q}-6\beta_0\Big)$$
$$+\frac{\beta_{0,Q}\zeta_3}{12}\Big([\beta_{0,Q}-2\beta_0]\gamma_{gg}^{(0)}+2[\beta_0+6\beta_{0,Q}]\beta_{0,Q}-4\beta_0^2\Big)$$

$$
\begin{aligned}
&-\frac{\hat{\gamma}_{qg}^{(0)}\zeta_2}{16}\Big(\gamma_{gq}^{(1)}+\hat{\gamma}_{gq}^{(1)}\Big)+\frac{\beta_{0,Q}\zeta_2}{8}\Big(\hat{\gamma}_{gg}^{(1)}-2\gamma_{gg}^{(1)}-2\beta_1-2\beta_{1,Q}\Big)\\
&+\frac{\delta m_1^{(-1)}}{4}\Big(8a_{gg,Q}^{(2)}+24\delta m_1^{(0)}\beta_{0,Q}+8\delta m_1^{(1)}\beta_{0,Q}+\zeta_2\beta_{0,Q}\beta_0+9\zeta_2\beta_{0,Q}^2\Big)\\
&+\delta m_1^{(0)}\Big(\beta_{0,Q}\delta m_1^{(0)}+\hat{\gamma}_{gg}^{(1)}\Big)+\delta m_1^{(1)}\Big(\hat{\gamma}_{qg}^{(0)}\gamma_{gq}^{(0)}+2\beta_{0,Q}\gamma_{gg}^{(0)}\\
&+4\beta_{0,Q}\beta_0+8\beta_{0,Q}^2\Big)-2\delta m_2^{(0)}\beta_{0,Q}+a_{gg,Q}^{(3)}.
\end{aligned}
\tag{4.137}
$$

References

1. R. Hamberg, van W.L. Neerven, Nucl. Phys. **B379**, 143 (1992)
2. J.C. Collins, R.J. Scalise, Phys. Rev. **D50**, 4117 (1994) hep-ph/9403231
3. B.W. Harris, J. Smith, Phys. Rev. **D51**, 4550 (1995) hep-ph/9409405
4. K.G. Chetyrkin, B.A. Kniehl, M. Steinhauser, Nucl. Phys. **B814**, 231 (2009) hep-ph/0812.1337
5. G. 't Hooft, M.J.G. Veltman, Nucl. Phys. **B44**, 189 (1972)
6. J.F. Ashmore, Lett. Nuovo Cim. **4**, 289 (1972)
7. G.M. Cicuta, E. Montaldi, Nuovo Cim. Lett. **4**, 329 (1972)
8. C.G. Bollini, J.J. Giambiagi, Nuovo Cim. **B12**, 20 (1972)
9. J.C. Collins, *Renormalization*, (Cambridge University Press, Cambridge, 1984) pp. 380.
10. W. Pauli, F. Villars, Rev. Mod. Phys. **21**, 434 (1949)
11. E.R. Speer, J. Math. Phys. **15**, 1 (1974)
12. G. 't Hooft, Nucl. Phys. **B61**, 455 (1973)
13. W.A. Bardeen, A.J. Buras, D.W. Duke, T. Muta, Phys. Rev. **D18**, 3998 (1978)
14. Y. Matiounine, J. Smith, van W.L. Neerven, Phys. Rev. **D57**, 6701 (1998) hep-ph/9801224
15. R. Hamberg, Second order gluonic contributions to physical quantities, Ph.D. thesis, Leiden, 1991
16. M. Buza, Y. Matiounine, J. Smith, R. Migneron, van W.L. Neerven, Nucl. Phys. **B472**, 611 (1996) hep-ph/9601302
17. I. Bierenbaum, J. Blümlein, S. Klein, C. Schneider, Nucl. Phys. **B803**, 1 (2008) hep-ph/0803.0273
18. R. Tarrach, Nucl. Phys. **B183**, 384 (1981)
19. O. Nachtmann, W. Wetzel, Nucl. Phys. **B187**, 333 (1981)
20. N. Gray, D.J. Broadhurst, W. Gräfe, K. Schilcher, Z. Phys. **C48**, 673 (1990)
21. D.J. Broadhurst, N. Gray, K. Schilcher, Z. Phys. **C52**, 111 (1991)
22. J. Fleischer, F. Jegerlehner, O.V. Tarasov, O.L. Veretin, Nucl. Phys. **B539**, 671 (1999) hep-ph/9803493
23. D.J. Gross, F. Wilczek, Phys. Rev. Lett. **30**, 1343 (1973)
24. H.D. Politzer, Phys. Rev. Lett. **30**, 1346 (1973)
25. G. t' Hooft, (1972) unpublished
26. I.B. Khriplovich, Yad. Fiz. **10**, 409 (1969)
27. W.E. Caswell, Phys. Rev. Lett. **33**, 244 (1974)
28. D.R.T. Jones, Nucl. Phys. **B75**, 531 (1974)
29. L.F. Abbott, Nucl. Phys. **B185**, 189 (1981)
30. A. Rebhan, Z. Phys. **C30**, 309 (1986)
31. F. Jegerlehner, O.V. Tarasov, Nucl. Phys. **B549**, 481 (1999) hep-ph/9809485
32. F.J. Yndurain, *The theory of quark and gluon interactions*, 4th edn. (Springer, Berlin, 2006) p. 474.

33. I. Bierenbaum, J. Blümlein, S. Klein, Nucl. Phys. **B780**, 40 (2007) hep-ph/0703285
34. M. Buza, Y. Matiounine, J. Smith, van W.L. Neerven, Eur. Phys. J. **C1**, 301 (1998) hep-ph/9612398
35. I. Bierenbaum, J. Blümlein, S. Klein, Phys. Lett. **B672**, 401 (2009) hep-ph/0901.0669
36. I. Bierenbaum, J. Blümlein, S. Klein, Nucl.Phys. **B820**, 417 (2009) hep-ph/0904.3563
37. M. Steinhauser, Comput. Phys. Commun. **134**, 335 (2001) hep-ph/0009029
38. K.G. Chetyrkin, M. Steinhauser, Phys. Rev. Lett. **83**, 4001 (1999) hep-ph/9907509
39. K.G. Chetyrkin, M. Steinhauser, Nucl. Phys. **B573**, 617 (2000) hep-ph/9911434
40. D.J. Broadhurst, Z. Phys. **C54**, 599 (1992) Avdeev:1994db
41. L. Avdeev, J. Fleischer, S. Mikhailov, O. Tarasov, Phys. Lett. **B336**, 560 (1994) hep-ph/9406363
42. S. Laporta, E. Remiddi, Phys. Lett. **B379**, 283 (1996) hep-ph/9602417
43. D.J. Broadhurst, Eur. Phys. J. **C8**, 311 (1999) hep-th/9803091
44. R. Boughezal, J.B. Tausk, van der J.J. Bij, Nucl. Phys. **B713**, 278 (2005) hep-ph/0410216
45. S. Moch, J.A.M. Vermaseren, A. Vogt, Nucl. Phys. **B688**, 101 (2004) hep-ph/0403192
46. J. Blümlein, S. Klein, B. Tödtli, Phys. Rev. **D80**, 094010 (2009) hep-ph/0909.1547

Chapter 5
Representation in Different Renormalization Schemes

As outlined in Chap. 4, there are different obvious possibilities to choose a scheme for the renormalization of the mass and the coupling constant. Concerning the coupling constant, we intermediately worked in a MOM–scheme, which derives from the condition that the external gluon lines have to be kept on–shell. In the end, we transformed back to the $\overline{\text{MS}}$–description via., Eq. 4.55, since this is the commonly used renormalization scheme. If masses are involved, it is useful to renormalize them in the on–mass–shell–scheme, as it was done in the previous Section. In this scheme, one defines the renormalized mass m as the pole of the quark propagator. In this Section, we present the relations required to transform the renormalized results from Sect. 4.7 into the different, related schemes. In Sect. 5.1, we show how these scheme transformations affect the NLO results. Denoting the $\overline{\text{MS}}$–mass by $\overline{m}$, there are in addition to the $\{a^{\overline{\text{MS}}}, m\}$-scheme adopted in Sect. 4.7 the following schemes

$$\left\{a_s^{\text{MOM}}, m\right\}, \quad \left\{a_s^{\text{MOM}}, \overline{m}\right\}, \quad \left\{a_s^{\overline{\text{MS}}}, \overline{m}\right\}. \tag{5.1}$$

In case of mass renormalization in the $\overline{\text{MS}}$–scheme, Eq. 4.27 becomes

$$\hat{m} = Z_m^{\overline{\text{MS}}}\overline{m} = \overline{m}[1 + \hat{a}_s\delta\overline{m}_1 + \hat{a}_s^2\delta\overline{m}_2] + O(\hat{a}_s^3). \tag{5.2}$$

The corresponding coefficients read, [1],

$$\delta\overline{m}_1 = \frac{6}{\varepsilon}C_F \equiv \frac{\delta\overline{m}_1^{(-1)}}{\varepsilon}, \tag{5.3}$$

$$\begin{aligned}\delta\overline{m}_2 &= \frac{C_F}{\varepsilon^2}\left(18C_F - 22C_A + 8T_F(n_f+1)\right) + \frac{C_F}{\varepsilon}\left(\frac{3}{2}C_F + \frac{97}{6}C_A - \frac{10}{3}T_F(n_f+1)\right)\\ &\equiv \frac{\delta\overline{m}_2^{(-2)}}{\varepsilon^2} + \frac{\delta\overline{m}_2^{(-1)}}{\varepsilon}.\end{aligned} \tag{5.4}$$

S. Klein, *Charm Production in Deep Inelastic Scattering*, Springer Theses,
DOI: 10.1007/978-3-642-23286-2_5,

One notices that the following relations hold between the expansion coefficients in ε of the on–shell– and $\overline{\mathsf{MS}}$–terms

$$\delta\overline{m}_1^{(-1)} = \delta m_1^{(-1)}, \tag{5.5}$$

$$\delta\overline{m}_2^{(-2)} = \delta m_2^{(-2)}, \tag{5.6}$$

$$\delta\overline{m}_2^{(-1)} = \delta m_2^{(-1)} - \delta m_1^{(-1)}\delta m_1^{(0)} + 2\delta m_1^{(0)}(\beta_0 + \beta_{0,Q}). \tag{5.7}$$

One has to be careful, since the choice of this scheme also affects the renormalization constant of the coupling in the MOM–scheme. This is due to the fact that in Eq. 4.42 mass renormalization had been performed in the on–shell–scheme. Going through the same steps as in Eqs. 4.42–4.47, but using the $\overline{\mathsf{MS}}$–mass, we obtain for Z_g in the MOM–scheme.

$$\begin{aligned} Z_g^{\mathrm{MOM}^2}(\varepsilon, n_f+1, \mu^2, \overline{m}^2) = 1 &+ a_s^{\mathrm{MOM}}(\mu^2)\left[\frac{2}{\varepsilon}(\beta_0(n_f) + \beta_{0,Q} f(\varepsilon))\right] \\ &+ a_s^{\mathrm{MOM}^2}(\mu^2)\left[\frac{\beta_1(n_f)}{\varepsilon} + \frac{4}{\varepsilon^2}(\beta_0(n_f) + \beta_{0,Q} f(\varepsilon))^2 \right. \\ &+ \frac{2\beta_{0,Q}}{\varepsilon}\delta\overline{m}_1^{(-1)} f(\varepsilon) + \frac{1}{\varepsilon}\left(\frac{\overline{m}^2}{\mu^2}\right)^{\varepsilon} \\ &\left.\times\left(\overline{\beta}_{1,Q} + \varepsilon\overline{\beta}_{1,Q}^{(1)} + \varepsilon^2\overline{\beta}_{1,Q}^{(2)}\right)\right] + O(a_s^{\mathrm{MOM}^3}), \end{aligned} \tag{5.8}$$

where in the term $f(\varepsilon)$, cf. Eq. 4.44, the $\overline{\mathsf{MS}}$–mass has to be used. The coefficients differing from the on–shell–scheme in the above equation are given by, cf. Eqs. 4.50 and 4.51

$$\overline{\beta}_{1,Q} = \beta_{1,Q} - 2\beta_{0,Q}\delta m_1^{(-1)}, \tag{5.9}$$

$$\overline{\beta}_{1,Q}^{(1)} = \beta_{1,Q}^{(1)} - 2\beta_{0,Q}\delta m_1^{(0)}, \tag{5.10}$$

$$\overline{\beta}_{1,Q}^{(2)} = \beta_{1,Q}^{(2)} - \frac{\beta_{0,Q}}{4}\left(8\delta m_1^{(1)} + \delta m_1^{(-1)}\zeta_2\right). \tag{5.11}$$

The transformation formulas between the different schemes follow from the condition that the unrenormalized terms are equal.

In order to transform from the $\{a_s^{\overline{\mathrm{MS}}}, m\}$–scheme to the $\{a_s^{\mathrm{MOM}}, m\}$–scheme, the inverse of Eq. 4.55

$$a_s^{\overline{\rm MS}}(m^2) = a_s^{\rm MOM}\left[1 + \beta_{0,Q}\ln\left(\frac{m^2}{\mu^2}\right)a_s^{\rm MOM} + \left\{\beta_{0,Q}^2\ln^2\left(\frac{m^2}{\mu^2}\right) + \beta_{1,Q}\ln\left(\frac{m^2}{\mu^2}\right) + \beta_{1,Q}^{(1)}\right\}a_s^{\rm MOM\,2}\right] \tag{5.12}$$

is used. For the transformation to the $\{a_s^{\overline{\rm MS}}, \overline{m}\}$–scheme one obtains

$$\begin{aligned} m(a_s^{\overline{\rm MS}}) = \overline{m}(a_s^{\overline{\rm MS}})\Biggl(1 &+ \left\{-\frac{\delta m_1^{(-1)}}{2}\ln\left(\frac{\overline{m}^2}{\mu^2}\right) - \delta m_1^{(0)}\right\}a_s^{\overline{\rm MS}} \\ &+ \Biggl\{\frac{\delta m_1^{(-1)}}{8}\left[2\beta_0 + 2\beta_{0,Q} + \delta m_1^{(-1)}\right]\ln^2\left(\frac{\overline{m}^2}{\mu^2}\right) \\ &+ \frac{1}{2}\Bigl[-\delta m_1^{(0)}\left(2\beta_0 + 2\beta_{0,Q} - 3\delta m_1^{(-1)}\right) \\ &+ \delta m_1^{(-1)2} - 2\delta m_2^{(-1)}\Bigr]\ln\left(\frac{\overline{m}^2}{\mu^2}\right) \\ &+ \delta m_1^{(1)}\left[\delta m_1^{(-1)} - 2\beta_0 - 2\beta_{0,Q}\right] \\ &+ \delta m_1^{(0)}\left[\delta m_1^{(-1)} + \delta m_1^{(0)}\right] - \delta m_2^{(0)}S\Biggr\}a_s^{\overline{\rm MS}2}\Biggr). \end{aligned} \tag{5.13}$$

Finally, the transformation to the $\{a_s^{\rm MOM}, \overline{m}\}$ is achieved via.,

$$\begin{aligned} a_s^{\overline{\rm MS}}(m^2) = a_s^{\rm MOM}\Biggl[1 &+ \beta_{0,Q}\ln\left(\frac{\overline{m}^2}{\mu^2}\right)a_s^{\rm MOM} \\ &+ \Biggl\{\beta_{0,Q}^2\ln^2\left(\frac{\overline{m}^2}{\mu^2}\right) + \left(\beta_{1,Q} - \beta_{0,Q}\delta m_1^{(-1)}\right)\ln\left(\frac{\overline{m}^2}{\mu^2}\right) \\ &+ \beta_{1,Q}^{(1)} - 2\delta m_1^{(0)}\beta_{0,Q}\Biggr\}a_s^{\rm MOM\,2}\Biggr], \end{aligned} \tag{5.14}$$

and

$$\begin{aligned} m(a_s^{\overline{\rm MS}}) = \overline{m}(a_s^{\rm MOM})\Biggl(1 &+ \left\{-\frac{\delta m_1^{(-1)}}{2}\ln\left(\frac{\overline{m}^2}{\mu^2}\right) - \delta m_1^{(0)}\right\}a_s^{\rm MOM} \\ &+ \Biggl\{\frac{\delta m_1^{(-1)}}{8}\left[2\beta_0 - 2\beta_{0,Q} + \delta m_1^{(-1)}\right]\ln^2\left(\frac{\overline{m}^2}{\mu^2}\right) \\ &+ \frac{1}{2}\Bigl[-\delta m_1^{(0)}\left(2\beta_0 + 4\beta_{0,Q} - 3\delta m_1^{(-1)}\right) \end{aligned}$$

$$+ \delta m_1^{(-1)^2} - 2\delta m_2^{(-1)}\Big] \ln\left(\frac{\overline{m}^2}{\mu^2}\right)$$
$$+ \delta m_1^{(1)} \left[\delta m_1^{(-1)} - 2\beta_0 - 2\beta_{0,Q}\right]$$
$$+ \delta m_1^{(0)} \left[\delta m_1^{(-1)} + \delta m_1^{(0)}\right] - \delta m_2^{(0)}\Big\} a_s^{\mathrm{MOM}^2}\Big) \tag{5.15}$$

The expressions for the OMEs in different schemes are then obtained by inserting the relations (5.12)–(5.15) into the general expression (4.77) and expanding in the coupling constant.

5.1 Scheme Dependence at NLO

Finally, we would like to comment on how the factorization formulas for the heavy flavor Wilson coefficients, (3.26)–(3.30), have to be applied to obtain a complete description. Here, the renormalization of the coupling constant has to be carried out in the same way for all quantities contributing. The general factorization formula (3.16) holds only for completely inclusive quantities, including radiative corrections containing heavy quark loops, [2].

One has to distinguish one-particle irreducible and reducible diagrams, which both contribute in the calculation. We would like to remind the reader of the background of this aspect. If one evaluates the heavy-quark Wilson coefficients, diagrams of the type shown in Fig. 5.1 may appear as well. Diagram (a) contains a virtual heavy quark loop correction to the gluon propagator in the initial state and contributes to the terms $L_{g,i}$ and $H_{g,i}$, respectively, depending on whether a light or heavy quark pair is produced in the final state. Diagrams (b), (c) contribute to $L_{q,i}^{\mathsf{NS}}$ and contain radiative corrections to the gluon propagator due to heavy quarks as well. The latter diagrams contribute to $F_{(2,L)}(x, Q^2)$ in the inclusive case, but are absent in the semi–inclusive $Q\overline{Q}$– section. The same holds for Diagram (a) if a $q\overline{q}$–pair is produced. In Refs. [3, 4], the coupling constant was renormalized in the MOM–scheme by absorbing the contributions of diagram (a) into the coupling constant, as a consequence of which the term $L_{g,i}$ appears for the first time at $O(a_s^3)$. This can be made explicit by considering the complete gluonic Wilson coefficient up to $O(a_s^2)$, including one heavy quark, cf. Eqs. 3.28 and 3.30,

$$\begin{aligned}
&C_{g,(2,L)}(n_f) + L_{g,(2,L)}(n_f+1) + H_{g,(2,L)}(n_f+1)\\
&\quad = a_s^{\overline{\mathrm{MS}}}\left[A_{Qg}^{(1),\overline{\mathrm{MS}}}\delta_2 + C_{g,(2,L)}^{(1)}(n_f+1)\right]\\
&\qquad + a_s^{\overline{\mathrm{MS}}^2}\left[A_{Qg}^{(2),\overline{\mathrm{MS}}}\delta_2 + A_{Qg}^{(1),\overline{\mathrm{MS}}}C_{q,(2,L)}^{(1),\mathrm{NS}}(n_f+1)\right.\\
&\qquad\qquad \left. + A_{gg,Q}^{(1),\overline{\mathrm{MS}}}C_{g,(2,L)}^{(1)}(n_f+1) + C_{g,(2,L)}^{(2)}(n_f+1)\right].
\end{aligned} \tag{5.16}$$

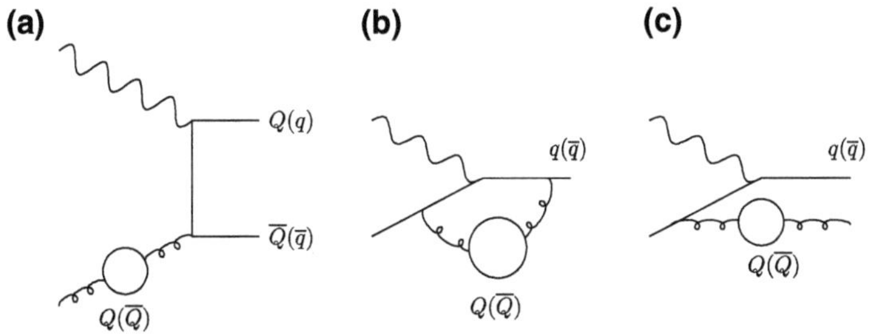

Fig. 5.1 $O(a_s^2)$ virtual heavy quark corrections

The above equation is given in the $\overline{\text{MS}}$–scheme , and the structure of the OMEs can be inferred from Eqs. 4.113 and 4.114. Here, diagram (a) gives a contribution, corresponding exactly to the color factor T_F^2. The transformation to the MOM–scheme for a_s, cf. Eqs. 4.55 and 4.56, yields

$$
\begin{aligned}
&C_{g,(2,L)}(n_f) + L_{g,(2,L)}(n_f+1) + H_{g,(2,L)}(n_f+1) \\
&= a_s^{\text{MOM}}\left[A_{Qg}^{(1),\overline{MS}}\delta_2 + C_{g,(2,L)}^{(1)}(n_f+1)\right] \\
&+ a_s^{MOM^2}\left[A_{Qg}^{(2),\overline{MS}}\delta_2 + \beta_{0,Q}\ln\left(\frac{m^2}{\mu^2}\right)A_{Qg}^{(1),\overline{MS}}\delta_2\right. \\
&\qquad + A_{Qg}^{(1),\overline{MS}}C_{q,(2,L)}^{(1),NS}(n_f+1) + A_{gg,Q}^{(1),\overline{\text{MS}}}C_{g,(2,L)}^{(1)}(n_f+1) \\
&\qquad \left. + \beta_{0,Q}\ln\left(\frac{m^2}{\mu^2}\right)C_{g,(2,L)}^{(1)}(n_f+1) + C_{g,(2,L)}^{(2)}(n_f+1)\right]. \qquad (5.17)
\end{aligned}
$$

By using the general structure of the renormalized OMEs, Eqs. 4.113, 4.114 and 4.135 , one notices that all contributions due to diagram (a) cancel in the MOM–scheme, i.e., the color factor T_F^2 does not occur at the two–loop level. Thus the factorization formula reads

$$
\begin{aligned}
&C_{g,(2,L)}(n_f) + L_{g,(2,L)}(n_f+1) + H_{g,(2,L)}(n_f+1) \\
&= a_s^{\text{MOM}}\left[A_{Qg}^{(1),\text{MOM}}\delta_2 + C_{g,(2,L)}^{(1)}(n_f+1)\right] \\
&+ a_s^{\text{MOM}^2}\left[A_{Qg}^{(2),\text{MOM}}\delta_2 + A_{Qg}^{(1),\text{MOM}}C_{q,(2,L)}^{(1),NS}(n_f+1) + C_{g,(2,L)}^{(2)}(n_f+1)\right]. \qquad (5.18)
\end{aligned}
$$

Splitting up Eq. 5.18 into $H_{g,i}$ and $L_{g,i}$, one observes that $L_{g,i}$ vanishes at $O(a_s^2)$ and the term $H_{g,i}$ is the one calculated in Ref. [5]. This is the asymptotic expression of the gluonic heavy flavor Wilson coefficient as calculated in Refs. [3, 4]. Note that the observed cancellation was due to the fact that the term $A_{gg,Q}^{(1)}$ receives only contributions from the heavy quark loops of the gluon–self energy, which also enters into the definition of the MOM–scheme. It is not clear whether this can be achieved at the three–loop level as well, i.e., transforming the general inclusive factorization formula (3.16) in such a way that only the contributions due to heavy flavors in the final state remain. Therefore one should use these asymptotic expressions only for completely inclusive analyzes, where heavy and light flavors are treated together. This approach has also been adopted in Ref. [2] for the renormalization of the massive OMEs, which was performed in the $\overline{\mathsf{MS}}$–scheme and not in the MOM–scheme, as previously in Ref. [5]. The radiative corrections in the NS–case can be treated in the same manner. Here the scheme transformation affects only the light Wilson coefficients and not the OMEs at the two–loop level. In the $\overline{\mathsf{MS}}$–scheme, one obtains the following asymptotic expression up to $O(a_s^2)$ from Eqs. 3.21 and 3.26.

$$C_{q,(2,L)}^{\rm NS}(n_f) + L_{q,(2,L)}^{\rm NS}(n_f+1) = 1 + a_s^{\overline{\rm MS}} C_{q,(2,L)}^{(1),\rm NS}(n_f+1)$$
$$+ a_s^{\overline{\rm MS}^2}\left[A_{qq,Q}^{(2),\rm NS,\overline{MS}}(n_f+1)\delta_2 + C_{q,(2,L)}^{(2),\rm NS}(n_f+1)\right]. \tag{5.19}$$

Transformation to the MOM-scheme yields

$$C_{q,(2,L)}^{\rm NS}(n_f) + L_{q,(2,L)}^{\rm NS}(n_f+1) = 1 + a_s^{\rm MOM} C_{q,(2,L)}^{(1),\rm NS}(n_f+1) + a_s^{\rm MOM^2}$$
$$\times\left[A_{qq,Q}^{(2),\rm NS,MOM}(n_f+1)\delta_2 + \beta_{0,Q}\ln\left(\frac{m^2}{\mu^2}\right)C_{q,(2,L)}^{(1),\rm NS}(n_f+1) + C_{q,(2,L)}^{(2),\rm NS}(n_f+1)\right] \tag{5.20}$$

Note that $A_{qq,Q}^{(2),\mathsf{NS}}$, Eq. 4.95, is not affected by this scheme transformation. As is obvious from Fig. 5.1, the logarithmic term in Eq. 5.20 can therefore only be attributed to the massless Wilson coefficient. Separating the light from the heavy part one obtains

$$L_{q,(2,L)}^{(2),\rm NS,MOM}(n_f+1) = A_{qq,Q}^{(2),\rm NS,MOM}(n_f+1)\delta_2 + \beta_{0,Q}\ln\left(\frac{m^2}{\mu^2}\right)$$
$$\times C_{q,(2,L)}^{(1),\rm NS}(n_f+1) + \hat{C}_{q,(2,L)}^{(2),\rm NS}(n_f). \tag{5.21}$$

This provides the same results as Eqs. 4.23–4.29 of Ref. [5]. These are the asymptotic expressions of the NS heavy flavor Wilson coefficients from Refs. [3, 4], where only the case of $Q\overline{Q}$–production in the final state has been considered. Hence the logarithmic term in Eq. 5.21 just cancels the contributions due to diagrams (b), (c) in Fig. 5.1.

References

1. R. Tarrach, Nucl. Phys **B183**, 384 (1981)
2. M. Buza, Y. Matiounine, J. Smith, W.L. van Neerven, Eur. Phys. J. C **1**, 301 (1998), hep-ph/9612398

3. E. Laenen, S. Riemersma, J. Smith, W.L. van Neerven, Nucl. Phys. B **392**, 162, 229 (1993)
4. S. Riemersma, J. Smith, W.L. van Neerven, Phys. Lett. B **347**, 143 (1995), hep-ph/9411431
5. M. Buza, Y. Matiounine, J. Smith, R. Migneron, W.L. van Neerven, Nucl. Phys. B **472**, 611 (1996), hep-ph/9601302

Chapter 6
Calculation of the Massive Operator Matrix Elements up to $O(a_s^2\varepsilon)$

The quarkonic 2-loop massive OMEs $A_{Qg}^{(2)}$, $A_{Qq}^{(2),\rm PS}$ and $A_{qq,Q}^{(2)}$ have been calculated for the first time in Ref. [1] to construct asymptotic expressions for the NLO heavy flavor Wilson Coefficients in the limit $Q^2 \gg m^2$, cf. Sect. 3.2. The corresponding gluonic OMEs $A_{gg,Q}^{(2)}$ and $A_{gq,Q}^{(2)}$ were calculated in Ref. [2], where they were used within a VFNS description of heavy flavors in high-energy scattering processes, see Sect. 3.3. In these calculations, the integration-by-parts technique, [3], has been applied to reduce the number of propagators occurring in the momentum integrals. Subsequently, the integrals were calculated in z-space, which led to a variety of multiple integrals of logarithms, partially with complicated arguments. The final results were given in terms of polylogarithms and Nielsen-integrals, see Appendix C.4. The quarkonic terms have been confirmed in Ref. [4], cf. also [5], where a different approach was followed. The calculation was performed in Mellin-N space and by avoiding the integration-by-parts technique. Using representations in terms of generalized hypergeometric functions, the integrals could be expressed in terms of multiple finite and infinite sums with one free parameter, N. The advantage of this approach is that the evaluation of these sums can be automatized using various techniques, simplifying the calculation. The final result is then obtained in Mellin-space in terms of nested harmonic sums or Z-sums, cf. [6, 7] and Appendix C.4. An additional simplification was found since the final result, e.g., for $A_{Qg}^{(2)}$ can be expressed in terms of two basic harmonic sums only, using algebraic, [8], and structural relations, [9, 10], between them. This is another example of an observation which has been made for many different single scale quantities in high-energy physics, namely that the Mellin-space representation is better suited to the problem than the Z-space representation.

As has been outlined in Chap. 4 , the $O(\varepsilon)$-terms of the unrenormalized 2-loop massive OMEs are needed in the renormalization of the 3-loop contributions. In this section, we calculate these terms based on the approach advocated in Ref. [4], which is a new result, [11, 12]. Additionally, we re-calculate the gluonic OMEs up to the

S. Klein, *Charm Production in Deep Inelastic Scattering*, Springer Theses,
DOI: 10.1007/978-3-642-23286-2_6, © Springer-Verlag Berlin Heidelberg 2012

Fig. 6.1 Examples for 2-loop diagrams contributing to the massive OMEs. *Thick lines* heavy quarks, *curly lines* gluons, *full lines* quarks

constant term in ε for the first time, cf. [2, 11]. Example diagrams for each OME are shown in Fig. 6.1.

In Sect. 6.1, we explain how the integrals are obtained in terms of finite and infinite sums using representations in terms of generalized hypergeometric functions, cf. [13–15] and Appendix C.2. For the calculation of these sums we mainly used the MATHEMATICA-based program Sigma, [16, 17], which is discussed in Sect. 6.2. The results are presented in Sect. 6.3. Additionally, we make several remarks on the MOM-scheme, which has to be adopted intermediately for the renormalization of the coupling constant, cf. Sect. 4.4. In Sect. 6.4, different checks of the results are presented.

6.1 Representation in Terms of Hypergeometric Functions

All diagrams contributing to the massive OMEs are shown in Figures 1–4 in Ref. [1] and in Figures 3,4 in Ref. [2], respectively. They represent 2-point functions with on-shell external momentum p, $p^2 = 0$. They are expressed in two parameters, the heavy quark mass m and the Mellin-parameter N. Since the mass can be factored out of the integrals, the problem effectively contains a single scale. The parameter N represents the spin of the composite operators, (2.86–2.88), and enters the calculation via the Feynman-rules for these objects, cf. Appendix B.

Since the external momentum does not appear in the final result, the corresponding scalar integrals reduce to massive tadpoles if one sets $N = 0$. In order to explain our method, we consider first the massive 2-loop tadpole shown in Fig. 6.2, from which all OMEs can be derived at this order, by attaching two outer legs and inserting the composite operator in all possible ways, i.e., both on the lines and on the vertices.

In Fig. 6.2, the wavy line is massless and the full lines are massive. Here ν_i labels the power of the propagator. We adopt the convention $\nu_{i\ldots j} \equiv \nu_i + \ldots + \nu_j$ etc. The corresponding dimensionless momentum integral reads in Minkowski-space

$$I_1 = \int\int \frac{d^D k_1 d^D k_2}{(4\pi)^{4D}} \frac{(4\pi)^4(-1)^{\nu_{123}-1}(m^2)^{\nu_{123}-D}}{(k_1^2 - m^2)^{\nu_1}(k_1^2 - k_2^2)^{\nu_2}(k_2^2 - m^2)^{\nu_3}}, \tag{6.1}$$

where we have attached a factor $(4\pi)^4(-1)^{\nu_{123}-1}$ for convenience. Using standard Feynman-parametrization and Eq. 4.3 for momentum integration, one obtains the following Feynman-parameter integral

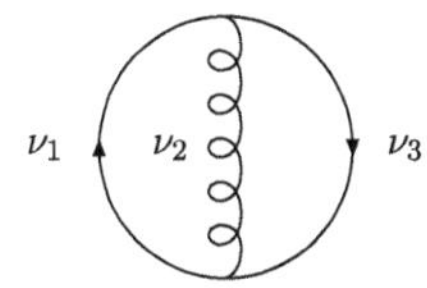

Fig. 6.2 Basic 2-loop massive tadpole

$$I_1 = \Gamma\begin{bmatrix} \nu_{123}-4-\varepsilon \\ \nu_1,\nu_2,\nu_3 \end{bmatrix} \iint_0^1 dxdy \frac{x^{1+\varepsilon/2-\nu_2}(1-x)^{\nu_{23}-3-\varepsilon/2}y^{\nu_3-1}(1-y)^{\nu_{12}-3-\varepsilon/2}}{(4\pi)^\varepsilon(1-xy)^{\nu_{123}-4-\varepsilon}}, \tag{6.2}$$

which belongs to the class of the hypergeometric function ${}_3F_2$ with argument $z = 1$, see Appendix C.2. Applying Eq. C.19, one obtains

$$I_1 = S_\varepsilon^2 \exp\left(\sum_{i=2}^{\infty}\frac{\zeta_i}{i}\varepsilon^i\right)\Gamma\begin{bmatrix} \nu_{123}-4-\varepsilon, 2+\varepsilon/2-\nu_2, \nu_{23}-2-\varepsilon/2, \nu_{12}-2-\varepsilon/2 \\ 1-\varepsilon, \nu_1,\nu_2,\nu_3,\nu_{123}-2-\varepsilon/2 \end{bmatrix}$$
$$\times\, {}_3F_2\left[\begin{matrix} \nu_{123}-4-\varepsilon, 2+\varepsilon/2-\nu_2, \nu_3 \\ \nu_3, \nu_{123}-2-\varepsilon/2 \end{matrix}; 1\right], \tag{6.3}$$

where we have used Eq. 4.8. The term ν_3 in the argument of the ${}_3F_2$ cancels between nominator and denominator and thus one can use Gauss's theorem, Eq. C.16, to write the result in terms of Γ-functions

$$I_1 = \Gamma\begin{bmatrix} \nu_{123}-4-\varepsilon, 2+\varepsilon/2-\nu_2, \nu_{12}-2-\varepsilon/2, \nu_{23}-2-\varepsilon/2 \\ 1-\varepsilon, 2+\varepsilon/2, \nu_1,\nu_3,\nu_{123}+\nu_2-4-\varepsilon \end{bmatrix} S_\varepsilon^2 \exp\left(\sum_{i=2}^{\infty}\frac{\zeta_i}{i}\varepsilon^i\right). \tag{6.4}$$

This calculation is of course trivial and Eq. 6.4 can be easily checked using MATAD, cf. Ref. [18] and Sect. 7.2. Next, let us consider the case of arbitrary moments in presence of the complete numerator structure. Since the final result contains the factor $(\Delta.p)^N$, one cannot set p to zero anymore. This increases the number of propagators and hence the number of Feynman-parameters in Eq. 6.2. Additionally, the terms $(\Delta.q)^A$ in the integral lead to polynomials in the Feynman-parameters to a symbolic power in the integral, which can not be integrated trivially. Hence neither Eq. C.19 nor Gauss's theorem can be applied anymore in the general case.

However, the structure of the integral in Eq. 6.2 does not change. For any diagram deriving from the 2-loop tadpole, a general integral of the type

$$I_2 = C_2 \iint_0^1 dxdy \frac{x^a(1-x)^b y^c(1-y)^d}{(1-xy)^e}\int_0^1 dz_1 \ldots \int_0^1 dz_i \mathsf{P}(x,y,z_1\ldots z_i,N) \tag{6.5}$$

is obtained. Here P is a rational function of x,y and possibly more parameters $z_1\ldots z_i$. N denotes the Mellin-parameter and occurs in some exponents. Note that operator insertions with more than two legs give rise to additional finite sums in P, see Appendix B. For fixed values of N, one can expand P and the integral I_2 turns into a finite sum over integrals of the type I_1. The terms ν_i in these integrals might have been shifted by integers, but after expanding in ε, the one-fold infinite sum can be performed, e.g., using the FORM-based code Summer, [7].

To illustrate the sophistication occurring once one keeps the complete dependence on N in an example, we consider the scalar integral contributing to $A_{Qg}^{(2)}$ shown in Fig. 6.1. After momentum integration, it reads

$$I_3 = \frac{(\Delta p)^{N-2}\Gamma(1-\varepsilon)}{(4\pi)^{4+\varepsilon}(m^2)^{1-\varepsilon}} \iiiint dudzdydx \frac{(1-u)^{-\varepsilon/2} z^{-\varepsilon/2}(1-z)^{\varepsilon/2-1}}{(1-u+uz)^{1-\varepsilon}(x-y)} \times \Big[(zyu + x(1-zu))^{N-1} - ((1-u)x+uy)^{N-1} \Big], \quad (6.6)$$

where we have performed the finite sum already, which stems from the operator insertion. Here and below, the Feynman-parameter integrals are carried out over the respective unit-cube. This integral is of the type of Eq. 6.5 and the term *x-y* in the denominator cancels for fixed values of *N*. Due to the operator insertion on an internal vertex, it is one of the more involved integrals in the 2-loop case. For almost all other integrals, all but two parameters can be integrated automatically, leaving only a single infinite sum of the type of Eq. 6.3 with *N* appearing in the parameters of the hypergeometric function, cf. e.g. [4, 5, 19]. In order to render this example calculable, suitable variable transformations, as, e.g., given in Ref. [20], are applied, [4, 5]. Thus one arrives at the following double sum

$$\begin{aligned} I_3 &= \frac{S_\varepsilon^2 (\Delta p)^{N-2}}{(4\pi)^4 (m^2)^{1-\varepsilon}} \exp\left\{ \sum_{l=2}^{\infty} \frac{\zeta_l}{l}\varepsilon^l \right\} \frac{2\pi}{N\sin(\frac{\pi}{2}\varepsilon)} \sum_{j=1}^{N} \left\{ \binom{N}{j}(-1)^j + \delta_{j,N} \right\} \\ &\quad \times \left\{ \frac{\Gamma(j)\Gamma\left(j+1-\frac{\varepsilon}{2}\right)}{\Gamma(j+2-\varepsilon)\Gamma(j+1+\frac{\varepsilon}{2})} - \frac{B\left(1-\frac{\varepsilon}{2}, 1+j\right)}{j} {}_3F_2\left[\begin{matrix} 1-\varepsilon, \frac{\varepsilon}{2}, j+1 \\ 1, j+2-\frac{\varepsilon}{2} \end{matrix}; 1 \right] \right\} \\ &= \frac{S_\varepsilon^2 (\Delta p)^{N-2}}{(4\pi)^4 (m^2)^{1-\varepsilon}} \left\{ I_3^{(0)} + I_3^{(1)}\varepsilon + O(\varepsilon^2) \right\}. \end{aligned} \quad (6.7)$$

Note that in our approach no expansion in ε is needed until a sum-representation of the kind of Eq. 6.7 is obtained. Having performed the momentum integrations, the expressions of almost all diagrams were given in terms of single generalized hypergeometric series ${}_3F_2$ at $z=1$, with possibly additional finite summations. These infinite sums could then be safely expanded in ε, leading to different kinds of sums depending on the Mellin-parameter *N*. The summands are typically products of harmonic sums with different arguments, weighted by summation parameters and contain hypergeometric terms,[1] like binomials or Beta-function factors *B(N,i)*, cf. Eq. C.9. Here *i* is a summation-index. In the most difficult cases, double sums as in Eq. 6.7 or even triple sums were obtained, which had to be treated accordingly. In general, these sums can be expressed in terms of nested harmonic sums and ζ-values. Note that sums containing Beta-functions with different arguments, e.g. $B(i,i)$, $B(N+i,i)$,

[1] *f(k)* is hypergeometric in *k* iff $f(k+1)/f(k) = g(k)$ for some fixed rational function *g(k)*.

usually do not lead to harmonic sums in the final result. Some of these sums can be performed by the existing packages [7, 21, 22]. However, there exists so far no automatic computer program to calculate sums which contain Beta-function factors of the type *B(N,i)* and single harmonic sums in the summand. These sums can be calculated applying analytic methods, as integral representations, and general summation methods, as encoded in the Sigma package [23–26, 16, 17]. In the next section, we will present details on this.

Before finishing this section, we give the result in terms of harmonic sums for the double sum in Eq. 6.7 applying these summation methods. The $O(\varepsilon^0)$ of Eq. 6.7 is needed for the constant term $a_{Qg}^{(2)}$, cf. Refs. [4, 19]. The linear term in ε reads

$$I_3^{(1)} = \frac{1}{N}\left[-2S_{2,1} + 2S_3 + \frac{4N+1}{N}S_2 - \frac{S_1^2}{N} - \frac{4}{N}S_1\right], \tag{6.8}$$

where we adopt the notation to take harmonic sums at argument N, if not stated otherwise.

6.2 Difference Equations and Infinite Summation

Single scale quantities in renormalizable quantum field theories are most simply represented in terms of nested harmonic sums, cf. [6, 7] and Appendix C.4, which holds at least up to 3-loop order for massless Yang–Mills theories and for a wide class of different processes. This includes the anomalous dimensions and massless Wilson coefficients for unpolarized and polarized space- and time-like processes to 3-loop order, the Wilson coefficients for the Drell-Yan process and pseudoscalar and scalar Higgs–boson production in hadron scattering in the heavy quark mass limit, as well as the soft- and virtual corrections to Bhabha scattering in the on-mass-shell-scheme to 2-loop order, cf. [27–34]. The corresponding Feynman-parameter integrals are such that nested harmonic sums appear in a natural way, working in Mellin space, [9, 10]. Single scale massive quantities at 2 loops, like the unpolarized and polarized heavy-flavor Wilson coefficients in the region $Q^2 \gg m^2$ as considered in this thesis, belong also to this class, [1, 4, 19, 35–40]. Finite harmonic sums obey algebraic, cf. [8], and structural relations, [9], which can be used to obtain simplified expressions and both shorten the calculations and yield compact final results. These representations have to be mapped to momentum-fraction space to use the respective quantities in experimental analyzes. This is obtained by an Mellin inverse transform which requires the analytic continuation of the harmonic sums w.r.t. the Mellin index $N \in \mathbb{C}$, [9, 10, 41, 42].

Calculating the massive OMEs in Mellin space, new types of infinite sums occur if compared to massless calculations. In the latter case, summation algorithms as Summer, [7], Nestedsums, [21], and Xsummer, [22], may be used to calculate the respective sums. Summer and Xsummer are based on FORM, while Nestedsums is based on GiNaC, [43]. The new sums which emerge in [4, 12, 37, 19, 39, 40] can be calculated in different ways. In Ref. [4, 37], we chose analytic methods and in the former reference all sums are given which are needed to calculate the constant term of the massive OMEs. Few of these sums can be calculated using general theorems, as Gauss' theorem, (C.16), Dixon's theorem, [13], or summation tables in the literature, cf. [7, 44, 45] (Blümlein, Collection of Polylog-Integrals, unpublished).

In order to calculate the gluonic OMEs as well as the $O(\varepsilon)$-terms, many new sums had to be evaluated. For this we adopted a more systematic technique based on difference equations, which are the discrete equivalent of differential equations, cf. [46, 47]. This is a promising approach, since it allowed us to obtain all sums needed automatically and it may be applied to entirely different single-scale processes as well. It is based on applying general summation algorithms in computer algebra. A first method is Gosper's telescoping algorithm, [48], for hypergeometric terms. For practical applications, Zeilberger's extension of Gosper's algorithm to creative telescoping, [49, 50], can be considered as the breakthrough in symbolic summation. The recent summation package Sigma, [23–26, 16, 17], written in MATHEMATICA opens up completely new possibilities in symbolic summation. Based on Karr's $\Pi\Sigma$-difference fields, [51], and further refinements, [23–25, 52], the package contains summation algorithms, [53], that allow to solve not only hypergeometric sums, like Gosper's and Zeilberger's algorithms, but also sums involving indefinite nested sums. In this algebraic setting, one can represent completely algorithmically indefinite nested sums and products without introducing any algebraic relations between them. Note that this general class of expressions covers as special cases the harmonic sums or generalized nested harmonic sums, cf. [54–57]. Given such an optimal representation, by introducing as less sums as possible, various summation principles are available in Sigma. In this work, we applied the following strategy which has been generalized from the hypergeometric case, [50, 58], to the $\Pi\Sigma$-field setting.

1. Given a definite sum that involves an extra parameter N, we compute a recurrence relation in N that is fulfilled by the input sum. The underlying difference field algorithms exploit Zeilberger's creative telescoping principle, [50, 58].
2. Then we solve the derived recurrence in terms of the so-called d'Alembertian solutions, [50, 58]. Since this class covers the harmonic sums, we find all solutions in terms of harmonic sums.
3. Taking the initial values of the original input sum, we can combine the solutions found from step 2 in order to arrive at a closed representation in terms of harmonic sums.

In the following, we give some examples on how Sigma works. A few typical sums we had to calculate are listed in Appendix D and a complete set of sums needed to calculate the 2-Loop OMEs up to $O(\varepsilon)$ can be found in Appendix B of Refs. [4, 12]. Note that in this calculation also more well-known sums are occurring which can, e.g., be easily solved using Summer.

6.2.1 The Sigma-Approach

As a first example we consider the sum

$$T_1(N) \equiv \sum_{i=1}^{\infty} \frac{B(N,i)}{i+N+2} S_1(i) S_1(N+i). \tag{6.9}$$

We treat the upper bound of the sum as a finite integer, i.e., we consider the truncated version

$$T_1(a,N) \equiv \sum_{i=1}^{a} \frac{B(N,i)}{i+N+2} S_1(i) S_1(N+i),$$

for $a \in \mathbb{N}$. Given this sum as input, we apply Sigma's creative telescoping algorithm and find a recurrence for $T_1(a,N)$ of the form

$$c_0(N)T(a,N) + \dots c_d(N)T(a,N+d) = q(a,N) \tag{6.10}$$

with order $d=4$. Here, the $c_i(N)$ and $q(a,N)$ are known functions of N and a. Finally, we perform the limit $a \to \infty$ and we end up at the recurrence

$$\begin{aligned}
&-N(N+1)(N+2)^2\{4N^5+68N^4+455N^3+1494N^2+2402N+1510\}T_1(N)\\
&-(N+1)(N+2)(N+3)\{16N^5+260N^4+1660N^3+5188N^2+7912N+4699\}\\
&\times T_1(N+1)+(N+2)(N+4)(2N+5)\{4N^6+74N^5+542N^4+1978N^3+3680N^2\\
&+3103N+767\}T_1(N+2)+(N+4)(N+5)\{16N^6+276N^5+1928N^4+6968N^3\\
&+13716N^2+13929N+5707\}T_1(N+3)-(N+4)(N+5)^2(N+6)\{4N^5+48N^4\\
&+223N^3+497N^2+527N+211\}T_1(N+4)=P_1(N)+P_2(N)S_1(N)
\end{aligned}$$

where

$$\begin{aligned}
P_1(N) = \Big(&32N^{18}+1232N^{17}+21512N^{16}+223472N^{15}+1514464N^{14}\\
&+6806114N^{13}+18666770N^{12}+15297623N^{11}-116877645N^{10}\\
&-641458913N^9-1826931522N^8-3507205291N^7\\
&-4825457477N^6-4839106893N^5-3535231014N^4\\
&-1860247616N^3-684064448N^2-160164480N-17395200\Big)\\
\Big/ \Big(&N^3(N+1)^3(N+2)^3(N+3)^2(N+4)(N+5)\Big)
\end{aligned}$$

and

$$
\begin{aligned}
P_2(N) = & -4\big(4N^{14} + 150N^{13} + 2610N^{12} + 27717N^{11} + 199197N^{10} \\
& + 1017704N^9 + 3786588N^8 + 10355813N^7 + 20779613N^6 \\
& + 30225025N^5 + 31132328N^4 + 21872237N^3 + 9912442N^2 \\
& + 2672360N + 362400\big) \\
& /\big(N^2(N+1)^2(N+2)^2(N+3)(N+4)(N+5)\big).
\end{aligned}
$$

In the next step, we apply Sigma's recurrence solver to the computed recurrence and find the four linearly independent solutions

$$
\begin{aligned}
h_1(N) &= \frac{1}{N+2}, \quad h_2(N) = \frac{(-1)^N}{N(N+1)(N+2)}, \\
h_3(N) &= \frac{S_1(N)}{N+2}, \quad h_4(N) = (-1)^N \frac{(1+(N+1)S_1(N))}{N(N+1)^2(N+2)},
\end{aligned}
$$

of the homogeneous version of the recurrence and the particular solution

$$
\begin{aligned}
p(N) = & \frac{2(-1)^N}{N(N+1)(N+2)}\Big[2S_{-2,1}(N) - 3S_{-3}(N) - 2S_{-2}(N)S_1(N) - \zeta_2 S_1(N) \\
& - \zeta_3 - \frac{2S_{-2}(N)+\zeta_2}{N+1}\Big] - 2\frac{S_3(N)-\zeta_3}{N+2} - \frac{S_2(N)-\zeta_2}{N+2}S_1(N) \\
& + \frac{2+7N+7N^2+5N^3+N^4}{N^3(N+1)^3(N+2)}S_1(N) + 2\frac{2+7N+9N^2+4N^3+N^4}{N^4(N+1)^3(N+2)}
\end{aligned}
$$

of the recurrence itself. Finally, we look for constants $c_1, \ldots, c_4$ such that

$$
T_1(N) = c_1 h_1(N) + c_2 h_2(N) + c_3 h_3(N) + c_4 h_4(N) + p(N).
$$

The calculation of the necessary initial values for $N = 0, 1, 2, 3$ does not pose a problem for Sigma and we conclude that $c_1 = c_2 = c_3 = c_4 = 0$. Hence the final result reads

$$
\begin{aligned}
T_1(N) = & \frac{2(-1)^N}{N(N+1)(N+2)}\Big[2S_{-2,1}(N) - 3S_{-3}(N) - 2S_{-2}(N)S_1(N) - \zeta_2 S_1(N) \\
& - \zeta_3 - \frac{2S_{-2}(N)+\zeta_2}{N+1}\Big] - 2\frac{S_3(N)-\zeta_3}{N+2} \\
& - \frac{S_2(N)-\zeta_2}{N+2}S_1(N) + \frac{2+7N+7N^2+5N^3+N^4}{N^3(N+1)^3(N+2)}S_1(N) \\
& + 2\frac{2+7N+9N^2+4N^3+N^4}{N^4(N+1)^3(N+2)}.
\end{aligned}
\tag{6.11}
$$

Using more refined algorithms of Sigma, see e.g. [59, 60], even a first order difference equation can be obtained

$$
\begin{aligned}
&(N+2)T_1(N) - (N+3)T_1(N+1) \\
&= 2\frac{(-1)^N}{N(N+2)}\Bigg(-\frac{3N+4}{(N+1)(N+2)}(\zeta_2 + 2S_{-2}(N)) - 2\zeta_3 - 2S_{-3}(N) \\
&\qquad - 2\zeta_2 S_1(N) - 4S_{1,-2}(N)\Bigg) \\
&\quad + \frac{N^6 + 8N^5 + 31N^4 + 66N^3 + 88N^2 + 64N + 16}{N^3(N+1)^2(N+2)^3} S_1(N) \\
&\quad + \frac{S_2(N) - \zeta_2}{N+1} + 2\frac{N^5 + 5N^4 + 21N^3 + 38N^2 + 28N + 8}{N^4(N+1)^2(N+2)^2}.
\end{aligned}
\tag{6.12}
$$

However, in deriving Eq. 6.12, use had to be made of further sums of less complexity, which had to be calculated separately. As above, we can easily solve the recurrence and obtain again the result (6.11). Here and in the following we applied various algebraic relations between harmonic sums to obtain a simplification of our results, cf. [8].

6.2.2 Alternative Approaches

As a second example we consider the sum

$$
T_2(N) \equiv \sum_{i=1}^{\infty} \frac{S_1^2(i+N)}{i^2}, \tag{6.13}
$$

which does not contain a Beta-function. In a first attempt, we proceed as in the first example $T_1(N)$. The naive application of Sigma yields a fifth order difference equation, which is clearly too complex for this sum. However, similar to the situation $T_1(N)$, Sigma can reduce it to a third order relation which reads

$$
\begin{aligned}
&T_2(N)(N+1)^2 - T_2(N+1)(3N^2 + 10N + 9) \\
&+ T_2(N+2)(3N^2 + 14N + 17) - T_2(N+3)(N+3)^2 \\
&= \frac{6N^5 + 48N^4 + 143N^3 + 186N^2 + 81N - 12}{(N+1)^2(N+2)^3(N+3)^2} - 2\frac{2N^2 + 7N + 7}{(N+1)(N+2)^2(N+3)} S_1(N) \\
&+ \frac{-2N^6 - 24N^5 - 116N^4 - 288N^3 - 386N^2 - 264N - 72}{(N+1)^2(N+2)^3(N+3)^2}\zeta_2.
\end{aligned}
\tag{6.14}
$$

Solving this recurrence relation in terms of harmonic sums gives a closed form, see (6.20) below. Still (6.14) represents a rather involved way to solve the problem. It is of advantage to map the numerator $S_1^2(i+N)$ into a linear representation, which can be achieved using Euler's relation

$$S_a^2(N) = 2S_{a,a}(N) - S_{2a}(N), a > 0. \tag{6.15}$$

This is realized in Summer by the basis-command for general-type harmonic sums,

$$T_2(N) = \sum_{i=1}^{\infty} \frac{2S_{1,1}(i+N) - S_2(i+N)}{i^2}. \tag{6.16}$$

As outlined in Ref. [7], sums of this type can be evaluated by considering the difference

$$D_2(j) = T_2(j) - T_2(j-1) = 2\sum_{i=1}^{\infty} \frac{S_1(j+i)}{i} - \sum_{i=1}^{\infty} \frac{1}{i^2(j+i)^2}. \tag{6.17}$$

The solution is then obtained by summing (6.17) to

$$T_2(N) = \sum_{j=1}^{N} D_2(j) + T_2(0). \tag{6.18}$$

The sums in Eq. 6.17 are now calculable trivially or are of less complexity than the original sum. In the case considered here, only the first sum on the left hand side is not trivial. However, after partial fractioning, one can repeat the same procedure, resulting into another difference equation, which is now easily solved. Thus using this technique, the solution of Eq. 6.13 can be obtained by summing two first order difference equations or solving a second order one. The above procedure is well known and some of the summation-algorithms of Summer are based on it. As a consequence, infinite sums with an arbitrary number of harmonic sums with the same argument can be performed using this package. Note that sums containing harmonic sums with different arguments, see e.g Eq. 6.21, can in principle be summed automatically using the same approach. However, this feature is not yet built into Summer. A third way to obtain the sum (6.13) consists of using integral representations for harmonic sums, [6]. One finds

$$\begin{aligned} T_2(N) &= 2\sum_{i=1}^{\infty} \int_0^1 dx \, \frac{x^{i+N}}{i^2} \left(\frac{\ln(1-x)}{1-x}\right)_+ - \sum_{i=1}^{\infty} \left(\int_0^1 dx \, \frac{x^{i+N}}{i^2} \frac{\ln(x)}{1-x} + \frac{\zeta_2}{i^2}\right) \\ &= 2\mathbf{M}\left[\left(\frac{\ln(1-x)}{1-x}\right)_+ \mathbf{Li}_2(x)\right](N+1) \\ &\quad - \left(\mathbf{M}\left[\frac{\ln(x)}{1-x}\mathbf{Li}_2(x)\right](N+1) + \zeta_2^2\right). \end{aligned} \tag{6.19}$$

Here the Mellin-transform is defined in Eq. 2.65. Equation 6.19 can then be easily calculated since the corresponding Mellin-transforms are well-known, [6]. Either of these three methods above lead to

$$T_2(N) = \frac{17}{10}\zeta_2^2 + 4S_1(N)\zeta_3 + S_1^2(N)\zeta_2 - S_2(N)\zeta_2 - 2S_1(N)S_{2,1}(N) - S_{2,2}(N). \tag{6.20}$$

As a third example we would like to evaluate the sum

$$T_3(N) = \sum_{i=1}^{\infty} \frac{S_1^2(i+N)S_1(i)}{i}. \tag{6.21}$$

Note that (6.21) is divergent. In order to treat this divergence, the symbol σ_1, cf. Eq. C.35, is used. The application of Sigma to this sum yields a fourth order difference equation

$$\begin{aligned}
&(N+1)^2(N+2)T_2(N) - (N+2)\left(4N^2+15N+15\right)T_2(N+1) \\
&+ (2N+5)\left(3N^2+15N+20\right)T_2(N+2) - (N+3)\left(4N^2+25N+40\right)T_2(N+3) \\
&+ (N+3)(N+4)^2T_2(N+4) \\
&= \frac{6N^5+73N^4+329N^3+684N^2+645N+215}{(N+1)^2(N+2)^2(N+3)^2} \\
&+ \frac{6N^2+19N+9}{(N+1)(N+2)(N+3)}S_1(N),
\end{aligned} \tag{6.22}$$

which can be solved. As in the foregoing example the better way to calculate the sum is to first change $S_1^2(i+N)$ into a linear basis representation

$$T_3(N) = \sum_{i=1}^{\infty} \frac{2S_{1,1}(i+N) - S_2(i+N)}{i} S_1(i). \tag{6.23}$$

One may now calculate $T_3(N)$ using telescoping for the difference

$$D_3(j) = T_3(j) - T_3(j-1) = 2\sum_{i=1}^{\infty} \frac{S_1(i+j)S_1(i)}{i(i+j)} - \sum_{i=1}^{\infty} \frac{S_1(i)}{i(i+j)^2}, \tag{6.24}$$

with

$$T_3(N) = \sum_{j=1}^{N} D_2(j) + T_3(0). \tag{6.25}$$

One finally obtains

$$\begin{aligned}
T_3(N) = &\frac{\sigma_1^4}{4} + \frac{43}{20}\zeta_2^2 + 5S_1(N)\zeta_3 + \frac{3S_1^2(N) - S_2(N)}{2}\zeta_2 \\
&- 2S_1(N)S_{2,1}(N) + S_1^2(N)S_2(N) + S_1(N)S_3(N) \\
&- \frac{S_2^2(N)}{4} + \frac{S_1^4(N)}{4}.
\end{aligned} \tag{6.26}$$

6.3 Results

For the singlet contributions, we leave out an overall factor

$$\frac{1+(-1)^N}{2} \tag{6.27}$$

in the following. This factor emerges naturally in our calculation and is due to the fact that in the light-cone expansion, only even values of N contribute to F_2 and F_L, cf. Sect. 2.3. Additionally, we do not choose a linear representation in terms of harmonic sums as was done in Refs. [28–30], since these are non-minimal w.r.t. to the corresponding quasi-shuffle algebra, [61, 62]. Due to this a much smaller number of harmonic sums contributes. Remainder terms can be expressed in polynomials $P_i(N)$. Single harmonic sums with negative index are expressed in terms of the function $\beta(N+1)$, cf. Appendix C.4. For completeness, we also give all pole terms and the constant terms of the quarkonic OMEs. The latter have been obtained before in Refs. [1, 4]. The pole terms can be expressed via the LO–, [63–65], and the fermionic parts of the NLO, [66–72], anomalous dimensions and the 1-loop β-function, [73, 74], (t'Hooft, 1972, unpublished) [75].

We first consider the matrix element $A_{Qg}^{(2)}$, which is the most complex of the 2-loop OMEs. For the calculation we used the projector given in Eq. 4.22 and therefore have to include diagrams with external ghost lines as well. The 1-loop result is straightforward to calculate and has already been given in Eqs. 4.111, 4.113. As explained in Chap. 4, we perform the calculation accounting for 1-particle reducible diagrams. Hence the 1-loop massive gluon self-energy term, Eq. 4.84, contributes. The unrenormalized 2-loop OME is then given in terms of 1-particle irreducible and reducible contributions by

$$\hat{\hat{A}}_{Qg}^{(2)} = \hat{\hat{A}}_{Qg}^{(2),irr} - \hat{\hat{A}}_{Qg}^{(1)} \hat{\Pi}^{(1)}\left(0, \frac{\hat{m}^2}{\mu^2}\right). \tag{6.28}$$

Using the techniques described in the previous sections, the pole-terms predicted by renormalization in Eq. 4.112 are obtained, which have been given in Refs. [1, 4] before. Here, the contributing 1-loop anomalous dimensions are

$$\gamma_{qq}^{(0)} = 4C_F\left\{2S_1 - \frac{3N^2+3N+2}{2N(N+1)}\right\}, \tag{6.29}$$

$$\hat{\gamma}_{qg}^{(0)} = -8T_F\frac{N^2+N+2}{N(N+1)(N+2)}, \tag{6.30}$$

$$\gamma_{gg}^{(0)} = 8C_A\left\{S_1 - \frac{2(N^2+N+1)}{(N-1)N(N+1)(N+2)}\right\} - 2\beta_0, \tag{6.31}$$

and the 2-loop contribution reads

$$\hat{\gamma}_{qg}^{(1)} = 8C_F T_F \left\{ 2\frac{N^2+N+2}{N(N+1)(N+2)} \left[S_2 - S_1^2 \right] + \frac{4}{N^2} S_1 - \frac{P_1}{N^3(N+1)^3(N+2)} \right\}$$
$$+ 16 C_A T_F \left\{ \frac{N^2+N+2}{N(N+1)(N+2)} \left[S_2 + S_1^2 - 2\beta' - \zeta_2 \right] \right.$$
$$\left. - \frac{4(2N+3)S_1}{(N+1)^2(N+2)^2} - \frac{P_2}{(N-1)N^3(N+1)^3(N+2)^3} \right\}, \tag{6.32}$$

$$P_1 = 5N^6 + 15N^5 + 36N^4 + 51N^3 + 25N^2 + 8N + 4,$$
$$P_2 = N^9 + 6N^8 + 15N^7 + 25N^6 + 36N^5 + 85N^4 + 128N^3$$
$$+ 104N^2 + 64N + 16. \tag{6.33}$$

These terms agree with the literature and provide a strong check on the calculation. The constant term in ε in Eq. 4.112 is determined after mass renormalization, [1, 4, 5].

$$a_{Qg}^{(2)} = T_F C_F \left\{ \frac{4(N^2+N+2)}{3N(N+1)(N+2)} \left(4S_3 - 3S_2 S_1 - S_1^3 - 6S_1\zeta_2 \right) + 4\frac{3N+2}{N^2(N+2)} S_1^2 \right.$$
$$+ 4\frac{N^4+17N^3+17N^2-5N-2}{N^2(N+1)^2(N+2)} S_2 + 2\frac{(3N^2+3N+2)(N^2+N+2)}{N^2(N+1)^2(N+2)} \zeta_2$$
$$\left. + 4\frac{N^4-N^3-20N^2-10N-4}{N^2(N+1)^2(N+2)} S_1 + \frac{2P_3}{N^4(N+1)^4(N+2)} \right\}$$
$$+ T_F C_A \left\{ \frac{2(N^2+N+2)}{3N(N+1)(N+2)} \left(-24S_{-2,1} + 6\beta'' + 16S_3 - 24\beta' S_1 + 18 S_2 S_1 + 2S_1^3 - 9\zeta_3 \right) \right.$$
$$- 16\frac{N^2-N-4}{(N+1)^2(N+2)^2}\beta' - 4\frac{7N^5+21N^4+13N^3+21N^2+18N+16}{(N-1)N^2(N+1)^2(N+2)^2} S_2$$
$$- 4\frac{N^3+8N^2+11N+2}{N(N+1)^2(N+2)^2} S_1^2 - 8\frac{N^4-2N^3+5N^2+2N+2}{(N-1)N^2(N+1)^2(N+2)} \zeta_2$$
$$\left. - \frac{4P_4}{N(N+1)^3(N+2)^3} S_1 + \frac{4P_5}{(N-1)N^4(N+1)^4(N+2)^4} \right\}, \tag{6.34}$$

where the polynomials in Eq. 6.34 are given by

$$P_3 = 12N^8 + 52N^7 + 132N^6 + 216N^5 + 191N^4$$
$$+ 54N^3 - 25N^2 - 20N - 4, \tag{6.35}$$

$$P_4 = N^6 + 8N^5 + 23N^4 + 54N^3 + 94N^2 + 72N + 8, \tag{6.36}$$

$$
\begin{aligned}
P_5 = {} & 2N^{12} + 20N^{11} + 86N^{10} + 192N^9 + 199N^8 - N^7 - 297N^6 \\
& - 495N^5 - 514N^4 - 488N^3 - 416N^2 - 176N - 32.
\end{aligned} \tag{6.37}
$$

The newly calculated $O(\varepsilon)$ contribution to $A_{Qg}^{(2)}$, [12], reads after mass renormalization

$$
\begin{aligned}
\overline{a}_{Qg}^{(2)} = {} & T_F C_F \left\{ \frac{N^2+N+2}{N(N+1)(N+2)} \left(16S_{2,1,1} - 8S_{3,1} - 8S_{2,1}S_1 + 3S_4 - \frac{4}{3}S_3S_1 - \frac{1}{2}S_2^2 \right.\right. \\
& \left. - S_2S_1^2 - \frac{1}{6}S_1^4 + 2\zeta_2S_2 - 2\zeta_2S_1^2 - \frac{8}{3}\zeta_3S_1 \right) - 8\frac{N^2-3N-2}{N^2(N+1)(N+2)}S_{2,1} \\
& + \frac{2}{3}\frac{3N+2}{N^2(N+2)}S_1^3 + \frac{2}{3}\frac{3N^4+48N^3+43N^2-22N-8}{N^2(N+1)^2(N+2)}S_3 \\
& + 2\frac{3N+2}{N^2(N+2)}S_2S_1 + 4\frac{S_1}{N^2}\zeta_2 + \frac{2}{3}\frac{(N^2+N+2)(3N^2+3N+2)}{N^2(N+1)^2(N+2)}\zeta_3 \\
& + \frac{P_6}{N^3(N+1)^3(N+2)}S_2 + \frac{N^4-5N^3-32N^2-18N-4}{N^2(N+1)^2(N+2)}S_1^2 \\
& - 2\frac{2N^5-2N^4-11N^3-19N^2-44N-12}{N^2(N+1)^3(N+2)}S_1 \\
& - \frac{5N^6+15N^5+36N^4+51N^3+25N^2+8N+4}{N^3(N+1)^3(N+2)}\zeta_2 \\
& \left. - \frac{P_7}{N^5(N+1)^5(N+2)} \right\} \\
& + T_F C_A \left\{ \frac{N^2+N+2}{N(N+1)(N+2)} \left(16S_{-2,1,1} - 4S_{2,1,1} - 8S_{-3,1} - 8S_{-2,2} \right.\right. \\
& - 4S_{3,1} - \frac{2}{3}\beta''' + 9S_4 - 16S_{-2,1}S_1 + \frac{40}{3}S_1S_3 + 4\beta''S_1 - 8\beta'S_2 + \frac{1}{2}S_2^2 \\
& \left. - 8\beta'S_1^2 + 5S_1^2S_2 + \frac{1}{6}S_1^4 - \frac{10}{3}S_1\zeta_3 - 2S_2\zeta_2 - 2S_1^2\zeta_2 - 4\beta'\zeta_2 - \frac{17}{5}\zeta_2^2 \right) \\
& - 8\frac{N^2+N-1}{(N+1)^2(N+2)^2}\zeta_2S_1 + \frac{4(N^2-N-4)}{(N+1)^2(N+2)^2}\left(-4S_{-2,1} + \beta'' - 4\beta'S_1\right) \\
& - \frac{2}{3}\frac{N^3+8N^2+11N+2}{N(N+1)^2(N+2)^2}S_1^3 - \frac{16}{3}\frac{N^5+10N^4+9N^3+3N^2+7N+6}{(N-1)N^2(N+1)^2(N+2)^2}S_3 \\
& + 8\frac{N^4+2N^3+7N^2+22N+20}{(N+1)^3(N+2)^3}\beta' + 2\frac{3N^3-12N^2-27N-2}{N(N+1)^2(N+2)^2}S_2S_1 \\
& - \frac{2}{3}\frac{9N^5-10N^4-11N^3+68N^2+24N+16}{(N-1)N^2(N+1)^2(N+2)^2}\zeta_3 \\
& - \frac{P_8S_2}{(N-1)N^3(N+1)^3(N+2)^3} - \frac{P_{10}S_1^2}{N(N+1)^3(N+2)^3} \\
& + \frac{2P_{11}S_1}{N(N+1)^4(N+2)^4}, \\
& \left. - \frac{2P_9\zeta_2}{(N-1)N^3(N+1)^3(N+2)^2} - \frac{2P_{12}}{(N-1)N^5(N+1)^5(N+2)^5} \right\}
\end{aligned} \tag{6.38}
$$

with the polynomials

$$P_6 = 3N^6 + 30N^5 + 15N^4 - 64N^3 - 56N^2 - 20N - 8, \tag{6.39}$$

$$P_7 = 24N^{10} + 136N^9 + 395N^8 + 704N^7 + 739N^6 + 407N^5 + 87N^4 + 27N^3 + 45N^2 + 24N + 4, \tag{6.40}$$

$$P_8 = N^9 + 21N^8 + 85N^7 + 105N^6 + 42N^5 + 290N^4 + 600N^3 + 456N^2 + 256N + 64, \tag{6.41}$$

$$P_9 = (N^3 + 3N^2 + 12N + 4)(N^5 - N^4 + 5N^2 + N + 2), \tag{6.42}$$

$$P_{10} = N^6 + 6N^5 + 7N^4 + 4N^3 + 18N^2 + 16N - 8, \tag{6.43}$$

$$P_{11} = 2N^8 + 22N^7 + 117N^6 + 386N^5 + 759N^4 + 810N^3 + 396N^2 + 72N + 32, \tag{6.44}$$

$$P_{12} = 4N^{15} + 50N^{14} + 267N^{13} + 765N^{12} + 1183N^{11} + 682N^{10} - 826N^9 - 1858N^8 - 1116N^7 + 457N^6 + 1500N^5 + 2268N^4 + 2400N^3 + 1392N^2 + 448N + 64. \tag{6.45}$$

Note that the terms $\propto \zeta_3$ in Eq. 6.34 and $\propto \zeta_2^2$ in Eq. 6.38 are only due to the representation using the $\beta^{(k)}$-functions and are absent in representations using harmonic sums. The results for the individual diagrams contributing to $A_{Qg}^{(2)}$ can be found up to $O(\varepsilon^0)$ in Ref. [4] and at $O(\varepsilon)$ in Ref. [12].

Since harmonic sums appear in a wide variety of applications, it is interesting to study the pattern in which they emerge. In Table 6.1, we list the harmonic sums contributing to each individual diagram.[2] The β-function and their derivatives can be traced back to the single non-alternating harmonic sums, allowing for half-integer arguments, cf. [6] and Appendix C.4. Therefore, all single harmonic sums form an equivalence class being represented by the sum S_1, from which the other single harmonic sums are easily derived through differentiation and half-integer relations Additionally, we have already made use of the algebraic relations, [8], between harmonic sums in deriving Eqs. 6.34, 6.38. Moreover, the sums $S_{-2,2}$ and $S_{3,1}$ obey structural relations to other harmonic sums, i.e., they lie in corresponding equivalence classes and may be obtained by either rational argument relations and/or differentiation w.r.t. N. Reference to these equivalence classes is useful since the representation of these sums for $N \varepsilon \mathbb{C}$ needs not to be derived newly, except of straightforward differentiations. All functions involved are meromorphic, with poles

[2] Cf. Ref. [1] for the labeling of the diagrams.

Table 6.1 Complexity of the results for the individual diagrams contributing to $A_{Qg}^{(2)}$

Diagram	S_1	S_2	S_3	S_4	S_{-2}	S_{-3}	S_{-4}	$S_{2,1}$	$S_{-2,1}$	$S_{-2,2}$	$S_{3,1}$	$S_{-3,1}$	$S_{2,1,1}$	$S_{-2,1,1}$
A		+	+											
B	+	+	+	+				+			+		+	
C		+	+											
D	+	+	+					+						
E	+	+	+					+						
F	+	+	+	+				+					+	
G	+	+	+					+						
H	+	+	+					+						
I	+	+	+	+	+	+	+	+	+	+	+	+	+	+
J		+	+											
K		+	+											
L	+	+	+	+				+			+		+	
M		+	+											
N	+	+	+	+	+	+	+	+	+	+	+	+	+	+
O	+	+	+	+				+			+		+	
P	+	+	+	+				+			+		+	
S		+	+											
T		+	+											

at the non-negative integers. Thus the $O(\varepsilon^0)$-term depends on two basic functions only, S_1 and $S_{-2,1}$.[3] This has to be compared to the Z-space representation used in Ref. [1], in which 48 different functions were needed. As shown in [6], various of these functions have Mellin transforms containing triple sums, which do not occur in our approach even on the level of individual diagrams. Thus the method applied here allowed to compactify the representation of the heavy flavor matrix elements and Wilson coefficients significantly.

The $O(\varepsilon)$-term consists of six basic functions only, which are given by

$$\begin{aligned} &\{S_1, S_2, S_3, S_4, S_{-2}, S_{-3}, S_{-4}\}, S_{2,1}, S_{-2,1}, S_{-3,1}, S_{2,1,1}, S_{-2,1,1}, \\ &\qquad S_{-2,2} : \text{depends on } S_{-2,1}, S_{-3,1} \\ &\qquad S_{3,1} \quad : \text{depends on } S_{2,1}. \end{aligned} \tag{6.46}$$

The absence of harmonic sums containing $\{-1\}$ as index was noted before for all other classes of space- and time-like anomalous dimensions and Wilson coefficients, including those for other hard processes having been calculated so far, cf. [27, 32–34]. This can not be seen if one applies the Z-space representation or the linear representation in Mellin-space, [76].

Analytic continuation, e.g., for $S_{-2,1}$ proceeds via the equality,

$$\mathbf{M}\left[\frac{\mathrm{Li}_2(x)}{1+x}\right](N+1) - \zeta_2\beta(N+1) = (-1)^{N+1}\left[S_{-2,1}(N) + \frac{5}{8}\zeta_3\right] \tag{6.47}$$

[3] The associated Mellin transform to this sum has been discussed in Ref. [68] first.

with similar representations for the remaining sums, [6].[4]

As discussed in [35], the result for $a_{Qg}^{(2)}$ agrees with that in Z-space given in Ref. [1]. However, there is a difference concerning the complete renormalized expression for $A_{Qg}^{(2)}$. This is due to the scheme-dependence for the renormalization of the coupling constant, which has been described in Sects. 4.4, 5.1 and emerges for the first time at $O(a_s^2)$. Comparing Eq. 4.114 for the renormalized result in the $\overline{\mathsf{MS}}$-scheme for the coupling constant with the transformation formula to the MOM-scheme, Eq. 5.12, this difference is given by

$$A_{Qg}^{(2),\overline{\mathsf{MS}}} = A_{Qg}^{(2),\mathrm{MOM}} - \beta_{0,Q}\frac{\hat{\gamma}_{qg}^{(0)}}{2}\ln^2\left(\frac{m^2}{\mu^2}\right). \tag{6.48}$$

As an example, the second moment of the massive OME up to 2-loops reads in the $\overline{\mathsf{MS}}$-scheme for coupling constant renormalization

$$\begin{aligned} A_{Qg}^{\overline{\mathsf{MS}}} &= a_s^{\overline{\mathsf{MS}}}\left\{-\frac{4}{3}T_F\ln\left(\frac{m^2}{\mu^2}\right)\right\} \\ &\quad + a_s^{\overline{\mathsf{MS}}^2}\left\{T_F\left[\frac{22}{9}C_A - \frac{16}{9}C_F - \frac{16}{9}T_F\right]\ln^2\left(\frac{m^2}{\mu^2}\right)\right. \\ &\qquad \left. + T_F\left[-\frac{70}{27}C_A - \frac{148}{27}C_F\right]\ln\left(\frac{m^2}{\mu^2}\right) - \frac{7}{9}C_AT_F + \frac{1352}{81}C_FT_F\right\}, \end{aligned} \tag{6.49}$$

and in the MOM-scheme

$$\begin{aligned} A_{Qg}^{\mathrm{MOM}} &= a_s^{\mathrm{MOM}}\left\{-\frac{4}{3}T_F\ln\left(\frac{m^2}{\mu^2}\right)\right\} + a_s^{\mathrm{MOM}^2}\left\{T_F\left[\frac{22}{9}C_A - \frac{16}{9}C_F\right]\ln^2\left(\frac{m^2}{\mu^2}\right)\right. \\ &\qquad \left. + T_F\left[-\frac{70}{27}C_A - \frac{148}{27}C_F\right]\ln\left(\frac{m^2}{\mu^2}\right) - \frac{7}{9}C_AT_F + \frac{1352}{81}C_FT_F\right\}. \end{aligned} \tag{6.50}$$

As one infers from the above formulas, this difference affects at the 2-loop level only the double logarithmic term and stems from the treatment of the 1-particle-reducible contributions. In Ref. [1], these contributions were absorbed into the coupling constant, applying the MOM-scheme. This was motivated by the need to eliminate the virtual contributions due to heavier quarks (b, t) and was also extended to the charm-quark, thus adopting the same renormalization scheme as has been used in Refs. [77–79] for the exact calculation of the heavy flavor contributions to the Wilson coefficients. Contrary, in Ref. [2], the $\overline{\mathsf{MS}}$-description was applied and the strong coupling constant depends on $n_f + 1$ flavors, cf. the discussion in Sect. 5.1.

The remaining massive OMEs are less complex than the term $A_{Qg}^{(2)}$ and depend only on single harmonic sums, i.e. on only one basic function, S_1. In the PS-case, the LO and NLO anomalous dimensions

---

4 Note that the argument of the Mellin-transform in Eq. 6.36, Ref. [35], should read $(N + 1)$.

$$\gamma_{gq}^{(0)} = -4C_F \frac{N^2+N+2}{(N-1)N(N+1)}, \tag{6.51}$$

$$\hat{\gamma}_{qq}^{(1),\mathrm{PS}} = -16C_F T_F \frac{5N^5+32N^4+49N^3+38N^2+28N+8}{(N-1)N^3(N+1)^3(N+2)^2} \tag{6.52}$$

contribute. The pole-terms are given by Eq. 4.100 and we obtain for the higher order terms in ε

$$a_{Qq}^{(2),\mathrm{PS}} = C_F T_F \left\{ -\frac{4(N^2+N+2)^2(2S_2+\zeta_2)}{(N-1)N^2(N+1)^2(N+2)} + \frac{4P_{13}}{(N-1)N^4(N+1)^4(N+2)^3} \right\}, \tag{6.53}$$

$$P_{13} = N^{10}+8N^9+29N^8+49N^7-11N^6-131N^5 - 161N^4-160N^3-168N^2-80N-16, \tag{6.54}$$

$$\overline{a}_{Qq}^{(2),\mathrm{PS}} = C_F T_F \left\{ -2\frac{(5N^3+7N^2+4N+4)(N^2+5N+2)}{(N-1)N^3(N+1)^3(N+2)^2}(2S_2+\zeta_2) - \frac{4(N^2+N+2)^2(3S_3+\zeta_3)}{3(N-1)N^2(N+1)^2(N+2)} + \frac{2P_{14}}{(N-1)N^5(N+1)^5(N+2)^4} \right\}, \tag{6.55}$$

$$P_{14} = 5N^{11}+62N^{10}+252N^9+374N^8-400N^6+38N^7-473N^5 - 682N^4-904N^3-592N^2-208N-32. \tag{6.56}$$

Since the PS-OME emerges for the first time at $O(a_s^2)$, there is no difference between its representation in the MOM- and the $\overline{\mathsf{MS}}$-scheme. The renormalized OME $A_{Qq}^{(2)PS}$ is given in Eq. 4.101 and the second moment reads

$$A_{Qq}^{PS,\overline{\mathsf{MS}}} = a_s^{\overline{\mathsf{MS}}^2} \left\{ -\frac{16}{9}\ln^2\left(\frac{m^2}{\mu^2}\right) - \frac{80}{27}\ln\left(\frac{m^2}{\mu^2}\right) - 4 \right\} C_F T_F + O\left(a_s^{\overline{\mathsf{MS}}^3}\right). \tag{6.57}$$

The flavor non-singlet NLO anomalous dimension is given by

$$\hat{\gamma}_{qq}^{(1),\mathrm{NS}} = \frac{4C_F T_F}{3} \left\{ 8S_2 - \frac{40}{3}S_1 + \frac{3N^4+6N^3+47N^2+20N-12}{3N^2(N+1)^2} \right\}. \tag{6.58}$$

The unrenormalized OME is obtained from the 1-particle irreducible graphs and the contributions of heavy quark loops to the quark self-energy. The latter is given at $O(\hat{a}_s^2)$ in Eq. 4.87. One obtains

$$\hat{\hat{A}}_{qq,Q}^{(2),\mathrm{NS}} = \hat{\hat{A}}_{qq,Q}^{(2),\mathrm{NS,irred}} - \hat{\Sigma}^{(2)}(0, \frac{\hat{m}^2}{\mu^2}). \tag{6.59}$$

Our result is of the structure given in Eq. 4.93 and the higher order terms in ε read

$$a_{qq,Q}^{(2),\mathrm{NS}} = \frac{C_F T_F}{3}\Biggl\{-8S_3 - 8\zeta_2 S_1 + \frac{40}{3}S_2 + 2\frac{3N^2+3N+2}{N(N+1)}\zeta_2 - \frac{224}{9}S_1 + \frac{219N^6+657N^5+1193N^4+763N^3-40N^2-48N+72}{18N^3(N+1)^3}\Biggr\}, \tag{6.60}$$

$$\overline{a}_{qq,Q}^{(2),\mathrm{NS}} = \frac{C_F T_F}{3}\Biggl\{4S_4 + 4S_2\zeta_2 - \frac{8}{3}S_1\zeta_3 + \frac{112}{9}S_2 + \frac{3N^4+6N^3+47N^2+20N-12}{6N^2(N+1)^2}\zeta_2 - \frac{20}{3}S_1\zeta_2 - \frac{20}{3}S_3 - \frac{656}{27}S_1 + 2\frac{3N^2+3N+2}{3N(N+1)}\zeta_3 + \frac{P_{15}}{216N^4(N+1)^4}\Biggr\}, \tag{6.61}$$

$$P_{15} = 1551N^8 + 6204N^7 + 15338N^6 + 17868N^5 + 8319N^4 + 944N^3 + 528N^2 - 144N - 432. \tag{6.62}$$

The anomalous dimensions in Eqs. 6.51, 6.52, 6.58 agree with the literature. Eqs. 6.53, 6.60, cf. Ref. [4], were first given in Ref. [1] and agree with the results presented there. Eqs. 6.55, 6.61, [12], are new results of this thesis. As in the PS case, the NS OME emerges for the first time at $O(a_s^2)$. The corresponding renormalized OME $A_{qq,Q}^{(2),\mathrm{NS}}$ is given in Eq. 4.95 and the second moment reads

$$A_{qq,Q}^{NS,\overline{\mathsf{MS}}} = a_s^{\overline{\mathsf{MS}}^2}\Biggl\{-\frac{16}{9}\ln^2\left(\frac{m^2}{\mu^2}\right) - \frac{128}{27}\ln\left(\frac{m^2}{\mu^2}\right) - \frac{128}{27}\Biggr\} \times C_F T_F + O\left(a_s^{\overline{\mathsf{MS}}^3}\right). \tag{6.63}$$

Note that the first moment of the NS-OME vanishes, even on the unrenormalized level up to $O(\varepsilon)$. This provides a check on the results in Eqs. 6.60, 6.61, because this is required by fermion number conservation.

At this point an additional comment on the difference between the MOM and the $\overline{\mathsf{MS}}$-scheme is in order. The MOM-scheme was applied in Ref. [1] for two different purposes. The first one is described below Eq. 6.50. It was introduced to absorb the contributions of one-particle reducible diagrams and heavier quarks into the definition of the coupling constant. However, in case of $A_{Qg}^{(2)}$, renormalization in the MOM-scheme and the scheme transformation from the MOM-scheme to the $\overline{\mathsf{MS}}$-scheme accidentally commute. This means, that one could apply Eq. 4.110 in the $\overline{\mathsf{MS}}$-scheme, i.e., set

$$\delta a_{s,1}^{\mathrm{MOM}} = \delta a_{s,1}^{\overline{\mathrm{MS}}}(n_f + 1) \tag{6.64}$$

from the start and obtain Eq. 4.114 for the renormalized result. This is not the case for $A_{qq,Q}^{(2),\mathrm{NS}}$. As mentioned earlier, the scheme transformation does not have an effect on this term at 2-loop order. This means that Eq. 4.91 should yield the same renormalized result in the MOM- and in the $\overline{\mathrm{MS}}$-scheme. However, in the latter case, the difference of Z-factors does not contain the mass. Thus a term

$$\propto \frac{1}{\varepsilon} \ln\left(\frac{m^2}{\mu^2}\right), \tag{6.65}$$

which stems from the expansion of the unrenormalized result in Eq. 4.93, can not be subtracted. The reason for this is the following. As pointed out in Ref. [1], the term $\hat{A}_{qq,Q}^{(2),\mathrm{NS}}$ is only UV-divergent. However, this is only the case if one imposes the condition that the heavy quark contributions to the gluon self-energy vanishes for on-shell momentum of the gluon. This is exactly the condition we imposed for renormalization in the MOM-scheme, cf. Sect. 4.4. Hence in this case, the additional divergences absorbed into the coupling are of the collinear type, contrary to the term in $A_{Qg}^{(2)}$. By applying the transformation back to the $\overline{\mathrm{MS}}$-scheme, we treat these two different terms in a concise way. This is especially important at the three-loop level, since in this case both effects are observed for all OMEs and the renormalization would not be possible if not applying the MOM-scheme first.

Let us now turn to the gluonic OMEs $A_{gg,Q}^{(2)}$, $A_{gq,Q}^{(2)}$, which are not needed for the asymptotic 2-loop heavy flavor Wilson coefficients. They contribute, however, in the VFNS-description of heavy flavor parton densities, cf. Ref. [2] and Sect. 3.3. The 1-loop term $A_{gg,Q}^{(1)}$ has already been given in Eqs. 4.132, 4.135. In case of $A_{gg,Q}^{(2)}$, the part

$$\hat{\gamma}_{gg}^{(1)} = 8C_F T_F \frac{N^8 + 4N^7 + 8N^6 + 6N^5 - 3N^4 - 22N^3 - 10N^2 - 8N - 8}{(N-1)N^3(N+1)^3(N+2)} + \frac{32C_A T_F}{9}\left\{-5S_1 + \frac{3N^6 + 9N^5 + 22N^4 + 29N^3 + 41N^2 + 28N + 6}{(N-1)N^2(N+1)^2(N+2)}\right\} \tag{6.66}$$

of the 2-loop anomalous dimension is additionally needed. As for $A_{Qg}^{(2)}$, the massive parts of the gluon self-energy contribute, Eqs. 4.84, 4.85. The unrenormalized OME at the 2-loop level is then given in terms of reducible and irreducible contributions via

$$\hat{\hat{A}}_{gg,Q}^{(2)} = \hat{\hat{A}}_{gg,Q}^{(2),\mathrm{irred}} - \hat{\hat{A}}_{gg,Q}^{(1)} \hat{\Pi}^{(1)}\left(0, \frac{\hat{m}^2}{\mu^2}\right) - \hat{\Pi}^{(2)}\left(0, \frac{\hat{m}^2}{\mu^2}\right). \tag{6.67}$$

In the unrenormalized result, we observe the same pole structure as predicted in Eq. 4.133. The constant and $O(\varepsilon)$ contributions $a_{gg,Q}^{(2)}$ and $\bar{a}_{gg,Q}^{(2)}$ are

$$a_{gg,Q}^{(2)} = T_F C_A \left\{ -\frac{8}{3}\zeta_2 S_1 + \frac{16(N^2+N+1)\zeta_2}{3(N-1)N(N+1)(N+2)} - 4\frac{56N+47}{27(N+1)}S_1 \right.$$
$$\left. + \frac{2P_{16}}{27(N-1)N^3(N+1)^3(N+2)} \right\}$$
$$+ T_F C_F \left\{ \frac{4(N^2+N+2)^2\zeta_2}{(N-1)N^2(N+1)^2(N+2)} - \frac{P_{17}}{(N-1)N^4(N+1)^4(N+2)} \right\}, \tag{6.68}$$

$$\overline{a}_{gg,Q}^{(2)} = T_F C_A \left\{ -\frac{8}{9}\zeta_3 S_1 - \frac{20}{9}\zeta_2 S_1 + \frac{16(N^2+N+1)}{9(N-1)N(N+1)(N+2)}\zeta_3 + \frac{2N+1}{3(N+1)}S_2 \right.$$
$$- \frac{S_1^2}{3(N+1)} - 2\frac{328N^4+256N^3-247N^2-175N+54}{81(N-1)N(N+1)^2}S_1$$
$$\left. + \frac{4P_{18}\zeta_2}{9(N-1)N^2(N+1)^2(N+2)} + \frac{P_{19}}{81(N-1)N^4(N+1)^4(N+2)} \right\}$$
$$+ T_F C_F \left\{ \frac{4(N^2+N+2)^2\zeta_3}{3(N-1)N^2(N+1)^2(N+2)} + \frac{P_{20}\zeta_2}{(N-1)N^3(N+1)^3(N+2)} \right.$$
$$\left. + \frac{P_{21}}{4(N-1)N^5(N+1)^5(N+2)} \right\}, \tag{6.69}$$

$$P_{16} = 15N^8 + 60N^7 + 572N^6 + 1470N^5 + 2135N^4$$
$$+ 1794N^3 + 722N^2 - 24N - 72, \tag{6.70}$$

$$P_{17} = 15N^{10} + 75N^9 + 112N^8 + 14N^7 - 61N^6 + 107N^5$$
$$+ 170N^4 + 36N^3 - 36N^2 - 32N - 16, \tag{6.71}$$

$$P_{18} = 3N^6 + 9N^5 + 22N^4 + 29N^3 + 41N^2 + 28N + 6, \tag{6.72}$$

$$P_{19} = 3N^{10} + 15N^9 + 3316N^8 + 12778N^7 + 22951N^6 + 23815N^5$$
$$+ 14212N^4 + 3556N^3 - 30N^2 + 288N + 216, \tag{6.73}$$

$$P_{20} = N^8 + 4N^7 + 8N^6 + 6N^5 - 3N^4 - 22N^3 - 10N^2 - 8N - 8, \tag{6.74}$$

$$P_{21} = 31N^{12} + 186N^{11} + 435N^{10} + 438N^9 - 123N^8 - 1170N^7$$
$$- 1527N^6 - 654N^5 + 88N^4 - 136N^2 - 96N - 32. \tag{6.75}$$

We agree with the result for $a_{gg,Q}^{(2)}$ given in [2], which is presented in Eq. 6.68. The new term $\overline{a}_{gg,Q}^{(2)}$, Eq. 6.69, contributes to all OMEs $A_{ij}^{(3)}$ through renormalization.

The renormalized OME is then given by Eq. 4.136. Since this OME already emerges at LO, the $O(a_s^2)$ term changes replacing the MOM– by the $\overline{\text{MS}}$-scheme. The second moment in the $\overline{\text{MS}}$-scheme reads

$$\begin{aligned} A_{gg,Q}^{\overline{\text{MS}}} &= a_s^{\overline{\text{MS}}}\left\{\frac{4}{3}T_F\ln\left(\frac{m^2}{\mu^2}\right)\right\} \\ &+ a_s^{\overline{\text{MS}}^2}\left\{T_F\left[-\frac{22}{9}C_A+\frac{16}{9}C_F+\frac{16}{9}T_F\right]\ln^2\left(\frac{m^2}{\mu^2}\right)\right. \\ &\left.+T_F\left[\frac{70}{27}C_A+\frac{148}{27}C_F\right]\ln\left(\frac{m^2}{\mu^2}\right)+\frac{7}{9}C_AT_F-\frac{1352}{81}C_FT_F\right\} \\ &+ O\left(a_s^{\overline{\text{MS}}^3}\right). \end{aligned} \tag{6.76}$$

In the MOM-scheme it is given by

$$\begin{aligned} A_{gg,Q}^{\text{MOM}} &= a_s^{\text{MOM}}\left\{\frac{4}{3}T_F\ln\left(\frac{m^2}{\mu^2}\right)\right\} \\ &+ a_s^{\text{MOM}^2}\left\{T_F\left[-\frac{22}{9}C_A+\frac{16}{9}C_F\right]\ln^2\left(\frac{m^2}{\mu^2}\right)\right. \\ &\left.+T_F\left[\frac{70}{27}C_A+\frac{148}{27}C_F\right]\ln\left(\frac{m^2}{\mu^2}\right)+\frac{7}{9}C_AT_F-\frac{1352}{81}C_FT_F\right\} \\ &+ O\left(a_s^{\text{MOM}^3}\right). \end{aligned} \tag{6.77}$$

The difference between the schemes reads

$$A_{gg,Q}^{(2),\overline{\text{MS}}} = A_{gg,Q}^{(2),\text{MOM}} + \beta_{0,Q}^2\ln^2\left(\frac{m^2}{\mu^2}\right). \tag{6.78}$$

The need for applying intermediately the MOM-scheme for renormalization becomes obvious again for the term $A_{gg,Q}^{(2)}$. As in the NS-case, renormalization in the $\overline{\text{MS}}$-scheme for the coupling constant does not cancel all singularities. The remaining term is $A_{gq,Q}^{(2)}$, which emerges for the first time at $O(a_s^2)$ and the same result is obtained in the $\overline{\text{MS}}$- and MOM-schemes. The corresponding NLO anomalous dimension is given by

$$\hat{\gamma}_{gq}^{(1)} = \frac{32C_FT_F}{3}\left\{-\frac{(N^2+N+2)S_1}{(N-1)N(N+1)}+\frac{8N^3+13N^2+27N+16}{3(N-1)N(N+1)^2}\right\}. \tag{6.79}$$

Again, we obtain the pole terms as predicted in Eq. 4.123. The constant and $O(\varepsilon)$ contributions $a_{gq,Q}^{(2)}$ and $\overline{a}_{gq,Q}^{(2)}$ then read

$$a_{gq,Q}^{(2)} = T_F C_F \left\{ \frac{4}{3} \frac{N^2+N+2}{(N-1)N(N+1)} \left(2\zeta_2 + S_2 + S_1^2\right) - \frac{8}{9} \frac{8N^3+13N^2+27N+16}{(N-1)N(N+1)^2} S_1 + \frac{8}{27} \frac{P_{22}}{(N-1)N(N+1)^3} \right\}, \tag{6.80}$$

$$\overline{a}_{gq,Q}^{(2)} = T_F C_F \left\{ \frac{2}{9} \frac{N^2+N+2}{(N-1)N(N+1)} \left(-2S_3 - 3S_2S_1 - S_1^3 + 4\zeta_3 - 6\zeta_2 S_1\right) + \frac{2}{9} \frac{8N^3+13N^2+27N+16}{(N-1)N(N+1)^2} \left(2\zeta_2 + S_2 + S_1^2\right) - \frac{4}{27} \frac{P_{22} S_1}{(N-1)N(N+1)^3} + \frac{4}{81} \frac{P_{23}}{(N-1)N(N+1)^4} \right\}, \tag{6.81}$$

with

$$P_{22} = 43N^4 + 105N^3 + 224N^2 + 230N + 86 \tag{6.82}$$

$$P_{23} = 248N^5 + 863N^4 + 1927N^3 + 2582N^2 + 1820N + 496. \tag{6.83}$$

The second moment of the renormalized result, cf. Eq. 4.125, reads

$$A_{gq,Q}^{\overline{\mathsf{MS}}} = a_s^{\overline{\mathsf{MS}}^2} \left\{ \frac{32}{9} \ln^2\left(\frac{m^2}{\mu^2}\right) + \frac{208}{27} \ln\left(\frac{m^2}{\mu^2}\right) + \frac{236}{27} \right\} C_F T_F + O\left(a_s^{\overline{\mathsf{MS}}^3}\right). \tag{6.84}$$

We agree with the result for $a_{gq,Q}^{(2)}$ given in [2], which is presented in (6.80).

Let us summarize so far. In this section, we newly calculated the $O(\varepsilon)$ terms of the 2-loop massive OMEs. We additionally recalculated for the first time the terms $a_{gg,Q}^{(2)}$, Eq. 6.68, and $a_{gq,Q}^{(2)}$, Eq. 6.80, which were given in Ref. [2] and find full agreement. For completeness, we showed as well the terms $a_{qq,Q}^{(2),\mathrm{NS}}$, $a_{Qq}^{(2),\mathrm{PS}}$ and $a_{Qg}^{(2)}$, which have been calculated for the first time in Ref. [1] and were recalculated in Refs. [4, 5]. The latter terms contribute to the heavy flavor Wilson coefficients in deeply inelastic scattering to the non power-suppressed contributions at $O(a_s^2)$. In the renormalization of the heavy flavor Wilson coefficients to 3-loop order, all these terms contribute together with lower order single pole terms. The $O(a_s^2\varepsilon)$ contributions form parts of the constant terms of the 3-loop heavy flavor unpolarized operator matrix elements needed to describe the 3-loop heavy flavor Wilson coefficients in the region $Q^2 \gg m^2$.

The mathematical structure of our results is as follows. The terms $\overline{a}_{ij}^{(2)}$ can be expressed in terms of polynomials of the basic nested harmonic sums up to weight

$w = 4$ and derivatives thereof. They belong to the complexity-class of the general two-loop Wilson coefficients or hard scattering cross sections in massless QED and QCD and are described by six basic functions and their derivatives in Mellin space. Their analytic continuation to complex values of N is known in explicit form. The package Sigma, [23–26, 16, 17], proved to be a useful tool to solve the sums occurring in the present problem and was extended accordingly by its author.

6.4 Checks on the Calculation

There are several checks which we can use for our results. First of all, the terms up to $O(\varepsilon^0)$ have been calculated in Refs. [1, 2] and we agree with all unrenormalized results. As described in Sects. 6.1, 6.2, we keep the complete ε-dependence until we expand the summand of the finite or infinite sums, which serves as a consistency check on the $O(\varepsilon)$ results.

Another test is provided by the sum rules in Eqs. 3.37, 3.38 for $N = 2$, which are fulfilled by the renormalized OMEs presented here and in Refs. [1, 2]. These rules are obeyed regardless of the renormalization scheme. We observe that they hold on the unrenormalized level as well, even up to $O(\varepsilon)$.

For the term $A_{Qg}^{(2)}$, we evaluated fixed moments of N for the contributing unrenormalized diagrams using the Mellin–Barnes method, [80–85], cf. also Appendix C.3. Here, we used an extension of a method developed for massless propagators in Ref. [86] to massive on-shell operator matrix elements, [87, 19, 40]. The Mellin–Barnes integrals are then evaluated numerically using the package MB, [88]. Using this method, we calculated the even moments $N = 2, 4, 6, 8$ and agree with the corresponding fixed moments of our all-N result.[5]

For the first moment of the Abelian part of the unrenormalized term $\hat{\hat{A}}_{Qg}^{(2)}$, there exists even another check. After analytic continuation from the even values of N to $N \varepsilon \mathbb{C}$ is performed, one may consider the limit $N \rightarrow 1$. In this procedure the term $(1 + (-1)^N)/2$ equals to 1. At $O(a_s^2)$ the terms $\propto T_F C_A$ contain $1/z$ contributions in momentum fraction space and their first moment diverges. For the other contributions to the unrenormalized operator matrix element, after mass renormalization to 2-loop order, the first moment is related to the Abelian part of the transverse contribution to the gluon propagator $\Pi_V(p^2, m^2)|_{p^2\, =\, 0}$, except the term $\propto T_F^2$ which results from wave function renormalization. Thiswas shown in [1] up to the constant term in ε. One obtains

$$\hat{\Pi}_V(p^2, m^2) = \hat{a}_s T_F \hat{\Pi}_V^{(1)}(p^2, m^2) + \hat{a}_s^2 C_F T_F \hat{\Pi}_V^{(2)}(p^2, m^2) + O(\hat{a}_s^3), \quad (6.85)$$

with

[5] In Table 2 of Ref. [12], the moments $N = 2$ and $N = 6$ for the more difficult 2-loop diagrams are presented.

$$\lim_{p^2 \to 0} \hat{\Pi}_V^{(1)}(p^2, m^2) = \frac{1}{2} \hat{\hat{A}}_{Qg}^{(1), N=1} \tag{6.86}$$

$$\lim_{p^2 \to 0} \hat{\Pi}_V^{(2)}(p^2, m^2) = \frac{1}{2} \hat{\hat{A}}_{Qg}^{(2), N=1}|_{C_F}. \tag{6.87}$$

Here, we extend the relation to the linear terms in ε. For the first moment the double pole contributions in ε vanish in Eq. 6.87. We compare with the corresponding QED-expression for the photon-propagator, $\Pi_T^{V,(k)}$, which has been obtained in Ref. [89]. Due to the transition from QED to QCD, the relative color factor at the 2–loop level has to be adjusted to $1/4 = 1/(C_F C_A)$. After asymptotic expansion in m^2/p^2, the comparison can be performed up to the linear term in ε. One obtains

$$\lim_{p^2 \to 0} \frac{1}{p^2} \hat{\Pi}_T^{V,(1)}(p^2, m^2) = \frac{1}{2T_F} \hat{\hat{A}}_{Qg}^{(1), N=1} = -\left(\frac{m^2}{\mu^2}\right)^{\varepsilon/2} \left[\frac{8}{3\varepsilon} + \frac{\varepsilon}{3}\zeta_2\right] \tag{6.88}$$

$$\lim_{p^2 \to 0} \frac{1}{p^2} \hat{\Pi}_T^{V,(2)}(p^2, m^2) = \frac{1}{2T_F C_F} \hat{\hat{A}}_{Qg}^{(2), N=1}|_{C_F} = \left(\frac{m^2}{\mu^2}\right)^{\varepsilon} \left[-\frac{4}{\varepsilon} + 15 - \left(\frac{31}{4} + \zeta_2\right)\varepsilon\right]. \tag{6.89}$$

Additionally, we notice that the renormalized results do not anymore contain ζ_2-terms. The renormalized terms in Eqs. 4.95, 4.101, 4.114, 4.125, 4.136 contain expressions proportional to ζ_2 in the non-logarithmic contributions, which just cancel the corresponding ζ_2-terms in $a_{ij}^{(2)}$, cf. Eqs. 6.34, 6.53, 6.60, 6.68, 6.80. For explicit examples of this cancellation, one may compare the second moments of the renormalized OMEs presented in Eqs. 6.49, 6.50, 6.57, 6.63, 6.76, 6.77, 6.84. The latter provides no stringent test, but is in accordance with general observations made in higher loop calculations, namely that even ζ-values cancel for massless calculations in even dimensions in the renormalized results if presented in the $\overline{\text{MS}}$-scheme, (Broadhurst, 2009, private communication). In the present work, this observation holds for the ζ_2-terms in a single-scale massive calculation as well.

The most powerful test is provided by the FORM-based program MATAD, [18], which we used to calculate fixed moments of the 2-loop OMEs up to $O(\varepsilon)$. The setup is the same as in the 3-loop case and is explained in the next section. At the 2-loop level we worked in general R_ξ-gauges and explicitly observe the cancellation of the gauge parameter. For the terms $A_{Qg}^{(2)}$, $A_{gg}^{(2)}$ we used both projection operators given in Eqs. 4.22, 4.23, which serves as another consistency check. In the singlet case, we calculated the even moments $N = 2, 4, \ldots, 12$ and found full agreement with the results presented in this section up to $O(\varepsilon)$. The same holds in the non-singlet case, where we calculated the odd moments as well, $N = 1, 2, 3, \ldots, 12$.

References

1. M. Buza, Y. Matiounine, J. Smith, R. Migneron, W.L. van Neerven, Nucl. Phys. **B472**, 611 (1996), hep-ph/9601302
2. M. Buza, Y. Matiounine, J. Smith, W.L. van Neerven, Eur. Phys. J. **C1**, 301 (1998), hep-ph/9612398
3. K.G. Chetyrkin, A.L. Kataev, F.V. Tkachov, Nucl. Phys. **B174**, 345 (1980)
4. I. Bierenbaum, J. Blümlein, S. Klein, Nucl. Phys. **B780**, 40 (2007), hep-ph/0703285
5. S. Klein, Diploma Thesis, University of Potsdam (2006)
6. J. Blümlein, S. Kurth, Phys. Rev. **D60**, 014018 (1999), hep-ph/9810241
7. J.A.M. Vermaseren, Int. J. Mod. Phys. **A14**, 2037 (1999), hep-ph/9806280
8. J. Blümlein, Comput. Phys. Commun. **159**, 19 (2004), hep-ph/0311046
9. J. Blümlein, Comput. Phys. Commun. **180**, 2218 (2009), hep-ph/0901.3106
10. J. Blümlein, Structural Relations of Harmonic Sums and Mellin Transforms at Weight w=6, in *Proceedings of the Workshop "Motives, Quantum Field Theory, and Pseudodifferential Operators, June (2008)"* (Clay Institute, Boston University, 2009), math-ph/0901.0837
11. I. Bierenbaum, J. Blümlein, S. Klein, Phys. Lett. **B672**, 401 (2009), hep-ph/0901.0669
12. I. Bierenbaum, J. Blümlein, S. Klein, C. Schneider, Nucl. Phys. **B803**, 1 (2008), hep-ph/0803.0273
13. L. Slater, *Generalized Hypergeometric Functions*, (Cambridge University Press, Cambridge, 1966) pp. 273.
14. W. Bailey, *Generalized Hypergeometric Series*, (Cambridge University Press, Cambridge, 1935) pp. 108.
15. G. Andrews, R. Askey, R. Roy, *Special Functions, Encyclopedia of Mathematics and its Applications, vol. 71*, (Cambridge University Press, Cambridge, 2001) pp. 663.
16. C. Schneider, J. Diffr. Equ. Appl. **11**(9), 799 (2005)
17. C. Schneider, Sém. Lothar. Combin. **56** (2007) Article B56b and Habilitation Thesis, JKU Linz (2007)
18. M. Steinhauser, Comput. Phys. Commun. **134**, 335 (2001), hep-ph/0009029
19. I. Bierenbaum, J. Blümlein, S. Klein, Phys. Lett. **B648**, 195 (2007), hep-ph/0702265
20. R. Hamberg, Second order gluonic contributions to physical quantities, Ph.D. Thesis, Leiden, 1991
21. S. Weinzierl, Comput. Phys. Commun. **145**, 357 (2002), math-ph/0201011
22. S. Moch, P. Uwer, Comput. Phys. Commun. **174**, 759 (2006), math-ph/0508008
23. C. Schneider, Ann. Comb. **9**(1), 75 (2005)
24. C. Schneider, in *Proceedings ISSAC'05*, p. 285, ACM Press, 2005
25. C. Schneider, in *Proceedings FPSAC'07*, p. 1, 2007
26. C. Schneider, J. Algebra Appl. **6**(3), 415 (2007)
27. M. Dittmar et al., Parton distributions: summary report for the HERA—LHC workshop, (2005), hep-ph/0511119
28. J.A.M. Vermaseren, A. Vogt, S. Moch, Nucl. Phys. **B724**, 3 (2005), hep-ph/0504242
29. S. Moch, J.A.M. Vermaseren, A. Vogt, Nucl. Phys. **B688**, 101 (2004), hep-ph/0403192
30. A. Vogt, S. Moch, J.A.M. Vermaseren, Nucl. Phys. **B691**, 129 (2004), hep-ph/0404111
31. J. Blümlein, S. Klein, PoS ACAT 2007, 084 (2007), hep-ph/0706.2426
32. J. Blümlein, Nucl. Phys. Proc. Suppl. **135**, 225 (2004), hep-ph/0407044
33. J. Blümlein, V. Ravindran, Nucl. Phys. **B716**, 128 (2005), hep-ph/0501178
34. J. Blümlein, V. Ravindran, Nucl. Phys. **B749**, 1 (2006), hep-ph/0604019
35. J. Blümlein, A. De Freitas, W.L. van Neerven, S. Klein, Nucl. Phys. **B755**, 272 (2006), hep-ph/0608024
36. J. Blümlein, S. Klein, B. Tödtli, Phys. Rev. **D80**, 094010 (2009), hep-ph/0909.1547

37. I. Bierenbaum, J. Blümlein, S. Klein, in *Proceedings of 15th International Workshop On Deep-Inelastic Scattering and Related Subjects (DIS2007)*, ed. by G. Grindhammer, K. Sachs. Two-loop massive operator matrix elements for polarized and unpolarized deep-inelastic scattering (16–20 April 2007, Munich), Vol. 2, p. 821, hep-ph/0706.2738
38. M. Buza, Y. Matiounine, J. Smith, W.L. van Neerven, Nucl. Phys. **B485**, 420 (1997), hep-ph/9608342
39. I. Bierenbaum, J. Blümlein, S. Klein, to appear
40. I. Bierenbaum, J. Blümlein, S. Klein, Nucl. Phys. Proc. Suppl. **160**, 85 (2006), hep-ph/0607300
41. J. Blümlein, Comput. Phys. Commun. **133**, 76 (2000), hep-ph/0003100
42. J. Blümlein, S.-O. Moch, Phys. Lett. **B614**, 53 (2005), hep-ph/0503188
43. C.W. Bauer, A. Frink, R. Kreckel, Introduction to the GiNaC framework for symbolic computation within the C++ programming language, (2000), cs-sc/0004015
44. A. Devoto, D.W. Duke, Riv. Nuovo. Cim. **7**(6), 1 (1984)
45. J. Blümlein, H. Kawamura, Nucl. Phys. **B708**, 467 (2005), hep-ph/0409289
46. N. Nörlund, *Vorlesungen über Differenzenrechnung*, (Springer, Berlin, 1924) pp. 551.
47. L. Milne-Thomson, *The Calculus of finite Differences*, (MacMillan, London, 1951) pp. 273.
48. R. Gosper, Proc. Nat. Acad. Sci. USA **75**, 40 (1978)
49. D. Zeilberger, J. Symb. Comput. **11**, 195 (1991)
50. M. Petkovšek, H.S. Wilf, D. Zeilberger, *A=B*. (AK Peters, Wellesley, 1996)
51. M. Karr, J. ACM **28**, 305 (1981)
52. C. Schneider, J. Symb. Comput. (2008). doi:10.1016/j.jsc.2008.01.001.
53. C. Schneider, in *Proceedings of ISSAC'04*, ACM Press, 2004, p. 282
54. J.M. Borwein, D.M. Bradley, D.J. Broadhurst, P. Lisonek, Trans. Am. Math. Soc. **353**, 907 (2001), math/9910045
55. A. Goncharov, Math. Res. Lett. **5**, 497 (1998)
56. M.P. Hoang Ngoc Minh, van der J. Hoeven, Discr. Math. **225**, 217 (2000)
57. S. Moch, P. Uwer, S. Weinzierl, J. Math. Phys. **43**, 3363 (2002), hep-ph/0110083
58. D. Zeilberger, J. Symb. Comput. **11**, 195 (1991)
59. P. Paule, C. Schneider, Adv. Appl. Math. **31**(2), 359 (2003)
60. K. Driver, H. Prodinger, C. Schneider, A. Weideman, Ramanujan J. **12**(3), 299 (2006)
61. M.E. Hoffman, J. Algebra **194**, 477 (1997)
62. M.E. Hoffman, Nucl. Phys. Proc. Suppl. **135**, 215 (2004), math/0406589
63. D.J. Gross, F. Wilczek, Phys. Rev. **D8**, 3633 (1973)
64. D.J. Gross, F. Wilczek, Phys. Rev. **D9**, 980 (1974)
65. H. Georgi, H.D. Politzer, Phys. Rev. **D9**, 416 (1974)
66. E.G. Floratos, D.A. Ross, C.T. Sachrajda, Nucl. Phys. **B129**, 66 (1977); [Erratum-ibid.] **B139**, 545 (1978)
67. E.G. Floratos, D.A. Ross, C.T. Sachrajda, Nucl. Phys. **B152**, 493 (1979)
68. A. Gonzalez-Arroyo, C. Lopez, F.J. Yndurain, Nucl. Phys. **B153**, 161 (1979)
69. A. Gonzalez-Arroyo, C. Lopez, Nucl. Phys. **B166**, 429 (1980)
70. G. Curci, W. Furmanski, R. Petronzio, Nucl. Phys. **B175**, 27 (1980)
71. W. Furmanski, R. Petronzio, Phys. Lett. **B97**, 437 (1980)
72. R. Hamberg, van W.L. Neerven, Nucl. Phys. **B379**, 143 (1992)
73. D.J. Gross, F. Wilczek, Phys. Rev. Lett. **30**, 1343 (1973)
74. H.D. Politzer, Phys. Rev. Lett. **30**, 1346 (1973)
75. I.B. Khriplovich, Yad. Fiz. **10**, 409 (1969)
76. S. Moch, J.A.M. Vermaseren, Nucl. Phys. **B573**, 853 (2000), hep-ph/9912355
77. E. Laenen, S. Riemersma, J. Smith, W.L. van Neerven, Nucl. Phys. **B392**, 162 (1993)
78. E. Laenen, S. Riemersma, J. Smith, W.L. van Neerven, Nucl. Phys. **B392**, 229 (1993)
79. S. Riemersma, J. Smith, W.L. van Neerven, Phys. Lett. **B347**, 143 (1995), hep-ph/9411431
80. E. Barnes, Proc. Lond. Math. Soc. **6**(2), 141 (1908)
81. E. Barnes, Quart. J. Math. **41**, 136 (1910)
82. H. Mellin, Math. Ann. **68**, 305 (1910)

83. E. Whittaker, G. Watson, *A Course of Modern Analysis* (Cambridge University Press, Cambridge, 1927) p. 616, reprinted 1996
84. E. Titchmarsh, *Introduction to the Theory of Fourier Integrals* (Calendron Press, Oxford, 1937; 2nd Edn. 1948)
85. R. Paris, D.D. Kaminski, *Asymptotics and Mellin-Barnes Integrals*, (Cambridge University Press, Cambridge, 2001) pp. 438.
86. I. Bierenbaum, S. Weinzierl, Eur. Phys. J. **C32**, 67 (2003), hep-ph/0308311
87. I. Bierenbaum, J. Blümlein, S. Klein, Two-loop massive operator matrix elements for polarized and unpolarized deep-inelastic scattering, PoS ACAT 2007, 070 (2007)
88. M. Czakon, Comput. Phys. Commun. **175**, 559 (2006), hep-ph/0511200
89. A. Djouadi, P. Gambino, Phys. Rev. **D49**, 3499 (1994), hep-ph/9309298

Chapter 7
Calculation of Moments at $O(a_s^3)$

Sebastian Klein

In this Chapter, we describe the computation of the 3-loop corrections to the massive operator matrix elements in detail, cf. [1]. Typical Feynman diagrams contributing for the different processes are shown in Fig. 7.1, where ⊗ denotes the corresponding composite operator insertions, cf. Appendix B. The generation of these diagrams with the FORTRAN-based program QGRAF [2], is described in Sect. 7.1 along with the subsequent steps to prepare the input for the FORM-based program MATAD [3]. The latter allows the calculation of massive tadpole integrals in D dimensions up to three loops and relies on the MINCER algorithm [4, 5]. The use of MATAD and the projection onto fixed moments are explained in Sect. 7.2. Finally, we present our results for the fixed moments of the 3-loop OMEs and the fermionic contributions to the anomalous dimensions in Sect. 7.3. The calculation is mainly performed using FORM programs while in a few cases codes have also been written in MAPLE.

7.1 Generation of Diagrams

QGRAF is a quite general program to generate Feynman diagrams and allows to specify various kinds of particles and interactions. Our main issue is to generate diagrams which contain composite operator insertions, cf. (2.86–2.88) and Appendix B, as special vertices. To give an example, let us consider the contributions to $A_{Qg}^{(1)}$. Within the light-cone expansion, Sect. 2.3, this term derives from the Born diagrams squared of the photon–gluon fusion process shown in Fig. 7.2, cf. Sect. 3.1 and Fig. 3.2.

After expanding these diagrams with respect to the virtuality of the photon, the mass effects are given by the diagrams in Fig. 7.3. These are obtained by contracting the lines between the external photons.

Thus, one may think of the operator insertion as being coupled to two external particles, an incoming and an outgoing one, which carry the same momentum.

Center of Physics , Institute for Theoretical Particle Physics , Aachen Germany

S. Klein, *Charm Production in Deep Inelastic Scattering*, Springer Theses,
DOI: 10.1007/978-3-642-23286-2_7, © Springer-Verlag Berlin Heidelberg 2012

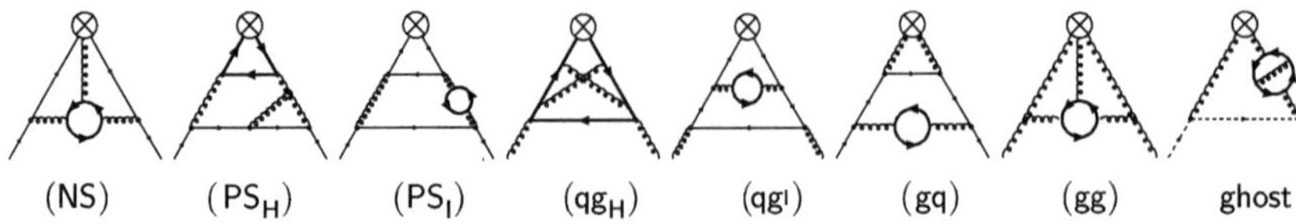

Fig. 7.1 Examples for 3-loop diagrams contributing to the massive operator matrix elements: *NS* non-singlet, $PS_{H,l}$ pure-singlet, singlet $qg_{H,\,l}$, gq, gg and *ghost* contributions. Here the coupling of the gauge boson to a heavy or light *fermion line* is labeled by H and l, respectively. *Thick lines* heavy quarks, *curly lines* gluons, *full lines* quarks, *dashed lines* ghosts

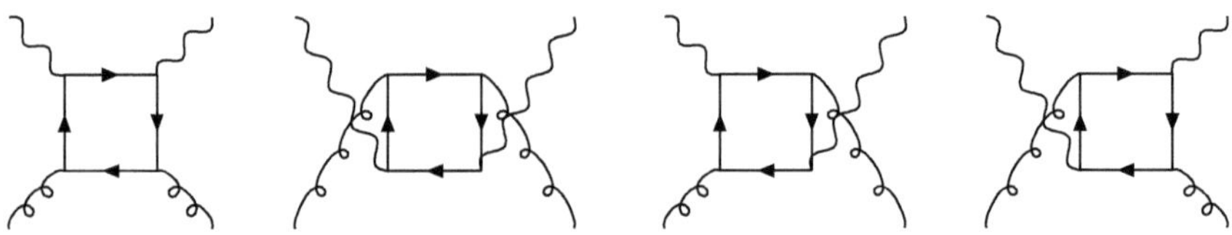

Fig. 7.2 Diagrams contributing to $H_{g,(2,\ L)}^{(1)}$ via the optical theorem. *Wavy lines* photons; *curly lines* gluons; *full lines* quarks

Fig. 7.3 Diagrams contributing to $A_{Qg}^{(1)}$

Fig. 7.4 Generation of the operator insertion

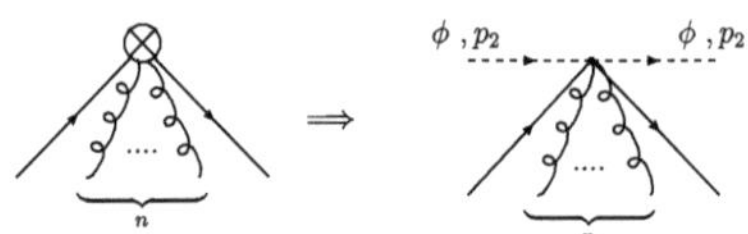

Therefore, one defines in the model file of QGRAF vertices which resemble the operator insertions in this manner, using a scalar field ϕ, which shall not propagate in order to ensure that there is only one of these vertices for each diagram. For the quarkonic operators, one defines the vertices

$$\phi + \phi + q + \overline{q} + ng, \quad 0 \leq n \leq 3, \tag{7.1}$$

which is illustrated in Fig. 7.4.

The same procedure can be used for the purely gluonic interactions and one defines in this case

$$\phi + \phi + ng, \quad 0 \leq n \leq 4. \tag{7.2}$$

The Green's functions we have to consider and their relation to the respective OMEs were given in Eqs. 4.18–4.21. The number of diagrams we obtain contributing to each OME is shown in Table 7.1.

Table 7.1 Number of diagrams contributing to the 3-loop massive OMEs

Term	#	Term	#	Term	#	Term	#
$A_{Qg}^{(3)}$	1358	$A_{qg,Q}^{(3)}$	140	$A_{Qq}^{(3),\mathsf{PS}}$	125	$A_{qq,Q}^{(3),\mathsf{PS}}$	8
$A_{qq,Q}^{(3),\mathsf{NS}}$	129	$A_{gq,Q}^{(3)}$	89	$A_{gg,Q}^{(3)}$	886		

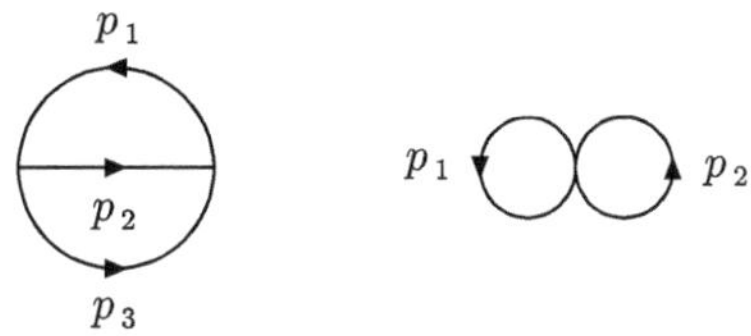

Fig. 7.5 2-Loop topologies for MATAD, indicating labeling of momenta

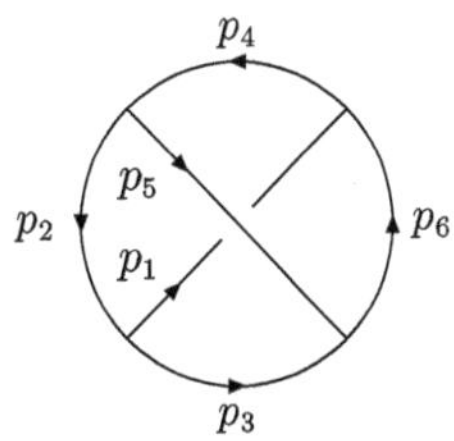

Fig. 7.6 Master 3-loop topology for MATAD, indicating labeling of momenta

The next step consists in rewriting the output provided by QGRAF in such a way, that the Feynman rules given in Appendix B can be inserted. Thus, one has to introduce Lorentz and color indices and align the fermion lines. Additionally, the integration momenta have to be written in such a way that MATAD can handle them. For the latter step, all information on the types of particles, the operator insertion and the external momentum are irrelevant, leading to only two basic topologies to be considered at the 2-loop level, which are shown in Fig. 7.5.

Note, that in the case at hand the topology on the right-hand side of Fig. 7.5 always yields zero after integration. At the 3-loop level, the master topology is given in Fig. 7.6.

From this topology, five types of diagrams are derived by shrinking various lines. These diagrams are shown in Fig. 7.7. Finally the projectors given in Eqs. 4.22, 4.24 are applied to project onto the scalar massive OMEs. We only use the physical projector (4.23) as a check for lower moments, since it causes a significant increase of the computation time. To calculate the color factor of each diagram, we use the program provided in Ref. [6] and for the calculation of fermion traces we use MATAD. Up to this point, all operations have been performed for general values of Mellin N and the dimensional parameter ε. The integrals do not contain any Lorentz or color indices anymore. In order to use MATAD, one now has to assign to N a specific value. Additionally, the unphysical momentum Δ has to be replaced by a suitable projector, which we define in the following section.

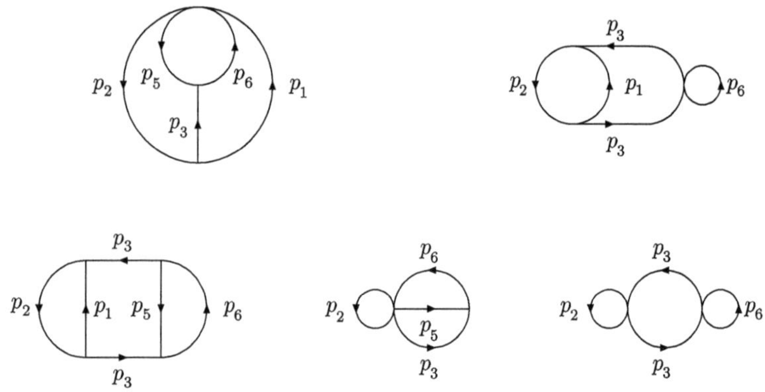

Fig. 7.7 Additional 3-loop topologies for MATAD

7.2 Calculation of Fixed 3-Loop Moments Using MATAD

We consider integrals of the type

$$I_l(p, m, n_1 \dots n_j) \equiv \int \frac{d^D k_1}{(2\pi)^D} \dots \int \frac{d^D k_l}{(2\pi)^D} (\Delta . q_1)^{n_1} \dots (\Delta . q_j)^{n_j} f(k_1 \dots k_l, p, m). \tag{7.3}$$

Here p denotes the external momentum, $p^2 = 0$, m is the heavy quark mass, and Δ is a light-like vector, $\Delta^2 = 0$. The momenta q_i are given by any linear combination of the loop momenta k_i and external momentum p. The exponents n_i are integers or possibly sums of integers, see the Feynman rules in Appendix B. Their sum is given by

$$\sum_{i=1}^{j} n_i = N. \tag{7.4}$$

The function f in Eq. 7.3 contains propagators, of which at least one is massive, dot-products of its arguments and powers of m. If one sets $N = 0$, (7.3) is given by

$$I_l(p, m, 0 \dots 0) = I_l(m) = \int \frac{d^D k_1}{(2\pi)^D} \dots \int \frac{d^D k_l}{(2\pi)^D} f(k_1 \dots k_l, m). \tag{7.5}$$

From $p^2 = 0$ it follows, that the result can not depend on p anymore. The above integral is a massive tadpole integral and thus of the type MATAD can process. Additionally, MATAD can calculate the integral up to a given order as a power series in p^2/m^2. Let us return to the general integral given in Eq. 7.3. One notes, that for fixed moments of N, each integral of this type splits up into one or more integrals of the same type with the n_i having fixed integer values. At this point, it is useful to recall that the auxiliary vector Δ has only been introduced to get rid of the trace terms of the expectation values of the composite operators and has no physical significance. By undoing the contraction with Δ, these trace terms appear again. Consider as an example

$$I_l(p,m,2,1) = \int \frac{d^D k_1}{(2\pi)^D} \cdots \int \frac{d^D k_l}{(2\pi)^D} (\Delta . q_1)^2 (\Delta . q_2) f(k_1 \dots k_l, p, m) \tag{7.6}$$

$$= \Delta^{\mu_1}\Delta^{\mu_2}\Delta^{\mu_3} \int \frac{d^D k_1}{(2\pi)^D} \cdots \int \frac{d^D k_l}{(2\pi)^D} q_{1,\mu_1} q_{1,\mu_2} q_{2,\mu_3} f(k_1 \dots k_l, p, m). \tag{7.7}$$

One notices that the way of distributing the indices in Eq. 7.7 is somewhat arbitrary, since after the contraction with the totally symmetric tensor $\Delta^{\mu_1}\Delta^{\mu_2}\Delta^{\mu_3}$ only the completely symmetric part of the corresponding tensor integral contributes. This is made explicit by distributing the indices among the q_i in all possible ways and dividing by the number of permutations one has used. Thus Eq. 7.7 is written as

$$I_l(p,m,2,1) = \Delta^{\mu_1}\Delta^{\mu_2}\Delta^{\mu_3} \frac{1}{3} \int \frac{d^D k_1}{(2\pi)^D} \cdots \int \frac{d^D k_l}{(2\pi)^D} (q_{1,\mu_2} q_{1,\mu_3} q_{2,\mu_1} + q_{1,\mu_1} q_{1,\mu_3} q_{2,\mu_2} + q_{1,\mu_1} q_{1,\mu_2} q_{2,\mu_3}) f(k_1 \dots k_l, p, m). \tag{7.8}$$

Generally speaking, the symmetrization of the tensor resulting from

$$\prod_{i=1}^{j} (\Delta \cdot q_1)^{n_i} \tag{7.9}$$

can be achieved by shuffling indices, [7–12], and dividing by the number of terms. The shuffle product is given by

$$C \left[\underbrace{(k_1, \dots, k_1)}_{n_1} \sqcup\!\sqcup \underbrace{(k_2, \dots, k_2)}_{n_2} \sqcup\!\sqcup \dots \sqcup\!\sqcup \underbrace{(k_I, \dots, k_I)}_{n_I} \right], \tag{7.10}$$

where C is the normalization constant

$$C = \binom{N}{n_1, \dots, n_I}^{-1}. \tag{7.11}$$

As an example, the symmetrization of

$$q_{1,\mu_1} q_{1,\mu_2} q_{2,\mu_3} \tag{7.12}$$

can be inferred from Eq. 7.8. After undoing the contraction with Δ in (7.3) and shuffling the indices, one may make the following ansatz for the result of this integral, which follows from the necessity of complete symmetry in the Lorentz indices

$$R_{\{\mu_1\ldots\mu_N\}} \equiv \sum_{j=1}^{[N/2]+1} A_j \Big(\prod_{k=1}^{j-1} g_{\{\mu_{2k}\mu_{2k-1}}\Big)\Big(\prod_{l=2j-1}^{N} p_{\mu_l\}}\Big). \tag{7.13}$$

In the above equation, [ ] denotes the Gauss-bracket and {} symmetrization with respect to the indices enclosed and dividing by the number of terms, as outlined above. The first few terms are then given by

$$R_0 \equiv 1, \tag{7.14}$$

$$R_{\{\mu_1\}} = A_1 p_{\mu_1}, \tag{7.15}$$

$$R_{\{\mu_1\mu_2\}} = A_1 p_{\mu_1} p_{\mu_2} + A_2 g_{\mu_1\mu_2}, \tag{7.16}$$

$$R_{\{\mu_1\mu_2\mu_3\}} = A_1 p_{\mu_1} p_{\mu_2} p_{\mu_3} + A_2 g_{\{\mu_1\mu_2} p_{\mu_3\}}. \tag{7.17}$$

The scalars A_j have in general different mass dimensions. By contracting again with Δ, all trace terms vanish and one obtains

$$I_l(p, m, n_1 \ldots n_j) = \Delta^{\mu_1} \ldots \Delta^{\mu_N} R_{\{\mu_1\ldots\mu_N\}} \tag{7.18}$$

$$= A_1 (\Delta \cdot p)^N \tag{7.19}$$

and thus the coefficient A_1 in Eq. 7.13 gives the desired result. To obtain it, one constructs a different projector, which is made up only of the external momentum p and the metric tensor. By making a general ansatz for this projector, applying it to Eq. 7.13 and demanding that the result shall be equal to A_1, the coefficients of the different Lorentz structures can be determined. The projector reads

$$\Pi_{\mu_1\ldots\mu_N} = F(N) \sum_{i=1}^{[N/2]+1} C(i, N) \left(\prod_{l=1}^{[N/2]-i+1} \frac{g_{\mu_{2l-1}\mu_{2l}}}{p^2}\right)\left(\prod_{k=2[N/2]-2i+3}^{N} \frac{p_{\mu_k}}{p^2}\right). \tag{7.20}$$

For the overall pre-factors $F(N)$ and the coefficients $C(i,N)$, one has to distinguish between even and odd values of N,

$$C^{odd}(k, N) = (-1)^{N/2+k+1/2} \frac{2^{2k-N/2-3/2}\Gamma(N+1)\Gamma(D/2+N/2+k-3/2)}{\Gamma(N/2-k+3/2)\Gamma(2k)\Gamma(D/2+N/2-1/2)}, \tag{7.21}$$

$$F^{odd}(N) = \frac{2^{3/2-N/2}\Gamma(D/2+1/2)}{(D-1)\Gamma(N/2+D/2-1)}, \tag{7.22}$$

$$C^{even}(k,N) = (-1)^{N/2+k+1}\frac{2^{2k-N/2-2}\Gamma(N+1)\Gamma(D/2+N/2-2+k)}{\Gamma(N/2-k+2)\Gamma(2k-1)\Gamma(D/2+N/2-1)}, \tag{7.23}$$

$$F^{even}(N) = \frac{2^{1-N/2}\Gamma(D/2+1/2)}{(D-1)\Gamma(N/2+D/2-1/2)}. \tag{7.24}$$

The projector obeys the normalization condition

$$\Pi_{\mu_1\ldots\mu_N} R^{\mu_1\ldots\mu_N} = A_1, \tag{7.25}$$

which implies

$$\Pi_{\mu_1\ldots\mu_N} p^{\mu_1}\ldots p^{\mu_N} = 1. \tag{7.26}$$

As an example for the above procedure, we consider the case $N = 3$,

$$\Pi_{\mu_1\mu_2\mu_3} = \frac{1}{D-1}\Big(-3\frac{g_{\mu_1\mu_2}p_{\mu_3}}{p^4} + (D+2)\frac{p_{\mu_1}p_{\mu_2}p_{\mu_3}}{p^6}\Big). \tag{7.27}$$

Applying this term to (7.8) yields

$$\begin{aligned} I_l(p,m,2,1) = \frac{1}{(D-1)p^6}\int\frac{d^Dk_1}{(2\pi)^D}\cdots\int\frac{d^Dk_l}{(2\pi)^D}\Big(-2p^2q_1.q_2p.q_1 \\ - p^2q_1^2p.q_2 + (D+2)(q_1.p)^2q_2.p\Big)f(k_1\ldots k_l,p,m). \end{aligned} \tag{7.28}$$

Up to 3-loop integrals of the type (7.28) can be calculated by MATAD as a Taylor series in p^2/m^2. It is important to keep p artificially off-shell until the end of the calculation. By construction, the overall result will not contain any term $\propto 1/p^2$, since the integral one starts with is free of such terms. Thus, at the end, these terms have to cancel. The remaining constant term in p^2 is the desired result.

The above projectors are similar to the harmonic projectors used in the MINCER-program, cf. [5, J.A.M. Vermaseren, The Form version of MINCER, unpublished]. These are, however, applied to the virtual forward Compton-amplitude to determine the anomalous dimensions and the moments of the massless Wilson coefficients up to 3-loop order.

The calculation was performed in Feynman gauge in general. Part of the calculation was carried out keeping the gauge parameter in R_ξ-gauges, in particular for the moments $N = 2, 4$ in the singlet case and for $N = 1, 2, 3, 4$ in the non-singlet case, yielding agreement with the results being obtained using Feynman-gauge. In addition, for the moments $N = 2, 4$ in the terms with external gluons, we applied the physical projector in Eq. 4.23, which serves as another verification of our results. The computation of the more complicated diagrams was performed on various 32/64 Gb machines using FORM and for part of the calculation TFORM, [13], was used. The complete calculation required about 250 CPU days.

7.3 Results

We calculated the unrenormalized operator matrix elements treating the 1PI-contributions explicitly. They contribute to $A_{Qg}^{(3)}$, $A_{gg,Q}^{(3)}$ and $A_{qq,Q}^{(3),\mathsf{NS}}$. One obtains the following representations

$$\begin{aligned}\hat{\hat{A}}_{Qg}^{(3)} &= \hat{\hat{A}}_{Qg}^{(3),\mathrm{irr}} - \hat{\hat{A}}_{Qg}^{(2),\mathrm{irr}} \hat{\Pi}^{(1)}\left(0, \frac{\hat{m}^2}{\mu^2}\right) - \hat{\hat{A}}_{Qg}^{(1)} \hat{\Pi}^{(2)}\left(0, \frac{\hat{m}^2}{\mu^2}\right) \\ &\quad + \hat{\hat{A}}_{Qg}^{(1)} \hat{\Pi}^{(1)}\left(0, \frac{\hat{m}^2}{\mu^2}\right) \hat{\Pi}^{(1)}\left(0, \frac{\hat{m}^2}{\mu^2}\right),\end{aligned} \tag{7.29}$$

$$\begin{aligned}\hat{\hat{A}}_{gg,Q}^{(3)} &= \hat{\hat{A}}_{gg,Q}^{(3),\mathrm{irr}} - \hat{\Pi}^{(3)}\left(0, \frac{\hat{m}^2}{\mu^2}\right) - \hat{\hat{A}}_{gg,Q}^{(2),\mathrm{irr}} \hat{\Pi}^{(1)}\left(0, \frac{\hat{m}^2}{\mu^2}\right) \\ &\quad - 2\hat{\hat{A}}_{gg,Q}^{(1)} \hat{\Pi}^{(2)}\left(0, \frac{\hat{m}^2}{\mu^2}\right) + \hat{\hat{A}}_{gg,Q}^{(1)} \hat{\Pi}^{(1)}\left(0, \frac{\hat{m}^2}{\mu^2}\right) \hat{\Pi}^{(1)}\left(0, \frac{\hat{m}^2}{\mu^2}\right),\end{aligned} \tag{7.30}$$

$$\hat{\hat{A}}_{qq,Q}^{(3),\mathsf{NS}} = \hat{\hat{A}}_{qq,Q}^{(3),\mathsf{NS},\mathrm{irr}} - \hat{\Sigma}^{(3)}\left(0, \frac{\hat{m}^2}{\mu^2}\right). \tag{7.31}$$

The self-energies are given in Eqs. (4.84–4.86, 4.88). The calculation of the one-particle irreducible 3-loop contributions is performed as described in the previous Section. [1] The amount of moments, which could be calculated, depended on the available computer resources w.r.t. memory and computational time, as well as the possible parallelization using TFORM. Increasing the Mellin moment from $N \rightarrow N+2$ demands both a factor of 6–8 larger memory and CPU time. We have calculated the even moments $N = 2, \ldots, 10$ for $A_{Qg}^{(3)}$, $A_{gg,Q}^{(3)}$, and $A_{qg,Q}^{(3)}$, for $A_{Qq}^{(3),\mathrm{PS}}$ up to $N = 12$, and for $A_{qq,Q}^{(3),\mathrm{NS}}$, $A_{qq,Q}^{(3),\mathrm{PS}}$, $A_{gq,Q}^{(3)}$ up to $N = 14$. In the NS-case, we also calculated the odd moments $N = 1, \ldots, 13$, which correspond to the NS^--terms.

7.3.1 *Anomalous Dimensions*

The pole terms of the unrenormalized OMEs emerging in the calculation agree with the general structure we presented in Eqs. 4.94, 4.103, 4.104, 4.116, 4.117, 4.124, 4.134. Using lower order renormalization coefficients and the constant terms of the 2-loop results, [16–19], allows to determine the fixed moments of the 2-loop anomalous dimensions and the contributions $\propto T_F$ of the 3-loop anomalous dimensions,

[1] Partial results of the calculation were presented in [14, 15].

cf. Appendix E. All our results agree with the results of Refs. [20–25]. The anomalous dimensions $\gamma_{qg^{(2)}}$ and $\gamma_{qq}^{(2),\mathsf{PS}}$ are obtained completely. The present calculation is fully independent both in the algorithms and codes compared to Refs. [21–24, 26] and thus provides a stringent check on these results.

7.3.2 The Constant Terms $a_{ij}^{(3)}(N)$:

The constant terms at $O(a_s^3)$, cf. Eqs. 4.94, 4.103, 4.104, 4.116, 4.117, 4.124, 4.134, are the new contributions to the non-logarithmic part of the 3-loop massive operator matrix elements, which can not be constructed by other renormalization constants calculated previously. They are given in Appendix F. All other contributions to the heavy flavor Wilson coefficients in the region $Q^2 \gg m^2$ are known for general values of N, cf. Sect. 4.7 and Chap. 6. The functions $a_{ij}^{(3)}(N)$ still contain coefficients $\propto \zeta_2$ and we will see below, under which circumstances these terms will contribute to the heavy flavor contributions to the deep-inelastic structure functions. The constant B_4, (4.89), emerges as in other massive single-scale calculations, [26–30].

7.3.3 Moments of the Constant Terms of the 3-loop Massive OMEs

The logarithmic terms of the renormalized 3-loop massive OMEs are determined by known renormalization constants and lower order contributions to the massive OMEs. They can be inferred from Eqs. 4.96, 4.105, 4.106, 4.118, 4.119, 4.126, 4.137. In the following, we consider as examples the non-logarithmic contributions to the second moments of the renormalized massive OMEs. We refer to coupling constant renormalization in the $\overline{\mathsf{MS}}$-scheme and compare the results performing the mass renormalization in the on-shell-scheme (m) and the $\overline{\mathsf{MS}}$-scheme ($\overline{m}$), cf. Chap. 5. For the matrix elements with external gluons, we obtain :

$$\begin{aligned}
A_{Qg}^{(3),\overline{\mathsf{MS}}}(\mu^2=m^2,2) = & T_F C_A^2\left(\frac{174055}{4374}-\frac{88}{9}B_4+72\zeta_4-\frac{29431}{324}\zeta_3\right)\\
&+T_F C_F C_A\left(-\frac{18002}{729}+\frac{208}{9}B_4-104\zeta_4+\frac{2186}{9}\zeta_3-\frac{64}{3}\zeta_2+64\zeta_2\ln(2)\right)\\
&+T_F C_F^2\left(-\frac{8879}{729}-\frac{64}{9}B_4+32\zeta_4-\frac{701}{81}\zeta_3+80\zeta_2-128\zeta_2\ln(2)\right)\\
&+T_F^2 C_A\left(-\frac{21586}{2187}+\frac{3605}{162}\zeta_3\right)+T_F^2 C_F\left(-\frac{55672}{729}+\frac{889}{81}\zeta_3+\frac{128}{3}\zeta_2\right)\\
&+n_f T_F^2 C_A\left(-\frac{7054}{2187}-\frac{704}{81}\zeta_3\right)+n_f T_F^2 C_F\left(-\frac{22526}{729}+\frac{1024}{81}\zeta_3-\frac{64}{3}\zeta_2\right).
\end{aligned} \tag{7.32}$$

$$A_{Qg}^{(3),\overline{\mathsf{MS}}}(\mu^2=\overline{m}^2,2)=T_FC_A^2\left(\frac{174055}{4374}-\frac{88}{9}B_4+72\zeta_4-\frac{29431}{324}\zeta_3\right)$$
$$+T_FC_FC_A\left(-\frac{123113}{729}+\frac{208}{9}B_4-104\zeta_4+\frac{2330}{9}\zeta_3\right)$$
$$+T_FC_F^2\left(-\frac{8042}{729}-\frac{64}{9}B_4+32\zeta_4-\frac{3293}{81}\zeta_3\right)$$
$$+T_F^2C_A\left(-\frac{21586}{2187}+\frac{3605}{162}\zeta_3\right)+T_F^2C_F\left(-\frac{9340}{729}+\frac{889}{81}\zeta_3\right)$$
$$+n_fT_F^2C_A\left(-\frac{7054}{2187}-\frac{704}{81}\zeta_3\right)+n_fT_F^2C_F\left(\frac{478}{729}+\frac{1024}{81}\zeta_3\right). \tag{7.33}$$

$$A_{qg,Q}^{(3),\overline{\mathsf{MS}}}(\mu^2=m^2,2)=n_fT_F^2C_A\left(\frac{64280}{2187}-\frac{704}{81}\zeta_3\right)+n_fT_F^2C_F\left(-\frac{7382}{729}+\frac{1024}{81}\zeta_3\right). \tag{7.34}$$

$$A_{gg,Q}^{(3),\overline{\mathsf{MS}}}(\mu^2=m^2,2)=T_FC_A^2\left(-\frac{174055}{4374}+\frac{88}{9}B_4-72\zeta_4+\frac{29431}{324}\zeta_3\right)$$
$$+T_FC_FC_A\left(\frac{18002}{729}-\frac{208}{9}B_4+104\zeta_4-\frac{2186}{9}\zeta_3+\frac{64}{3}\zeta_2-64\zeta_2\ln(2)\right)$$
$$+T_FC_F^2\left(\frac{8879}{729}+\frac{64}{9}B_4-32\zeta_4+\frac{701}{81}\zeta_3-80\zeta_2+128\zeta_2\ln(2)\right)$$
$$+T_F^2C_A\left(\frac{21586}{2187}-\frac{3605}{162}\zeta_3\right)+T_F^2C_F\left(\frac{55672}{729}-\frac{889}{81}\zeta_3-\frac{128}{3}\zeta_2\right)$$
$$+n_fT_F^2C_A\left(-\frac{57226}{2187}+\frac{1408}{81}\zeta_3\right)+n_fT_F^2C_F\left(\frac{29908}{729}-\frac{2048}{81}\zeta_3+\frac{64}{3}\zeta_2\right). \tag{7.35}$$

$$A_{gg,Q}^{(3),\overline{\mathsf{MS}}}(\mu^2=\overline{m}^2,2)=T_FC_A^2\left(-\frac{174055}{4374}+\frac{88}{9}B_4-72\zeta_4+\frac{29431}{324}\zeta_3\right)$$
$$+T_FC_FC_A\left(\frac{123113}{729}-\frac{208}{9}B_4+104\zeta_4-\frac{2330}{9}\zeta_3\right)$$
$$+T_FC_F^2\left(\frac{8042}{729}+\frac{64}{9}B_4-32\zeta_4+\frac{3293}{81}\zeta_3\right)$$
$$+T_F^2C_A\left(\frac{21586}{2187}-\frac{3605}{162}\zeta_3\right)+T_F^2C_F\left(\frac{9340}{729}-\frac{889}{81}\zeta_3\right)$$
$$+n_fT_F^2C_A\left(-\frac{57226}{2187}+\frac{1408}{81}\zeta_3\right)+n_fT_F^2C_F\left(\frac{6904}{729}-\frac{2048}{81}\zeta_3\right). \tag{7.36}$$

Comparing the operator matrix elements in case of the on-shell-scheme and $\overline{\mathsf{MS}}$-scheme, one notices that the terms $\ln(2)\zeta_2$ and ζ_2 are absent in the latter. The ζ_2 terms, which contribute to $a_{ij}^{(3)}$, are canceled by other contributions through renormalization. Although the present process is massive, this observation resembles the known result that ζ_2-terms do not contribute in space-like massless higher order calculations in even dimensions, [D. Broadhurst, private communication (2009)]. This behavior is found for all calculated moments. The occurring ζ_4-terms may partly cancel with those in the 3-loop light Wilson coefficients, [31]. Note, that Eq. 7.34 is not sensitive to mass renormalization due to the structure of the contributing diagrams.

An additional check is provided by the sum rule (3.38), which is fulfilled in all renormalization schemes and also on the unrenormalized level.

Unlike the operator matrix elements with external gluons, the second moments of the quarkonic OMEs emerge for the first time at $O(a_s^2)$. To 3-loop order, the renormalized quarkonic OMEs do not contain terms $\propto \zeta_2$. Due to their simpler structure, mass renormalization in the on-shell-scheme does not give rise to terms $\propto \zeta_2, \ln(2)\zeta_2$. Only the rational contribution in the color factor $\propto T_F C_F^2$ turns out to be different compared to the on-mass-shell-scheme and $A_{qq,Q}^{\mathsf{PS},(3)}$, (7.39), is not affected at all. This holds again for all moments we calculated. The non-logarithmic contributions are given by

$$A_{Qq}^{(3),\mathsf{PS},\overline{\mathsf{MS}}}(\mu^2 = m^2, 2) = T_F C_F C_A \left(\frac{830}{2187} + \frac{64}{9} B_4 - 64\zeta_4 + \frac{1280}{27}\zeta_3\right) + T_F C_F^2 \left(\frac{95638}{729} - \frac{128}{9} B_4 + 64\zeta_4 - \frac{9536}{81}\zeta_3\right) + T_F^2 C_F \left(\frac{53144}{2187} - \frac{3584}{81}\zeta_3\right) + n_f T_F^2 C_F \left(-\frac{34312}{2187} + \frac{1024}{81}\zeta_3\right). \tag{7.37}$$

$$A_{Qq}^{(3),\mathsf{PS},\overline{\mathsf{MS}}}(\mu^2 = \overline{m}^2, 2) = T_F C_F C_A \left(\frac{830}{2187} + \frac{64}{9} B_4 - 64\zeta_4 + \frac{1280}{27}\zeta_3\right) + T_F C_F^2 \left(\frac{78358}{729} - \frac{128}{9} B_4 + 64\zeta_4 - \frac{9536}{81}\zeta_3\right) + T_F^2 C_F \left(\frac{53144}{2187} - \frac{3584}{81}\zeta_3\right) + n_f T_F^2 C_F \left(-\frac{34312}{2187} + \frac{1024}{81}\zeta_3\right). \tag{7.38}$$

$$A_{qq,Q}^{(3),\mathsf{PS},\overline{\mathsf{MS}}}(\mu^2 = m^2, 2) = n_f T_F^2 C_F \left(-\frac{52168}{2187} + \frac{1024}{81}\zeta_3\right). \tag{7.39}$$

$$A_{qq,Q}^{(3),\mathsf{NS},\overline{\mathsf{MS}}}(\mu^2 = m^2, 2) = T_F C_F C_A\left(-\frac{101944}{2187} + \frac{64}{9}B_4 - 64\zeta_4 + \frac{4456}{81}\zeta_3\right)$$
$$+ T_F C_F^2\left(\frac{283964}{2187} - \frac{128}{9}B_4 + 64\zeta_4 - \frac{848}{9}\zeta_3\right)$$
$$+ T_F^2 C_F\left(\frac{25024}{2187} - \frac{1792}{81}\zeta_3\right)$$
$$+ n_f T_F^2 C_F\left(-\frac{46336}{2187} + \frac{1024}{81}\zeta_3\right). \quad (7.40)$$

$$A_{qq,Q}^{(3),\mathsf{NS},\overline{\mathsf{MS}}}(\mu^2 = \overline{m}^2, 2) = T_F C_F C_A\left(-\frac{101944}{2187} + \frac{64}{9}B_4 - 64\zeta_4 + \frac{4456}{81}\zeta_3\right)$$
$$+ T_F C_F^2\left(\frac{201020}{2187} - \frac{128}{9}B_4 + 64\zeta_4 - \frac{848}{9}\zeta_3\right)$$
$$+ T_F^2 C_F\left(\frac{25024}{2187} - \frac{1792}{81}\zeta_3\right)$$
$$+ n_f T_F^2 C_F\left(-\frac{46336}{2187} + \frac{1024}{81}\zeta_3\right). \quad (7.41)$$

$$A_{gq,Q}^{(3),\overline{\mathsf{MS}}}(\mu^2 = m^2, 2) = T_F C_F C_A\left(\frac{101114}{2187} - \frac{128}{9}B_4 + 128\zeta_4 - \frac{8296}{81}\zeta_3\right)$$
$$+ T_F C_F^2\left(-\frac{570878}{2187} + \frac{256}{9}B_4 - 128\zeta_4 + \frac{17168}{81}\zeta_3\right)$$
$$+ T_F^2 C_F\left(-\frac{26056}{729} + \frac{1792}{27}\zeta_3\right)$$
$$+ n_f T_F^2 C_F\left(\frac{44272}{729} - \frac{1024}{27}\zeta_3\right). \quad (7.42)$$

$$A_{gq,Q}^{(3),\overline{\mathsf{MS}}}(\mu^2 = \overline{m}^2, 2) = T_F C_F C_A\left(\frac{101114}{2187} - \frac{128}{9}B_4 + 128\zeta_4 - \frac{8296}{81}\zeta_3\right)$$
$$+ T_F C_F^2\left(-\frac{436094}{2187} + \frac{256}{9}B_4 - 128\zeta_4 + \frac{17168}{81}\zeta_3\right)$$
$$+ T_F^2 C_F\left(-\frac{26056}{729} + \frac{1792}{27}\zeta_3\right)$$
$$+ n_f T_F^2 C_F\left(\frac{44272}{729} - \frac{1024}{27}\zeta_3\right). \quad (7.43)$$

Finally, the sum rule (3.38) holds on the unrenormalized level, as well as for the renormalized expressions in all schemes considered.

FORM-codes for the constant terms $a_{ij}^{(3)}$, Appendix F, and the corresponding moments of the renormalized massive operator matrix elements, both for the mass renormalization carried out in the on-shell- and $\overline{\text{MS}}$-scheme, are attached to Ref. [1] and can be obtained upon request. Phenomenological studies of the 3-loop heavy flavor Wilson coefficients in the region $Q^2 \gg m^2$ will be given elsewhere, (J. Blümlein, S. Klein, in preparation).

References

1. I. Bierenbaum, J. Blümlein, S. Klein, Nucl.Phys. **B820**, 417 (2009) hep-ph/0904.3563
2. P. Nogueira, J. Comput. Phys. **105**, 279 (1993)
3. M. Steinhauser, Comput. Phys. Commun. **134**, 335 (2001) hep-ph/0009029
4. S.G. Gorishnii, S.A. Larin, L.R. Surguladze, F.V. Tkachov, Comput. Phys. Commun. **55**, 381 (1989)
5. S.A. Larin, F.V. Tkachov, J.A.M. Vermaseren, The FORM version of MINCER, NIKHEF-H-91-18 (1991)
6. van T. Ritbergen, A.N. Schellekens, J.A.M. Vermaseren, Int. J. Mod. Phys. **A14**, 41 (1999) hep-ph/9802376
7. J. Blümlein, S. Kurth, Phys. Rev. **D60**, 014018 (1999) hep-ph/9810241
8. J.A.M. Vermaseren, Int. J. Mod. Phys. **A14**, 2037 (1999) hep-ph/9806280
9. J. Blümlein, Comput. Phys. Commun. **159**, 19 (2004) hep-ph/0311046
10. J.M. Borwein, D.M. Bradley, D.J. Broadhurst, P. Lisonek, Trans. Am. Math. Soc. **353**, 907 (2001) math/9910045
11. S. Moch, P. Uwer, S. Weinzierl, J. Math. Phys. **43**, 3363 (2002) hep-ph/0110083
12. E. Remiddi, J.A.M. Vermaseren, Int. J. Mod. Phys. **A15**, 725 (2000) hep-ph/9905237
13. M. Tentyukov, J. Vermaseren, Comput. Phys. Commun. **181**, 1419 (2010) hep-ph/0702279
14. I. Bierenbaum, J. Blümlein, S. Klein, Nucl. Phys. Proc. Suppl. **183**, 162 (2008) hep-ph/0806.4613
15. I. Bierenbaum, J. Blümlein, S. Klein, PoS Confinement8, **185** (2008). hep-ph/0812.2427
16. M. Buza, Y. Matiounine, J. Smith, R. Migneron, van W.L. Neerven, Nucl. Phys. **B472**, 611 (1996) hep-ph/9601302
17. I. Bierenbaum, J. Blümlein, S. Klein, Nucl. Phys. **B780**, 40 (2007) hep-ph/0703285
18. M. Buza, Y. Matiounine, J. Smith, van W.L. Neerven, Eur. Phys. J. **C1**, 301 (1998) hep-ph/9612398
19. I. Bierenbaum, J. Blümlein, S. Klein, Phys. Lett. **B672**, 401 (2009) hep-ph/0901.0669
20. S.A. Larin, P. Nogueira, van T. Ritbergen, J.A.M. Vermaseren, Nucl. Phys. **B492**, 338 (1997) hep-ph/9605317
21. A. Retey, J.A.M. Vermaseren, Nucl. Phys. **B604**, 281 (2001) hep-ph/0007294
22. S. Moch, J.A.M. Vermaseren, A. Vogt, Nucl. Phys. **B688**, 101 (2004) hep-ph/0403192
23. A. Vogt, S. Moch, J.A.M. Vermaseren, Nucl. Phys. **B691**, 129 (2004) hep-ph/0404111
24. J.A. Gracey, Phys. Lett. **B322**, 141 (1994) hep-ph/9401214
25. S. Moch, J.A.M. Vermaseren, A. Vogt, Nucl. Phys. **B646**, 181 (2002) hep-ph/0209100
26. D.J. Broadhurst, Z. Phys. **C54**, 599 (1992)
27. L. Avdeev, J. Fleischer, S. Mikhailov, O. Tarasov, Phys. Lett. **B336**, 560 (1994) hep-ph/9406363
28. S. Laporta, E. Remiddi, Phys. Lett. **B379**, 283 (1996) hep-ph/9602417
29. D.J. Broadhurst, Eur. Phys. J. **C8**, 311 (1999) hep-th/9803091
30. R. Boughezal, J.B. Tausk, van der J.J. Bij, Nucl. Phys. **B713**, 278 (2005) hep-ph/0410216
31. J.A.M. Vermaseren, A. Vogt, S. Moch, Nucl. Phys. **B724**, 3 (2005) hep-ph/0504242

Chapter 8
Heavy Flavor Corrections to Polarized Deep-Inelastic Scattering

The composition of the proton spin in terms of partonic degrees of freedom has attracted much interest after the initial experimental finding, [1–5], that the polarization of the three light quarks alone does not add to the required value 1/2. Subsequently, the polarized proton structure functions have been measured in great detail by various experiments [6–19].[1] To determine the different contributions to the nucleon spin, both the flavor dependence as well as the contributions due to gluons and angular excitations at virtualities Q^2 in the perturbative region have to be studied in more detail in the future. As the nucleon spin contributions are related to the first moments of the respective distribution functions, it is desirable to measure to very small values of x at high energies, cf. [23–25].

A detailed treatment of the flavor structure requires the inclusion of heavy flavor. As in the unpolarized case, this contribution is driven by the gluon and sea-quark densities. Exclusive data on charm-quark pair production in polarized deep-inelastic scattering are available only in the region of very low photon virtualities at present [26, 27]. However, the inclusive measurement of the structure functions $g_1(x, Q^2)$ and $g_2(x, Q^2)$ contains the heavy flavor contributions for hadronic masses $W^2 \geq (2m + M)^2$.

The polarized heavy flavor Wilson coefficients are known to first order in the whole kinematic range [28–30]. In these references, numerical illustrations for the LO contributions were given as well, cf. also [31]. The polarized parton densities have been extracted from deep-inelastic scattering data in [32–37]. Unlike the case for photo-production, [38], the NLO Wilson coefficients have not been calculated for the whole kinematic domain, but only in the region $Q^2 \gg m^2$, [39], applying the same technique as described in Sect. 3.2. As outlined in the same section, the heavy flavor contributions to the structure function $F_2(x, Q^2)$ are very well described by the asymptotic representation for $Q^2/m^2 \gtrsim 10$, i.e., $Q^2 \gtrsim 22.5\,\mathrm{GeV}^2$, in case of charm. A similar approximation should hold in case of the polarized structure function $g_1(x, Q^2)$.

[1] For theoretical surveys see [20–22].

S. Klein, *Charm Production in Deep Inelastic Scattering*, Springer Theses,
DOI: 10.1007/978-3-642-23286-2_8,

In this chapter, we re-calculate for the first time the heavy flavor contributions to the longitudinally polarized structure function $g_1(x, Q^2)$ to $O(a_s^2)$ in the asymptotic region $Q^2 \gg m^2$ [39]. The corresponding contributions to the structure function $g_2(x, Q^2)$ can be obtained by using the Wandzura–Wilczek relation, [40], at the level of twist-2 operators, as has been shown in Refs. [26, 41–44] within the covariant parton model.

In the polarized case, the twist-2 heavy flavor Wilson coefficients factorize in the limit $Q^2 \gg m^2$ in the same way as in the unpolarized case, cf. Sect. 3.2 and [39]. The corresponding light flavor Wilson coefficients were obtained in Ref. [45]. We proceed by calculating the two-loop polarized massive quarkonic OMEs, as has been done in Ref. [39]. Additionally, we newly calculate the $O(\varepsilon)$ terms of these objects, which will be needed to evaluate the $O(a_s^3)$ corrections, cf. Chap. 4.

The calculation is performed in the same way as described in Chap. 6 and we therefore only discuss aspects that are specific to the polarized case. The notation for the heavy flavor Wilson coefficients is the same as in Eq. 3.2 and below, except that the index $(2, L)$ has to be replaced by (g_1, g_2). The polarized massive operator matrix elements are denoted by ΔA_{ij} and obey the same relations as in Chaps. 3 and 4, if one replaces the anomalous dimensions, cf. Eq. 2.107, 2.108, by their polarized counterparts, $\Delta\gamma_{ij}$.

The asymptotic heavy flavor corrections for polarized deeply inelastic scattering to $O(a_s^2)$, [39], were calculated in a specific scheme for the treatment of γ_5 in dimensional regularization. This was done in order to use the same scheme as has been applied in the calculation of the massless Wilson coefficients in [45]. Here, we refer to the version prior to an Erratum submitted in 2007, which connected the calculation to the $\overline{\mathsf{MS}}$-scheme. In this chapter we would like to compare to the results given in Ref. [39], which requires to apply the conventions used there.

In Sect. 8.1, we summarize main relations such as the differential cross sections for polarized deeply inelastic scattering and the leading order heavy flavor corrections. We give a brief outline on the representation of the asymptotic heavy flavor corrections at NLO. In Sects. 8.2.1–8.2.3, the contributions to the operator matrix elements $\Delta A_{qq,Q}^{(2),\mathsf{NS}}$, $\Delta A_{Qg}^{(2)}$ and $\Delta A_{Qq}^{\mathsf{PS},(2)}$ are calculated up to the linear terms in ε.

8.1 Polarized Scattering Cross Sections

We consider the process of deeply inelastic longitudinally polarized charged lepton scattering off longitudinally (L) or transversely (T) polarized nucleons in case of single photon exchange.[2] The differential scattering cross section is given by

$$\frac{d^3\sigma}{dxdyd\theta} = \frac{y\alpha^2}{Q^4} L^{\mu\nu} W_{\mu\nu}, \tag{8.1}$$

cf. [21, 44]. Here, θ is the azimuthal angle of the final state lepton. One may define an asymmetry between the differential cross sections for opposite nucleon polarization

[2] For the basic kinematics of DIS, see Sect. 2.1.

$$A(x,y,\theta)_{L,T} = \frac{d^3\sigma_{L,T}^{\rightarrow}}{dxdyd\theta} - \frac{d^3\sigma_{L,T}^{\leftarrow}}{dxdyd\theta}, \tag{8.2}$$

which projects onto the asymmetric part of both the leptonic and hadronic tensors, $L_{\mu\nu}^A$ and $W_{\mu\nu}^A$. The hadronic tensor is then expressed by two nucleon structure functions

$$W_{\mu\nu}^A = i\varepsilon_{\mu\nu\lambda\sigma}\left[\frac{q^\lambda S^\sigma}{P\cdot q}g_1(x,Q^2) + \frac{q^\lambda(P\cdot qS^\sigma - S\cdot qP^\sigma)}{(P\cdot q)^2}g_2(x,Q^2)\right]. \tag{8.3}$$

Here S denotes the nucleon's spin vector

$$\begin{aligned} S_L &= (0,0,0,M) \\ S_T &= M(0,\cos(\bar\theta),\sin(\bar\theta),0), \end{aligned} \tag{8.4}$$

with $\bar\theta$ a fixed angle in the plane transverse to the nucleon beam. $\varepsilon_{\mu\nu\lambda\sigma}$ is the Levi-Civita symbol. The asymmetries $A(x,y,\theta)_{L,T}$ read

$$A(x,y)_L = 4\lambda\frac{\alpha^2}{Q^2}\left[\left(2-y-\frac{2xyM^2}{s}\right)g_1(x,Q^2) + 4\frac{yxM^2}{s}g_2(x,Q^2)\right], \tag{8.5}$$

$$\begin{aligned} A(x,y,\bar\theta,\theta)_T &= -8\lambda\frac{\alpha^2}{Q^2}\sqrt{\frac{M^2}{s}}\sqrt{\frac{x}{y}\left[1-y-\frac{xyM^2}{S}\right]} \\ &\times\cos\left(\bar\theta-\theta\right)\left[yg_1\left(x,Q^2\right)+2g_2\left(x,Q^2\right)\right], \end{aligned} \tag{8.6}$$

where λ is the degree of polarization. In case of $A(x,y)_L$, the azimuthal angle was integrated out, since the differential cross section depends on it only through phase space.

The twist-2 heavy flavor contributions to the structure function $g_1(x,Q^2)$ are calculated using the collinear parton model. This is not possible in case of the structure function $g_2(x,Q^2)$. As has been shown in Ref. [31], the Wandzura–Wilczek relation holds for the gluonic heavy flavor contributions as well

$$g_2^{\tau=2}(x,Q^2) = -g_1^{\tau=2}(x,Q^2) + \int_x^1 \frac{dz}{z}g_1^{\tau=2}(z,Q^2), \tag{8.7}$$

from which $g_2(x,Q^2)$ can be calculated for twist $\tau=2$. At leading order the heavy flavor corrections are known for the whole kinematic region, [28–30],

$$g_1^{Q\overline{Q}}(x,Q^2,m^2) = 4e_Q^2 a_s\int_{ax}^1 \frac{dz}{z}H_{g,g_1}^{(1)}\left(\frac{x}{z},\frac{m^2}{Q^2}\right)\Delta G(z,n_f,Q^2) \tag{8.8}$$

and are of the same structure as in the unpolarized case, cf. Eq. 3.12. Here, ΔG is the polarized gluon density. The LO heavy flavor Wilson coefficient then reads

$$H_{g,g_1}^{(1)}\left(\tau, \frac{m^2}{Q^2}\right) = 4T_F\left[v(3-4\tau)+(1-2\tau)\ln\left(\frac{1-v}{1+v}\right)\right]. \tag{8.9}$$

The support of $H_{g,g_1}^{(1)}\left(\tau, m^2/Q^2\right)$ is $\tau \in [0, 1/a]$. As is well known, its first moment vanishes

$$\int_0^{1/a} d\tau H_{g_1}^{(1)}\left(\tau, \frac{m^2}{Q^2}\right) = 0, \tag{8.10}$$

which has a phenomenological implication on the heavy flavor contributions to polarized structure functions, resulting in an oscillatory profile [31]. The unpolarized heavy flavor Wilson coefficients, [46–55], do not obey a relation like (8.10) but exhibit a rising behavior towards smaller values of x.

At asymptotic values $Q^2 \gg m^2$ one obtains

$$H_{g,g_1}^{(1),\mathrm{as}}\left(\tau, \frac{m^2}{Q^2}\right) = 4T_F\left[(3-4\tau)-(1-2\tau)\ln\left(\frac{Q^2}{m^2}\frac{1-\tau}{\tau}\right)\right]. \tag{8.11}$$

The factor in front of the logarithmic term $\ln(Q^2/m^2)$ in (8.11) is the leading order splitting function $\Delta P_{qg}(\tau)$, [56–59],[3]

$$\Delta P_{qg}(\tau) = 8T_F\left[\tau^2-(1-\tau)^2\right] = 8T_F\left[2\tau-1\right]. \tag{8.12}$$

The sum-rule (8.10) also holds in the asymptotic case extending the range of integration to $\tau \in [0, 1]$,

$$\int_0^1 d\tau H_{g,g_1}^{(1),\mathrm{as}}\left(\tau, \frac{m^2}{Q^2}\right) = 0. \tag{8.13}$$

8.2 Polarized Massive Operator Matrix Elements

The asymptotic heavy flavor Wilson coefficients obey the same factorization relations in the limit $Q^2 \gg m^2$ as in the unpolarized case, Eqs. 3.21–3.25, if one replaces all quantities by their polarized counterparts.

---

[3] Early calculations of the leading order polarized singlet splitting functions in Refs. [57–59] still contained some errors.

The corresponding polarized twist-2 composite operators, cf. Eqs. 2.86–2.88, are given by

$$O^{\mathsf{NS}}_{q,r;\mu_1,\ldots,\mu_N} = i^{N-1}\mathbf{S}\left[\overline{\psi}\gamma_5\gamma_{\mu_1}D_{\mu_2}\ldots D_{\mu_N}\frac{\lambda_r}{2}\psi\right] - \text{trace terms}, \tag{8.14}$$

$$O^{\mathsf{S}}_{q;\mu_1,\ldots,\mu_N} = i^{N-1}\mathbf{S}[\overline{\psi}\gamma_5\gamma_{\mu_1}D_{\mu_2}\ldots D_{\mu_N}\psi] - \text{trace terms}, \tag{8.15}$$

$$O^{\mathsf{S}}_{g;\mu_1,\ldots,\mu_N} = 2i^{N-2}\mathbf{SSp}\left[\frac{1}{2}\varepsilon^{\mu_1\alpha\beta\gamma}F^a_{\beta\gamma}D^{\mu_2}\ldots D^{\mu_{N-1}}F^{\mu_N}_{\alpha,a}\right] - \text{trace terms}. \tag{8.16}$$

The Feynman rules needed are given in Appendix B. The polarized anomalous dimensions of these operators are defined in the same way as in Eqs. 2.107 and 2.108, as is the case for the polarized massive OMEs, cf. Eq. 3.17 and below.

In the subsequent investigation, we will follow Ref. [39] and calculate the quarkonic heavy quark contributions to $O(a_s^2)$. The diagrams contributing to the corresponding massive OMEs are the same as in the unpolarized case and are shown in Figs. 1–4 in Ref. [49]. The formal factorization relations for the heavy flavor Wilson coefficients can be inferred from Eqs. 3.26, 3.29, 3.30. Here, we perform the calculation in the MOM-scheme, cf. Sect. 5.1, to account for heavy quarks in the final state only. The same scheme has been adopted in Ref. [39]. Identifying $\mu^2 = Q^2$, the heavy flavor Wilson coefficients in the limit $Q^2 \gg m^2$ become, [39],

$$H^{(1)}_{g,g_1}\left(\frac{Q^2}{m^2}, N\right) = -\frac{1}{2}\Delta\hat{\gamma}^{(0)}_{qg}\ln\left(\frac{Q^2}{m^2}\right) + \hat{c}^{(1)}_{g,g_1}, \tag{8.17}$$

$$\begin{aligned} H^{(2)}_{g,g_1}\left(\frac{Q^2}{m^2}, N\right) = \Bigg\{ & \frac{1}{8}\Delta\hat{\gamma}^{(0)}_{qg}\left[\Delta\gamma^{(0)}_{qq} - \Delta\gamma^{(0)}_{gg} - 2\beta_0\right]\ln^2\left(\frac{Q^2}{m^2}\right) \\ & -\frac{1}{2}\left[\Delta\hat{\gamma}^{(1)}_{qg} + \Delta\hat{\gamma}^{(0)}_{qg}c^{(1)}_{q,g_1}\right]\ln\left(\frac{Q^2}{m^2}\right) \\ & +\left[\Delta\gamma^{(0)}_{gg} - \Delta\gamma^{(0)}_{qq} + 2\beta_0\right] \\ & \times\frac{\Delta\hat{\gamma}^{(0)}_{qg}\zeta_2}{8} + \hat{c}^{(2)}_{g,g_1} + \Delta a^{(2)}_{Qg}\Bigg\}, \end{aligned} \tag{8.18}$$

$$\begin{aligned} H^{(2),\mathsf{PS}}_{q,g_1}\left(\frac{Q^2}{m^2}, N\right) = \Bigg\{ & -\frac{1}{8}\Delta\hat{\gamma}^{(0)}_{qg}\Delta\gamma^{(0)}_{gq}\ln^2\left(\frac{Q^2}{m^2}\right) - \frac{1}{2}\Delta\hat{\gamma}^{(1),\mathsf{PS}}_{qq}\ln\left(\frac{Q^2}{m^2}\right) \\ & + \frac{\Delta\hat{\gamma}^{(0)}_{qg}\Delta\gamma^{(0)}_{gq}}{8}\zeta_2 + \hat{c}^{(2),\mathsf{PS}}_{q,g_1} + \Delta a^{(2),\mathsf{PS}}_{Qq}\Bigg\}, \end{aligned} \tag{8.19}$$

$$L_{q,g_1}^{(2),\mathsf{NS}}\left(\frac{Q^2}{m^2},N\right)=\left\{\frac{1}{4}\beta_{0,Q}\Delta\gamma_{qq}^{(0)}\ln^2\left(\frac{Q^2}{m^2}\right)-\left[\frac{1}{2}\Delta\hat{\gamma}_{qq}^{(1),\mathsf{NS}}+\beta_{0,Q}c_{q,g_1}^{(1)}\right]\right.$$
$$\left.\times\ln\left(\frac{Q^2}{m^2}\right)-\frac{1}{4}\beta_{0,Q}\zeta_2\Delta\gamma_{qq}^{(0)}+\hat{c}_{q,g_1,Q}^{(2),\mathsf{NS}}+\Delta a_{qq,Q}^{(2),\mathsf{NS}}\right\}. \quad (8.20)$$

$c_{i,g_1}^{(k)}$ are the kth order non-logarithmic terms of the polarized coefficient functions. As has been described in [39], the relations (8.18)–(8.20) hold if one uses the same scheme for the description of γ_5 in dimensional regularization for the massive OMEs and the light flavor Wilson coefficients. This is the case for the massive OMEs as calculated in [39], to which we refer, and the light flavor Wilson coefficients as calculated in Ref. [45].

8.2.1 $\Delta A_{qq,Q}^{(2),\mathsf{NS}}$

The non-singlet operator matrix element $\Delta A_{qq,Q}^{(2),\mathsf{NS}}$ has to be the same as in the unpolarized case due to the Ward-Takahashi identity [60, 61]. Since it is obtained as zero-momentum insertion on a graph for the transition $\langle p| \rightarrow |p\rangle$, one may write it equivalently in terms of the momentum derivative of the self-energy. The latter is independent of the operator insertion and yields therefore the same in case of $\not{\Delta}\Delta(\Delta\cdot p)^{N-1}$ and $\not{\Delta}\Delta\gamma_5(\Delta\cdot p)^{N-1}$. Hence, $\Delta A_{qq,Q}^{(2),\mathsf{NS}}$ reads, cf. Eq. (4.95),

$$\Delta A_{qq,Q}^{(2),\mathsf{NS}}\left(N,\frac{m^2}{\mu^2}\right)=A_{qq,Q}^{(2),\mathsf{NS}}\left(N,\frac{m^2}{\mu^2}\right)$$
$$=\frac{\beta_{0,Q}\gamma_{qq}^{(0)}}{4}\ln^2\left(\frac{m^2}{\mu^2}\right)+\frac{\hat{\gamma}_{qq}^{(1),\mathsf{NS}}}{2}$$
$$\times\ln\left(\frac{m^2}{\mu^2}\right)+a_{qq,Q}^{(2),\mathsf{NS}}-\frac{\gamma_{qq}^{(0)}}{4}\beta_{0,Q}\zeta_2, \quad (8.21)$$

where the constant term in ε of the unrenormalized result, Eq. 4.93, is given in Eq. 6.60 and the $O(\varepsilon)$-term in Eq. (6.61).

8.2.2 $\Delta A_{Qg}^{(2)}$

To calculate the OME ΔA_{Qg} up to $O(a_s^2)$, the Dirac-matrix γ_5 is represented in $D=4+\varepsilon$ dimensions via, [39, 62–66],

$$\not{\Delta}\gamma^5=\frac{i}{6}\varepsilon_{\mu\nu\rho\sigma}\Delta^\mu\gamma^\nu\gamma^\rho\gamma^\sigma. \quad (8.22)$$

The Levi-Civita symbol will be contracted later with a second Levi-Civita symbol emerging in the general expression for the Green's function, cf. Eq. 4.18,

$$\Delta \hat{G}^{ab}_{Q,\mu\nu} = \Delta \hat{\hat{A}}_{Qg}\left(\frac{\hat{m}^2}{\mu^2}, \varepsilon, N\right)\delta^{ab}(\Delta \cdot p)^{N-1}\varepsilon_{\mu\nu\alpha\beta}\Delta^\alpha p^\beta, \tag{8.23}$$

by using the following relation in D-dimensions, [67],

$$\begin{aligned} \varepsilon_{\mu\nu\rho\sigma}\varepsilon^{\alpha\lambda\tau\gamma} &= -\operatorname{Det}\left[g^{\beta}_{\omega}\right], \\ &\beta = \alpha, \lambda, \tau, \gamma, \quad \omega = \mu, \nu, \rho, \sigma. \end{aligned} \tag{8.24}$$

In particular, anti-symmetry relations of the Levi-Civita tensor or the relation $\gamma_5^2 = \mathbf{1}$, holding in four dimensions, are not used. The projector for the gluonic OME then reads

$$\Delta \hat{\hat{A}}_{Qg} = \frac{\delta^{ab}}{N_c^2 - 1}\frac{1}{(D-2)(D-3)}(\Delta \cdot p)^{-N-1}\varepsilon^{\mu\nu\rho\sigma}\Delta \hat{G}^{ab}_{Q,\mu\nu}\Delta_\rho p_\sigma. \tag{8.25}$$

In the following, we will present the results for the operator matrix element using the above prescription for γ_5. This representation allows a direct comparison to Ref. [39] despite the fact that in this scheme even some of the anomalous dimensions are not those of the $\overline{\text{MS}}$-scheme. We will discuss operator matrix elements for which only mass renormalization was carried out, cf. Sect. 4.3. Due to the crossing relations of the forward Compton amplitude corresponding to the polarized case, only odd moments contribute. Therefore the overall factor

$$\frac{1}{2}\left[1 - (-1)^N\right], \quad N \in \mathbb{N}, \tag{8.26}$$

is implied in the following. To obtain the results in x-space the analytic continuation to complex values of N can be performed starting from the odd integers. The $O(a_s)$ calculation is straightforward

$$\Delta \hat{\hat{A}}^{(1)}_{Qg} = \left(\frac{m^2}{\mu^2}\right)^{\varepsilon/2}\left[\frac{1}{\varepsilon} + \frac{\zeta_2}{8}\varepsilon^2 + \frac{\zeta_3}{24}\varepsilon\right]\Delta\hat{\gamma}^{(0)}_{qg} + O(\varepsilon^3) \tag{8.27}$$

$$= \left(\frac{m^2}{\mu^2}\right)^{\varepsilon/2}\left[\frac{1}{\varepsilon}\Delta\hat{\gamma}^{(0)}_{qg} + \Delta a^{(1)}_{Qg} + \varepsilon\Delta\overline{a}^{(1)}_{Qg} + \varepsilon^2\Delta\overline{\overline{a}}^{(1)}_{Qg}\right] + O(\varepsilon^3). \tag{8.28}$$

The matrix element contains the leading order anomalous dimension $\Delta\hat{\gamma}^{(0)}_{qg}$,

$$\Delta A^{(1)}_{Qg} = \frac{1}{2}\Delta\hat{\gamma}^{(0)}_{qg}\ln\left(\frac{m^2}{\mu^2}\right), \tag{8.29}$$

where

$$\Delta\hat{\gamma}_{qg}^{(0)} = -8T_F \frac{N-1}{N(N+1)}. \tag{8.30}$$

The leading order polarized Wilson coefficient $c_{g,g_1}^{(1)}$ reads, [30, 45, 68],

$$c_{g,g_1}^{(1)} = -4T_F \frac{N-1}{N(N+1)} \left[S_1 + \frac{N-1}{N}\right]. \tag{8.31}$$

The Mellin transform of Eq. 8.11 then yields the same expression as one obtains from Eq. (8.17)

$$H_{g,g_1}^{(1),\mathrm{as}}\left(N, \frac{m^2}{Q^2}\right) = \left[-\frac{1}{2}\Delta\hat{\gamma}_{qg}^{(0)} \ln\left(\frac{Q^2}{m^2}\right) + c_{g,g_1}^{(1)}\right], \tag{8.32}$$

for which the proportionality

$$H_{g,g_1}^{(1),\mathrm{as}}\left(N, \frac{m^2}{Q^2}\right) \propto (N-1) \tag{8.33}$$

holds, leading to a vanishing first moment.

At the two-loop level, we express the operator matrix element $\Delta\hat{\hat{A}}_{Qg}^{(2)}$, after mass renormalization, in terms of anomalous dimensions, cf. [49–55], by

$$\begin{aligned}\Delta\hat{\hat{A}}_{Qg}^{(2)} = &\left(\frac{m^2}{\mu^2}\right)^{\varepsilon} \left[\frac{\Delta\hat{\gamma}_{qg}^{(0)}}{2\varepsilon^2}\left\{\Delta\gamma_{qq}^{(0)} - \Delta\gamma_{gg}^{(0)} - 2\beta_0\right\} + \frac{\Delta\hat{\gamma}_{qg}^{\prime(1)}}{\varepsilon} + \Delta a_{Qg}^{\prime(2)}\right. \\ &\left. + \Delta\overline{a}_{Qg}^{\prime(2)}\varepsilon\right] - \frac{2}{\varepsilon}\beta_{0,Q}\left(\frac{m^2}{\mu^2}\right)^{\varepsilon/2}\left(1 + \frac{\varepsilon^2}{8}\zeta_2 + \frac{\varepsilon^3}{24}\zeta_3\right)\Delta\hat{\hat{A}}_{Qg}^{(1)} + O(\varepsilon^2),\end{aligned} \tag{8.34}$$

The remaining LO anomalous dimensions are

$$\Delta\gamma_{qq}^{(0)} = -C_F\left(-8S_1 + 2\frac{3N^2+3N+2}{N(N+1)}\right), \tag{8.35}$$

$$\Delta\gamma_{gg}^{(0)} = -C_A\left(-8S_1 + 2\frac{11N^2+11N+24}{3N(N+1)}\right) + \frac{8}{3}T_F n_f. \tag{8.36}$$

The renormalized expression in the MOM-scheme is given by

$$\begin{aligned}\Delta A_{Qg}^{\prime(2),\mathrm{MOM}} = &\frac{\Delta\hat{\gamma}_{qg}^{(0)}}{8}\left[\Delta\gamma_{qq}^{(0)} - \Delta\gamma_{gg}^{(0)} - 2\beta_0\right]\ln^2\left(\frac{m^2}{\mu^2}\right) + \frac{\hat{\gamma}_{qg}^{\prime(1)}}{2} \\ &\times \ln\left(\frac{m^2}{\mu^2}\right) + \left(\Delta\gamma_{gg}^{(0)} - \Delta\gamma_{qq}^{(0)} + 2\beta_0\right)\frac{\hat{\gamma}_{qg}^{(0)}}{8}\zeta_2 + \Delta a_{Qg}^{\prime(2)}.\end{aligned} \tag{8.37}$$

The LO anomalous dimensions which enter the double pole term in Eq. 8.34 and the $\ln^2(m^2/\mu^2)$ term in Eq. 8.37, respectively, are scheme-independent. This is not the case for the remaining terms, which depend on the particular scheme we adopted in Eqs. 8.42, 8.22 and are therefore denoted by a prime. The NLO anomalous dimension we obtain is given by

$$\begin{aligned}\Delta\hat{\gamma}_{qg}^{\prime(1)} =& -T_F C_F\Biggl(-16\frac{N-1}{N(N+1)}S_2 + 16\frac{N-1}{N(N+1)}S_1^2 - 32\frac{N-1}{N^2(N+1)}S_1 \\ &+ 8\frac{(N-1)(5N^4+10N^3+8N^2+7N+2)}{N^3(N+1)^3}\Biggr) \\ &+ T_F C_A\Biggl(32\frac{N-1}{N(N+1)}\beta' + 16\frac{N-1}{N(N+1)}S_2 + 16\frac{N-1}{N(N+1)}S_1^2 \\ &- 16\frac{N-1}{N(N+1)}\zeta_2 - \frac{64S_1}{N(N+1)^2} \\ &- 16\frac{N^5+N^4-4N^3+3N^2-7N-2}{N^3(N+1)^3}\Biggr). \qquad (8.38)\end{aligned}$$

It differs from the result in the $\overline{\text{MS}}$-scheme, [69–71], by a finite renormalization. This is due to the fact that we contracted the Levi-Civita symbols in D dimensions. The correct NLO splitting function is obtained by

$$\Delta\hat{\gamma}_{qg}^{(1)} = \Delta\hat{\gamma}_{qg}^{\prime(1)} + 64T_F C_F\frac{N-1}{N^2(N+1)^2}. \qquad (8.39)$$

In an earlier version of Ref. [45], $\Delta\hat{\gamma}_{qg}^{\prime(1)}$ was used as the anomalous dimension departing from the $\overline{\text{MS}}$ scheme. Therefore, in Ref. [39] the finite renormalization (8.39), as the corresponding one for $c_{g,g_1}^{(2)}$, [45], was not used for the calculation of $\Delta A_{Qg}^{(2)}$. For the higher order terms in ε in Eq. (8.34) we obtain

$$\begin{aligned}\Delta a_{Qg}^{\prime(2)} =& -T_F C_F\Biggl\{\frac{4(N-1)}{3N(N+1)}\left(-4S_3 + S_1^3 + 3S_1S_2 + 6S_1\zeta_2\right) \\ &- \frac{4(3N^2+3N-2)S_1^2}{N^2(N+1)(N+2)} - 4\frac{N^4+17N^3+43N^2+33N+2}{N^2(N+1)^2(N+2)}S_2 \\ &- 2\frac{(N-1)(3N^2+3N+2)}{N^2(N+1)^2}\zeta_2 - 4\frac{N^3-2N^2-22N-36}{N^2(N+1)(N+2)}S_1 \\ &+ \frac{2P_1}{N^4(N+1)^4(N+2)}\Biggr\} \\ &- T_F C_A\Biggl\{4\frac{N-1}{3N(N+1)}\Biggl(12\mathrm{M}\left[\frac{\mathrm{Li}_2(x)}{1+x}\right](N+1) + 3\beta'' - 8S_3 - S_1^3 - 9S_1S_2 \\ &- 12S_1\beta' - 12\beta\zeta_2 - 3\zeta_3\Biggr) - 16\frac{N-1}{N(N+1)^2}\beta'\end{aligned}$$

$$
\begin{aligned}
&+4\frac{N^2+4N+5}{N(N+1)^2(N+2)}S_1^2+4\frac{7N^3+24N^2+15N-16}{N^2(N+1)^2(N+2)}S_2\\
&+8\frac{(N-1)(N+2)}{N^2(N+1)^2}\zeta_2+4\frac{N^4+4N^3-N^2-10N+2}{N(N+1)^3(N+2)}S_1\\
&-\frac{4P_2}{N^4(N+1)^4(N+2)}\Bigg\},
\end{aligned}
\tag{8.40}
$$

$$
\begin{aligned}
\Delta\overline{a}_{Qg}^{\prime(2)} = T_FC_F\Bigg\{&\frac{N-1}{N(N+1)}\Bigg(16S_{2,1,1}-8S_{3,1}-8S_{2,1}S_1+3S_4-\frac{4}{3}S_3S_1-\frac{1}{2}S_2^2-\frac{1}{6}S_1^4\\
&-\frac{8}{3}S_1\zeta_3-S_2S_1^2+2S_2\zeta_2-2S_1^2\zeta_2\Bigg)-8\frac{S_{2,1}}{N^2}+\frac{3N^2+3N-2}{N^2(N+1)(N+2)}\\
&\times\left(2S_2S_1+\frac{2}{3}S_1^3\right)+2\frac{3N^4+48N^3+123N^2+98N+8}{3N^2(N+1)^2(2+N)}S_3\\
&+\frac{4(N-1)}{N^2(N+1)}S_1\zeta_2+\frac{2}{3}\frac{(N-1)(3N^2+3N+2)}{N^2(N+1)^2}\zeta_3\\
&+\frac{P_3S_2}{N^3(N+1)^3(N+2)}+\frac{N^3-6N^2-22N-36}{N^2(N+1)(N+2)}S_1^2\\
&+\frac{P_4\zeta_2}{N^3(N+1)^3}-2\frac{2N^4-4N^3-3N^2+20N+12}{N^2(N+1)^2(N+2)}S_1\\
&+\frac{P_5}{N^5(N+1)^5(N+2)}\Bigg\}\\
+T_FC_A\Bigg\{&\frac{N-1}{N(N+1)}\Bigg(16S_{-2,1,1}-4S_{2,1,1}-8S_{-3,1}-8S_{-2,2}-4S_{3,1}+\frac{2}{3}\beta'''\\
&-16S_{-2,1}S_1-4\beta''S_1+8\beta'S_2+8\beta'S_1^2+9S_4+\frac{40}{3}S_3S_1+\frac{1}{2}S_2^2\\
&+5S_2S_1^2+\frac{1}{6}S_1^4+4\zeta_2\beta'-2\zeta_2S_2-2\zeta_2S_1^2-\frac{10}{3}S_1\zeta_3-\frac{17}{5}\zeta_2^2\Bigg)\\
&-\frac{N-1}{N(N+1)^2}\left(16S_{-2,1}+4\beta''-16\beta'S_1\right)-\frac{16}{3}\frac{N^3+7N^2+8N-6}{N^2(N+1)^2(N+2)}S_3\\
&+\frac{2(3N^2-13)S_2S_1}{N(N+1)^2(N+2)}-\frac{2(N^2+4N+5)}{3N(N+1)^2(N+2)}S_1^3-\frac{8\zeta_2S_1}{(N+1)^2}\\
&-\frac{2}{3}\frac{(N-1)(9N+8)}{N^2(N+1)^2}\zeta_3-\frac{8(N^2+3)}{N(N+1)^3}\beta'-\frac{P_6S_2}{N^3(N+1)^3(N+2)}\\
&-\frac{N^4+2N^3-5N^2-12N+2}{N(N+1)^3(N+2)}S_1^2-\frac{2P_7\zeta_2}{N^3(N+1)^3}+\frac{2P_8S_1}{N(N+1)^4(N+2)}\\
&-\frac{2P_9}{N^5(N+1)^5(N+2)}\Bigg\},
\end{aligned}
\tag{8.41}
$$

with the polynomials

$$
P_1=4N^8+12N^7+4N^6-32N^5-55N^4-30N^3-3N^2-8N-4, \tag{8.42}
$$

$$
P_2=2N^8+10N^7+22N^6+36N^5+29N^4+4N^3+33N^2+12N+4, \tag{8.43}
$$

$$P_3 = 3N^6 + 30N^5 + 107N^4 + 124N^3 + 48N^2 + 20N + 8, \tag{8.44}$$

$$P_4 = (N-1)(7N^4 + 14N^3 + 4N^2 - 7N - 2), \tag{8.45}$$

$$P_5 = 8N^{10} + 24N^9 - 11N^8 - 160N^7 - 311N^6 - 275N^5 - 111N^4 - 7N^3 + 11N^2 + 12N + 4, \tag{8.46}$$

$$P_6 = N^6 + 18N^5 + 63N^4 + 84N^3 + 30N^2 - 64N - 16, \tag{8.47}$$

$$P_7 = N^5 - N^4 - 4N^3 - 3N^2 - 7N - 2, \tag{8.48}$$

$$P_8 = 2N^5 + 10N^4 + 29N^3 + 64N^2 + 67N + 8, \tag{8.49}$$

$$P_9 = 4N^{10} + 22N^9 + 45N^8 + 36N^7 - 11N^6 - 15N^5 + 25N^4 - 41N^3 - 21N^2 - 16N - 4. \tag{8.50}$$

The Mellin-transform in Eq. 8.40 is given in Eq. 6.47 in terms of harmonic sums. As a check, we calculated several lower moments ($N = 1, \ldots, 9$) of each individual diagram contributing to $A_{Qg}^{(2)}$ [4] using the Mellin-Barnes method [54, 72]. In Table 8.1, we present the numerical results we obtain for the moments $N = 3, 7$ of the individual diagrams. We agree with the results obtained for general values of N. The contributions from the individual diagrams are given in [73]. Our results up to $O(\varepsilon^0)$, Eqs. 8.34, 8.40, agree with the results presented in [39], which we thereby confirm for the first time. Eq. 8.41 is a new result.

In this calculation extensive use was made of the representation of the Feynman-parameter integrals in terms of generalized hypergeometric functions, cf. Chap. 6. The infinite sums, which occur in the polarized calculation, are widely the same as in the unpolarized case [50–55]. The structure of the result for the higher order terms in ε is completely the same as in the unpolarized case as well, see Eq. 6.34 and the following discussion. Especially, the structural relations between the finite harmonic sums, [74–77], allow to express $\Delta a_{Qg}^{\prime(2)}$ by only two basic Mellin transforms, S_1 and $S_{-2,1}$. This has to be compared to the 24 functions needed in Ref. [39] to express the constant term in z-space. Thus we reached a more compact representation. $\Delta \overline{a}_{Qg}^{\prime(2)}$ depends on the six sums $S_1(N)$, $S_{\pm 2,1}(N)$, $S_{-3,1}(N)$, $S_{\pm 2,1,1}(N)$, after applying the structural relations. The $O(\varepsilon^0)$ term has the same complexity as the two-loop anomalous dimensions, whereas the complexity of the $O(\varepsilon)$ term corresponds to the level observed for two-loop Wilson coefficients and other hard scattering processes which depend on a single scale, cf. [78–80].

---

4 These are shown in Fig. 2 of Ref. [49].

Table 8.1 Numerical values for moments of individual diagrams of $\Delta\hat{\hat{A}}_{Qg}^{(2)}$

Order		$1/\varepsilon^2$	$1/\varepsilon$	1	ε	ε^2
A	N=3	−0.22222	0.06481	−0.13343	−0.15367	−0.06208
	$N=7$	−0.03061	0.00409	−0.01669	−0.01900	−0.00639
B	$N=3$	4.44444	−1.07407	4.45579	0.515535	3.13754
	$N=7$	5.46122	0.74491	6.09646	2.97092	5.35587
C	$N=3$	1.33333	−8.14444	0.13303	−6.55515	−2.64601
	$N=7$	0.85714	−5.12329	0.14342	−4.10768	−1.59526
D	$N=3$	2.66666	−0.02222	2.19940	1.03927	1.69331
	$N=7$	1.71428	0.85340	1.78773	1.56227	1.80130
E	$N=3$	−2.66667	5	−2.27719	4.89957	0.73208
	$N=7$	−1.71429	2.97857	−1.3471	2.83548	0.44608
F	$N=3$	0	0.77777	−5.80092	−2.63560	−6.57334
	$N=7$	0	1.40105	−3.54227	−0.78565	−3.72466
L	$N=3$	−9.33333	0.25000	−8.83933	−3.25228	−6.84460
	$N=7$	−6.73878	−1.86855	−7.09938	−4.56051	−6.501
M	$N=3$	−0.22222	0.71296	−0.41198	0.69938	−0.11618
	$N=7$	−0.03061	0.11324	−0.05861	0.11969	−0.01207
N	$N=3$	−2.22222	1.26851	−1.37562	0.69748	−0.36030
	$N=7$	−3.19184	−0.50674	−3.39832	−1.7667	−2.97339

8.2.3 $\Delta A_{Qq}^{(2),\mathrm{PS}}$

The operator matrix element $\Delta A_{Qq}^{(2),\mathrm{PS}}$ is obtained from the diagrams shown in Fig. 3 of Ref. [49]. In this calculation, we did not adopt any specific scheme for γ_5, but calculated the corresponding integrals without performing any traces or (anti)commuting γ_5.

The result can then be represented in terms of three bi-spinor structures

$$C_1(\varepsilon) = \frac{1}{\Delta\cdot p} Tr\left\{\not\Delta \not p \gamma^\mu\gamma^\nu\gamma_5\right\}\not\Delta\gamma_\mu\gamma_\nu = \frac{24}{3+\varepsilon}\not\Delta\gamma_5 \tag{8.51}$$

$$C_2(\varepsilon) = Tr\left\{\not\Delta\gamma^\mu\gamma^\nu\gamma^\rho\gamma_5\right\}\gamma_\mu\gamma_\nu\gamma_\rho = 24\not\Delta\gamma_5 \tag{8.52}$$

$$C_3(\varepsilon) = \frac{1}{m^2} Tr\left\{\not\Delta \not p \gamma^\mu\gamma^\nu\gamma_5\right\}\not p\gamma_\mu\gamma_\nu. \tag{8.53}$$

These are placed between the states $\langle p|\ldots|p\rangle$, with

$$\not p|p\rangle = m_0|p\rangle \tag{8.54}$$

and m_0 the light quark mass. Therefore, the contribution due to $C_3(\varepsilon)$ vanishes in the limit $m_0 \to 0$. The results for $C_{1,2}(\varepsilon)$ in the r.h.s. of Eqs. 8.51, 8.52 can be obtained by applying the projector

$$\frac{-3}{2(D-1)(D-2)(D-3)}\mathrm{Tr}[\not p\gamma_5 C_i] \tag{8.55}$$

and performing the trace in D-dimensions using relations (8.22, 8.24). Note that the result in four-dimensions is recovered by setting $\varepsilon = 0$. One obtains from the truncated two-loop Green's function $\Delta G_{Qq}^{ij,(2)}$

$$\Delta G_{Qq}^{ij,(2)} = \Delta\hat{\hat{A}}_{Qq}^{(2),\mathrm{PS}}\not\Delta\gamma_5\delta^{ij}(\Delta\cdot p)^{N-1}, \tag{8.56}$$

the following result for the massive OME

$$\begin{aligned}\Delta\hat{\hat{A}}_{Qq}^{(2),\mathrm{PS}}\not\Delta\gamma_5 = \left(\frac{m^2}{\mu^2}\right)^{\varepsilon} C(\varepsilon)8(N+2)\Bigg\{&-\frac{1}{\varepsilon^2}\frac{2(N-1)}{N^2(N+1)^2}+\frac{1}{\varepsilon}\frac{N^3+2N+1}{N^3(N+1)^3}\\&-\frac{(N-1)(\zeta_2+2S_2)}{2N^2(N+1)^2}-\frac{4N^3-4N^2-3N-1}{2N^4(N+1)^4}\\&+\varepsilon\Bigg[\frac{(N^3+2N+1)(\zeta_2+2S_2)}{4N^3(N+1)^3}\\&-\frac{(N-1)(\zeta_3+3S_3)}{6N^2(N+1)^2}\\&+\frac{N^5-7N^4+6N^3+7N^2+4N+1}{4N^5(N+1)^5}\Bigg]\Bigg\}+O(\varepsilon^2),\end{aligned} \tag{8.57}$$

where

$$\begin{aligned}C(\varepsilon) &= \frac{C_1(\varepsilon)\cdot(N-1)+C_2(\varepsilon)}{8(N+2)} = \not\Delta\gamma_5\frac{3(N+2+\varepsilon)}{(N+2)(3+\varepsilon)}\\&= 1+\frac{N-1}{3(N+2)}\left(-\varepsilon+\frac{\varepsilon^2}{3}-\frac{\varepsilon^3}{9}\right)+O(\varepsilon^4).\end{aligned} \tag{8.58}$$

Comparing our result, Eq. 8.57, to the result obtained in [39], one notices that there the factor $C(\varepsilon)$ was calculated in four-dimensions, i.e. $C(\varepsilon) = 1$. Therefore, we do the same and obtain

$$\Delta\hat{\hat{A}}_{Qq}^{(2),\mathrm{PS}} = S_\varepsilon^2\left(\frac{m^2}{\mu^2}\right)^{\varepsilon}\left[-\frac{\Delta\hat{\gamma}_{qg}^{(0)}\Delta\gamma_{gq}^{(0)}}{2\varepsilon^2}+\frac{\Delta\hat{\gamma}_{qq}^{(1),\mathrm{PS}}}{2\varepsilon}+\Delta a_{Qq}^{(2),\mathrm{PS}}+\Delta\overline{a}_{Qq}^{(2),\mathrm{PS}}\varepsilon\right]. \tag{8.59}$$

with

$$\Delta\gamma_{gq}^{(0)} = -4C_F\frac{N+2}{N(N+1)}, \tag{8.60}$$

$$\Delta\hat{\gamma}_{qq}^{(1),\mathrm{PS}} = 16T_FC_F\frac{(N+2)(N^3+2N+1)}{N^3(N+1)^3}, \tag{8.61}$$

$$\Delta a_{Qq}^{(2),\mathsf{PS}} = -\frac{(N-1)(\zeta_2+2S_2)}{2N^2(N+1)^2} - \frac{4N^3-4N^2-3N-1}{2N^4(N+1)^4}, \tag{8.62}$$

$$\Delta \overline{a}_{Qq}^{(2),\mathsf{PS}} = -\frac{(N-1)(\zeta_3+3S_3)}{6N^2(N+1)^2}\frac{(N^3+2N+1)(\zeta_2+2S_2)}{4N^3(N+1)^3} + \frac{N^5-7N^4+6N^3+7N^2+4N+1}{4N^5(N+1)^5}. \tag{8.63}$$

Here, we agree up to $O(\varepsilon^0)$ with Ref. [39] and Eq. 8.63 is a new result. Note, that Eq. 8.61 is already the $\overline{\mathsf{MS}}$ anomalous dimension as obtained in Refs. [69–71]. Therefore any additional scheme dependence due to γ_5 can only be contained in the higher order terms in ε. As a comparison the anomalous dimension $\Delta\hat{\gamma}_{qq}^{\prime(1),\mathsf{PS}}$ which is obtained by calculating $C(\varepsilon)$ in D dimensions, is related to the $\overline{\mathsf{MS}}$ one by

$$\Delta\hat{\gamma}_{qq}^{(1),\mathsf{PS}} = \Delta\hat{\gamma}_{qq}^{\prime(1),\mathsf{PS}} - T_F C_F \frac{16(N-1)^2}{3N^2(N+1)^2}. \tag{8.64}$$

The renormalized result becomes

$$\Delta A_{Qq}^{(2),\mathsf{PS}} = -\frac{\Delta\hat{\gamma}_{qg}^{(0)}\Delta\gamma_{gq}^{(0)}}{8}\ln^2\left(\frac{m^2}{\mu^2}\right) + \frac{\Delta\hat{\gamma}_{qq}^{(1),\mathsf{PS}}}{2} \times \ln\left(\frac{m^2}{\mu^2}\right) + \Delta a_{Qq}^{(2),\mathsf{PS}} + \frac{\Delta\hat{\gamma}_{qg}^{(0)}\Delta\gamma_{gq}^{(0)}}{8}\zeta_2. \tag{8.65}$$

The results in this section constitute a partial step towards the calculation of the asymptotic heavy flavor contributions at $O(a_s^2)$ in the $\overline{\mathsf{MS}}$-scheme, thereby going beyond the results of Ref. [39]. The same holds for the $O(a_s^2\varepsilon)$-terms, which we calculated for the first time, using the same description for γ_5 as has been done in [39]. The correct finite renormalization to transform to the $\overline{\mathsf{MS}}$-scheme remains to be worked out and will be presented elsewhere [73].

References

1. M.J. Alguard et al., Phys. Rev. Lett. **37**, 1261 (1976)
2. M.J. Alguard et al., Phys. Rev. Lett. **41**, 70 (1978)
3. G. Baum et al., Phys. Rev. Lett. **51**, 1135 (1983)
4. EMC collaboration, J. Ashman et al., Phys. Lett. **B206**, 364 (1988)
5. EMC collaboration, J. Ashman et al., Nucl. Phys. **B328**, 1 (1989)
6. SMC collaboration, B. Adeva et al., Phys. Lett. **B302**, 533 (1993)
7. E 142 collaboration, P.L. Anthony et al., Phys. Rev. **D54**, 6620 (1996), hepex/ 9610007
8. HERMES collaboration, K. Ackerstaff et al., Phys. Lett. **B404**, 383 (1997), hepex/ 9703005
9. E 154 collaboration, K. Abe et al., Phys. Rev. Lett. **79**, 26 (1997), hep-ex/9705012
10. SMC collaboration, B. Adeva et al., Phys. Rev. **D58**, 112001 (1998)
11. E 143 collaboration, K. Abe et al., Phys. Rev. **D58**, 112003 (1998), hep-ph/9802357

12. HERMES collaboration, A. Airapetian et al., Phys. Lett. **B442**, 484 (1998), hepex/ 9807015
13. E 155 collaboration, P.L. Anthony et al., Phys. Lett. **B463**, 339 (1999), hepex/ 9904002
14. E 155 collaboration, P.L. Anthony et al., Phys. Lett. **B493**, 19 (2000), hepph/ 0007248
15. Jefferson Lab Hall A collaboration, X. Zheng et al., Phys. Rev. Lett. **92**, 012004 (2004), nucl-ex/0308011
16. HERMES collaboration, A. Airapetian et al., Phys. Rev. **D71**, 012003 (2005), hep-ex/0407032
17. COMPASS collaboration, E.S. Ageev et al., Phys. Lett. **B612**, 154 (2005), hepex/ 0501073
18. COMPASS collaboration, E.S. Ageev et al., Phys. Lett. **B647**, 330 (2007), hepex/ 0701014
19. HERMES collaboration, A. Airapetian et al., Phys. Rev. **D75**, 012007 (2007), hep-ex/0609039
20. E. Reya, *The spin structure of the nucleon*, ed. by P.M. Zerwas, H.A. Kastrup. QCD - 20 years later. Proceedings, Workshop, Aachen, Germany, June 9–13, 1992, vol. 1 (World Scientific, Singapore, 1993), p. 272
21. B. Lampe, E. Reya, Phys. Rept. **332**, 1 (2000), hep-ph/9810270
22. M. Burkardt, A. Miller, W.-D. Nowak, Rept. Prog. Phys. **73**, 016201 (2010), hep-ph/0812.2208
23. J. Blümlein, W.D. Nowak (eds.), *Prospects of spin physics at HERA*. Proceedings of Workshop, Zeuthen, Germany, August 28–31, 1995, DESY-95–200 (DESY, Hamburg, 1995), 387 p
24. J. Blümlein, A. De Roeck, T. Gehrmann, W.D. Nowak (eds.), *Deep inelastic scattering off polarized targets: theory meets experiment. Physics with polarized protons at HERA*. Proceedings, Workshops, SPIN'97, Zeuthen, Germany, September 1–5, 1997 and Hamburg, Germany, March-September 1997, DESY-97-200
25. J. Blümlein, On the measurability of the structure function g1(x, Q^2) in ep collisions at HERA (1995), hep-ph/9508387
26. K. Kurek, ΔG from COMPASS (2006), hep-ex/0607061
27. COMPASS collaboration, G. Brona, Measurement of the gluon polarisation at COMPASS (2007), hep-ex/0705.2372
28. A.D. Watson, Z. Phys. **C12**, 123 (1982)
29. M. Glück, E. Reya, W. Vogelsang, Nucl. Phys. **B351**, 579 (1991)
30. W. Vogelsang, Z. Phys. **C50**, 275 (1991)
31. J. Blümlein, V. Ravindran, W.L. van Neerven, Phys. Rev. **D68**, 114004 (2003), hep-ph/0304292
32. G. Altarelli, R.D. Ball, S. Forte, G. Ridolfi, Acta Phys. Polon. **B29**, 1145 (1998), hep-ph/9803237
33. M. Glück, E. Reya, M. Stratmann, W. Vogelsang, Phys. Rev. **D63**, 094005 (2001), hep-ph/0011215
34. J. Blümlein, H. Böttcher, Nucl. Phys. **B636**, 225 (2002), hep-ph/0203155
35. M. Hirai, S. Kumano, N. Saito, Phys. Rev. **D74**, 014015 (2006), hepph/ 0603213
36. E. Leader, A.V. Sidorov, D.B. Stamenov, Phys. Rev. **D75**, 074027 (2007), hep-ph/0612360
37. D. de Florian, R. Sassot, M. Stratmann, W. Vogelsang, Phys. Rev. **D80**, 034030 (2009), hep-ph/0904.3821
38. I. Bojak, M. Stratmann, Nucl. Phys. **B540**, 345 (1999), hep-ph/9807405
39. M. Buza, Y. Matiounine, J. Smith, W.L. van Neerven, Nucl. Phys. **B485**, 420 (1997), hep-ph/9608342
40. S. Wandzura, F. Wilczek, Phys. Lett. **B72**, 195 (1977)
41. J. Blümlein, N. Kochelev, Nucl. Phys. **B498**, 285 (1997), hep-ph/9612318
42. J.D. Jackson, G.G. Ross, R.G. Roberts, Phys. Lett. **B226**, 159 (1989)
43. R.G. Roberts, G.G. Ross, Phys. Lett. **B373**, 235 (1996), hep-ph/9601235
44. J. Blümlein, N. Kochelev, Phys. Lett. **B381**, 296 (1996), hep-ph/9603397
45. E.B. Zijlstra, W.L. van Neerven, Nucl. Phys. **B417**, 61 (1994); [Erratum-ibid.] **B426**, 245 (1994); [Erratum-ibid.] **B773**, 105 (2007)
46. E. Laenen, S. Riemersma, J. Smith, W.L. van Neerven, Nucl. Phys. **B392**, 162 (1993)
47. E. Laenen, S. Riemersma, J. Smith, W.L. van Neerven, Nucl. Phys. **B392**, 229 (1993)
48. S. Riemersma, J. Smith, W.L. van Neerven, Phys. Lett. **B347**, 143 (1995), hep-ph/9411431
49. M. Buza, Y. Matiounine, J. Smith, R. Migneron, W.L. van Neerven, Nucl. Phys. **B472**, 611 (1996), hep-ph/9601302

50. I. Bierenbaum, J. Blümlein, S. Klein, Nucl. Phys. **B780**, 40 (2007), hepph/ 0703285
51. I. Bierenbaum, J. Blümlein, S. Klein, Acta Phys. Polon. **B38**, 3543 (2007)
52. I. Bierenbaum, J. Blümlein, S. Klein, Pos RADCOR 2007, 034 (2007), hep-ph/0710.3348
53. I. Bierenbaum, J. Blümlein, S. Klein, Acta Phys. Polon. **B39**, 1531 (2008), hep-ph/0806.0451
54. I. Bierenbaum, J. Blümlein, S. Klein, Phys. Lett. **B648**, 195 (2007), hepph/ 0702265
55. I. Bierenbaum, J. Blümlein, S. Klein, Nucl. Phys. Proc. Suppl. **160**, 85 (2006), hep-ph/0607300
56. G. Altarelli, G. Parisi, Nucl. Phys. **B126**, 298 (1977)
57. H. Ito, Prog. Theor. Phys. **54**, 555 (1975)
58. K. Sasaki, Prog. Theor. Phys. **54**, 1816 (1975)
59. M.A. Ahmed, G.G. Ross, Nucl. Phys. **B111**, 441 (1976)
60. J.C. Ward, Phys. Rev. **78**, 182 (1950)
61. Y. Takahashi, Nuovo Cim. **6**, 371 (1957)
62. G. 't Hooft, M.J.G. Veltman, Nucl. Phys. **B44**, 189 (1972)
63. D.A. Akyeampong, R. Delbourgo, Nuovo Cim. **A17**, 578 (1973)
64. D.A. Akyeampong, R. Delbourgo, Nuovo Cim. **A18**, 94 (1973)
65. D.A. Akyeampong, R. Delbourgo, Nuovo Cim. **A19**, 219 (1974)
66. P. Breitenlohner, D. Maison, Commun. Math. Phys. **52**, 55 (1977)
67. C. Itzykson, J. Zuber, *Quantum Field Theory*, (McGraw-Hill, New York, 1980) pp. 705.
68. G.T. Bodwin, J.-W. Qiu, Phys. Rev. **D41**, 2755 (1990)
69. R. Mertig, W.L. van Neerven, Z. Phys. **C70**, 637 (1996), hep-ph/9506451
70. W. Vogelsang, Phys. Rev. **D54**, 2023 (1996), hep-ph/9512218
71. W. Vogelsang, Nucl. Phys. **B475**, 47 (1996), hep-ph/9603366
72. M. Czakon, Comput. Phys. Commun. **175**, 559 (2006), hep-ph/0511200
73. I. Bierenbaum, J. Blümlein, S. Klein, in preparation
74. J. Blümlein, S. Klein, PoS ACAT 2007, 084 (2007), hep-ph/0706.2426
75. J. Blümlein, Nucl. Phys. Proc. Suppl. **135**, 225 (2004), hep-ph/0407044
76. J. Blümlein, Comput. Phys. Commun. **180**, 2218 (2009), hep-ph/0901.3106
77. J. Blümlein, *Structural Relations of Harmonic Sums and Mellin Transforms at Weight w=6.* Proceedings of the Workshop Motives, Quantum Field Theory, and Pseudodifferential Operators, June (2008), (Clay Institute, Boston University, Boston, 2009), math-ph/0901.0837
78. M. Dittmar et al., Parton distributions: summary report for the HERA - LHC workshop (2005), hep-ph/0511119
79. J. Blümlein, V. Ravindran, Nucl. Phys. **B716**, 128 (2005), hep-ph/0501178
80. J. Blümlein, V. Ravindran, Nucl. Phys. **B749**, 1 (2006), hep-ph/0604019

Chapter 9
Heavy Flavor Contributions to Transversity

The transversity distribution $\Delta_T f(x, Q^2)$ is one of the three possible quarkonic twist-2 parton distributions besides the unpolarized and the longitudinally polarized quark distribution, $f(x, Q^2)$ and $\Delta f(x, Q^2)$, respectively. Unlike the latter distribution functions, it cannot be measured in inclusive deeply inelastic scattering in case of massless partons since it is chirally odd. However, it can be extracted from semi-inclusive deep-inelastic scattering (SIDIS) studying isolated meson production [1–4], and in the Drell–Yan process, [4–9], off transversely polarized targets.[1] Measurements of the transversity distribution in different polarized hard scattering processes are currently performed or in preparation [11–16]. In the past, phenomenological models for the transversity distribution were developed based on bag-like models, chiral models, light-cone models, spectator models, and non-perturbative QCD calculations, cf. Chap. 8 of Ref. [10]. The main characteristics of the transversity distributions are that they vanish by some power law both at small and large values of Bjorken-x and exhibit a shifted bell-like shape. Recent attempts to extract the distribution out of data were made in Refs. [17, 18]. The moments of the transversity distribution can be measured in lattice simulations, which help to constrain it ab initio, where first results were given in Refs. [19–24] (D. Renner, private communication, 2009). From these investigations there is evidence, that the up-quark distribution is positive while the down-quark distribution is negative, with first moments between $\{0.85, \ldots, 1.0\}$ and $\{-0.20, \ldots, -0.24\}$, respectively. This is in qualitative agreement with phenomenological fits.

Some of the processes which have been proposed to measure transversity contain $k_\perp-$ and higher twist effects, cf. [10]. We will limit our considerations to the class of purely twist-2 contributions, for which the formalism to calculate the heavy flavor corrections is established, cf. Chap. 3. As for the unpolarized flavor non-singlet contributions, we apply the factorization relation of the heavy flavor Wilson coefficient (3.21) in the region $Q^2 \gg m^2$.

As physical processes one may consider the SIDIS process $lN \rightarrow l'h + X$ off transversely polarized targets in which the transverse momentum of the produced

[1] For a review see Ref. [10].

S. Klein, *Charm Production in Deep Inelastic Scattering*, Springer Theses,
DOI: 10.1007/978-3-642-23286-2_9, © Springer-Verlag Berlin Heidelberg 2012

final state hadron h is integrated. The differential scattering cross section in case of single photon exchange reads

$$\frac{d^3\sigma}{dxdydz} = \frac{4\pi\alpha^2}{xyQ^2} \sum_{i=q,\overline{q}} e_i^2 x \left\{ \frac{1}{2}[1+(1-y)^2] F_i(x, Q^2) \tilde{D}_i(z, Q^2) - (1-y)|\mathbf{S}_\perp||\mathbf{S}_{h\perp}| \right.$$
$$\left. \times \cos\left(\phi_S + \phi_{S_h}\right) \Delta_T F_i(x, Q^2) \Delta_T \tilde{D}_i(z, Q^2) \right\}. \tag{9.1}$$

Here, in addition to the Bjorken variables x and y, the fragmentation variable z occurs. $\mathbf{S}_\perp$ and $\mathbf{S}_{h\perp}$ are the transverse spin vectors of the incoming nucleon N and the measured hadron h. The angles ϕ_{S,S_h} are measured in the plane perpendicular to the $\gamma^* N$ (z-) axis between the x-axis and the respective vector. The transversity distribution can be obtained from Eq. 9.1 for a transversely polarized hadron h by measuring its polarization.

The functions $F_i, \tilde{D}_i, \Delta_T F_i, \Delta_T \tilde{D}_i$ are given by

$$F_i(x, Q^2) = C_i(x, Q^2) \otimes f_i(x, Q^2) \tag{9.2}$$

$$\tilde{D}_i(z, Q^2) = \tilde{C}_i(z, Q^2) \otimes D_i(z, Q^2) \tag{9.3}$$

$$\Delta_T F_i(x, Q^2) = \Delta_T C_i(x, Q^2) \otimes \Delta_T f_i(x, Q^2) \tag{9.4}$$

$$\Delta_T \tilde{D}_i(z, Q^2) = \Delta_T \tilde{C}_i(z, Q^2) \otimes \Delta_T D_i(z, Q^2). \tag{9.5}$$

Here, $D_i, \Delta_T D_i$ are the fragmentation functions and $\tilde{C}_i$, $\Delta_T C_i$, $\Delta_T \tilde{C}_i$ are the corresponding space- and time-like Wilson coefficients. The functions C_i are the Wilson coefficients as have been considered in the unpolarized case, cf. Chaps. 2 and 3. The Wilson coefficient for transversity, $\Delta_T C_i(x, Q^2)$, contains light- and heavy-flavor contributions, cf. Eq. (3.3),

$$\Delta_T C_i(x, Q^2) = \Delta_T C_i(x, Q^2) + \Delta_T H_i(x, Q^2). \tag{9.6}$$

$\Delta_T C_i$ denotes the light flavor transversity Wilson coefficient and $\Delta_T H_i(x, Q^2)$ the heavy flavor part. We dropped arguments of the type n_f, m^2, μ^2 for brevity, since they can all be inferred from the discussion in Chap. 3.

Equation 9.1 holds for spin-1/2 hadrons in the final state, but the transversity distribution may also be measured in the leptoproduction process of spin-1 hadrons, [25]. In this case, the $\mathbf{P}_{h\perp}$-integrated Born cross section reads

$$\frac{d^3\sigma}{dxdydz} = \frac{4\pi\alpha^2}{xyQ^2} \sin\left(\phi_S + \phi_{S_{LT}}\right) |\mathbf{S}_\perp||S_{LT}|(1-y) \sum_{i=q,\overline{q}} e_i^2 x \Delta_T F_i(x, Q^2) \widehat{H}_{i,1,LT}(z, Q^2). \tag{9.7}$$

Here, the polarization state of a spin-1 particle is described by a tensor with five independent components, [26]. ϕ_{LT} denotes the azimuthal angle of $\vec{S}_{LT}$, with

$$|S_{LT}| = \sqrt{\left(S^x_{LT}\right)^2 + \left(S^y_{LT}\right)^2}. \tag{9.8}$$

$\widehat{H}_{a,1,LT}(z, Q^2)$ is a T- and chirally odd twist-2 fragmentation function at vanishing $k_\perp$. The process (9.7) has the advantage that the transverse polarization of the produced hadron can be measured from its decay products.

The transversity distribution can also be measured in the transversely polarized Drell–Yan process, see Refs. [27–29]. However, the SIDIS processes have the advantage that in high luminosity experiments the heavy flavor contributions can be tagged like in deep-inelastic scattering. This is not the case for the Drell–Yan process, where the heavy flavor effects appear as inclusive radiative corrections in the Wilson coefficients only.

The same argument as in Sect. 3.2 can be applied to obtain the heavy flavor Wilson coefficients for transversity in the asymptotic limit $Q^2 \gg m^2$. Since transversity is a NS quantity, the relation is the same as in the unpolarized NS case and reads up to $O(a_s^3)$, cf. Eq. (3.26),

$$\begin{aligned} \Delta_T H_q^{\text{Asym}}(n_f+1) = a_s^2 & \Big[\Delta_T A_{qq,Q}^{(2),\text{NS}}(n_f+1) + \Delta_T \hat{C}_q^{(2)}(n_f)\Big] \\ & + a_s^3 \Big[\Delta_T A_{qq,Q}^{(3),\text{NS}}(n_f+1) + \Delta_T A_{qq,Q}^{(2),\text{NS}}(n_f+1) \\ & \quad \times \Delta_T C_q^{(1)}(n_f+1) + \Delta_T \hat{C}_q^{(3)}(n_f)\Big]. \end{aligned} \tag{9.9}$$

The operator matrix elements $\Delta_T A_{qq,Q}^{(2,3),NS}$ are—as in the unpolarized case—universal and account for all mass contributions but power corrections. The respective asymptotic heavy flavor Wilson coefficients are obtained in combination with the light flavor process-dependent Wilson coefficients.[2] In the following, we will concentrate on the calculation of the massive operator matrix elements. The twist-2 local operator in case of transversity has a different Lorentz-structure compared to Eqs. 2.86–2.88 and is given by

$$O_{q,r;\mu,\mu_1,\ldots,\mu_N}^{\text{TR,NS}} = i^{N-1}\mathbf{S}[\overline{\psi}\sigma^{\mu\mu_1}D^{\mu_2},\ldots,D^{\mu_N}\frac{\lambda_r}{2}\psi] - \text{trace terms}, \tag{9.10}$$

with $\sigma^{\nu\mu} = (i/2)\left[\gamma_\nu\gamma_\mu - \gamma_\mu\gamma_\nu\right]$ and the definition of the massive operator matrix element is the same as in Sect. 3.2. Since (9.10) denotes a twist-2 flavor non-singlet operator, it does not mix with other operators. After multiplying with the external source J_N, cf. Eq. 4.10 and below, the Green's function in momentum space corresponding to the transversity operator between quarkonic states is given by

---

[2] Apparently, the light flavor Wilson coefficients for SIDIS were not yet calculated even at $O(a_s)$, although this calculation and the corresponding soft-exponentiation should be straightforward. For the transversely polarized Drell–Yan process the $O(a_s)$ light flavor Wilson coefficients were given in [28] and higher order terms due to soft resummation were derived in [29].

$$\overline{u}(p,s)G^{ij,\mathsf{TR,NS}}_{\mu,q,Q}\lambda_r u(p,s) = J_N\langle\overline{\Psi}_i(p)|O^{\mathsf{TR,NS}}_{q,r;\mu,\mu_1,\ldots,\mu_N}|\Psi^j(p)\rangle_Q. \tag{9.11}$$

It relates to the unrenormalized transversity OME via

$$\hat{G}^{ij,\mathsf{TR,NS}}_{\mu,q,Q} = \delta_{ij}(\Delta\cdot p)^{N-1}\left(\Delta_\rho\sigma^{\mu\rho}\Delta_T\hat{\hat{A}}^{\mathsf{NS}}_{qq,Q}\left(\frac{\hat{m}^2}{\mu^2},\varepsilon,N\right)\right.$$
$$\left.+c_1\Delta^\mu+c_2p^\mu+c_3\gamma^\mu\not{p}+c_4\not{\Delta}\not{p}\Delta^\mu+c_5\not{\Delta}\not{p}p^\mu\right). \tag{9.12}$$

The Feynman rules for the operators multiplied with the external source are given in Appendix B. The projection onto the massive OME is found to be

$$\Delta_T\hat{\hat{A}}^{\mathsf{NS}}_{qq,Q}\left(\frac{\hat{m}^2}{\mu^2},\varepsilon,N\right)$$
$$= -\frac{i\delta^{ij}}{4N_c(\Delta.p)^2(D-2)}\left\{\mathrm{Tr}\left[\not{\Delta}\not{p}p^\mu\hat{G}^{ij,\mathsf{TR,NS}}_{\mu,q,Q}\right]-\Delta.p\,\mathrm{Tr}\left[p^\mu\hat{G}^{ij,\mathsf{TR,NS}}_{\mu,q,Q}\right]\right.$$
$$\left.+i\Delta.p\,\mathrm{Tr}\left[\sigma_{\mu\rho}p^\rho\hat{G}^{ij,\mathsf{TR,NS}}_{\mu,q,Q}\right]\right\}. \tag{9.13}$$

Renormalization for transversity proceeds in the same manner as in the NS-case. The structure of the unrenormalized expressions at the 2- and 3-loop level are the same as shown in Eqs. 4.93, 4.94, if one inserts the respective transversity anomalous dimensions. The expansion coefficients of the renormalized OME then read up to $O(a_s^3)$ in the $\overline{\mathsf{MS}}$-scheme, cf. Eqs. (4.95, 4.96),

$$\Delta_T A^{(2),\mathsf{NS},\overline{\mathsf{MS}}}_{qq,Q} = \frac{\beta_{0,Q}\gamma^{(0),\mathsf{TR}}_{qq}}{4}\ln^2\left(\frac{m^2}{\mu^2}\right)+\frac{\hat{\gamma}^{(1),\mathsf{TR}}_{qq}}{2}\ln\left(\frac{m^2}{\mu^2}\right)+a^{(2),\mathsf{TR}}_{qq,Q}-\frac{\beta_{0,Q}\gamma^{(0),\mathsf{TR}}_{qq}}{4}\zeta_2, \tag{9.14}$$

$$\Delta_T A^{(3),\mathsf{NS},\overline{\mathsf{MS}}}_{qq,Q} = -\frac{\gamma^{(0),\mathsf{TR}}_{qq}\beta_{0,Q}}{6}\left(\beta_0+2\beta_{0,Q}\right)\ln^3\left(\frac{m^2}{\mu^2}\right)$$
$$+\frac{1}{4}\left\{2\gamma^{(1),\mathsf{TR}}_{qq}\beta_{0,Q}-2\hat{\gamma}^{(1),\mathsf{TR}}_{qq}\left(\beta_0+\beta_{0,Q}\right)+\beta_{1,Q}\gamma^{(0),\mathsf{TR}}_{qq}\right\}$$
$$\times\ln^2\left(\frac{m^2}{\mu^2}\right)+\frac{1}{2}\left\{\hat{\gamma}^{(2),\mathsf{TR}}_{qq}-\left(4a^{(2),\mathsf{TR}}_{qq,Q}-\zeta_2\beta_{0,Q}\gamma^{(0),\mathsf{TR}}_{qq}\right)\right.$$
$$\left.\times\left(\beta_0+\beta_{0,Q}\right)+\gamma^{(0),\mathsf{TR}}_{qq}\beta^{(1)}_{1,Q}\right\}$$

$$
\times \ln\left(\frac{m^2}{\mu^2}\right) + 4\overline{a}_{qq,Q}^{(2),\mathsf{TR}}(\beta_0 + \beta_{0,Q}) - \gamma_{qq}^{(0)}\beta_{1,Q}^{(2)}
- \frac{\gamma_{qq}^{(0),\mathsf{TR}}\beta_0\beta_{0,Q}\zeta_3}{6} - \frac{\gamma_{qq}^{(1),\mathsf{TR}}\beta_{0,Q}\zeta_2}{4}
+ 2\delta m_1^{(1)}\beta_{0,Q}\gamma_{qq}^{(0),\mathsf{TR}} + \delta m_1^{(0)}\hat{\gamma}_{qq}^{(1),\mathsf{TR}} + 2\delta m_1^{(-1)} a_{qq,Q}^{(2),\mathsf{TR}} + a_{qq,Q}^{(3),\mathsf{TR}}. \tag{9.15}
$$

Here, $\gamma_{qq}^{(k),\mathsf{TR}}$, $\{k = 0, 1, 2\}$, denote the transversity quark anomalous dimensions at $O(a_s^{k+1})$ and $a_{qq,Q}^{(2,3),\mathsf{TR}}$, $\overline{a}_{qq,Q}^{(2),\mathsf{TR}}$ are the constant and $O(\varepsilon)$ terms of the massive operator matrix element at 2- and 3-loop order, respectively, cf. the discussion in Chap. 4. At LO the transversity anomalous dimension was calculated in [5, 30–34],[3] and at NLO in [28, 36, 37].[4] At three-loop order the moments $N = 1, \ldots, 8$ are known, [41–44].

The 2-loop calculation for all N proceeds in the same way as described in Chap. 6. We also calculated the unprojected Green's function to check the projector (9.13). Fixed moments at the 2- and 3-loop level were calculated using MATAD as described in Chap. 7.

From the pole terms of the unrenormalized 2-loop OMEs, the leading and next-to-leading order anomalous dimensions $\gamma_{qq}^{(0),\mathsf{TR}}$ and $\hat{\gamma}_{qq}^{(1),\mathsf{TR}}$ can be determined. We obtain

$$
\gamma_{qq}^{(0),\mathsf{TR}} = 2C_F\left[-3 + 4S_1\right], \tag{9.16}
$$

and

$$
\hat{\gamma}_{qq}^{(1),\mathsf{TR}} = \frac{32}{9}C_F T_F\left[3S_2 - 5S_1 + \frac{3}{8}\right], \tag{9.17}
$$

confirming earlier results [28, 36, 37]. The finite and $O(\varepsilon)$ contributions are given by

$$
a_{qq,Q}^{(2),\mathsf{TR}} = C_F T_F\left\{-\frac{8}{3}S_3 + \frac{40}{9}S_2 - \left[\frac{224}{27} + \frac{8}{3}\zeta_2\right]S_1 + 2\zeta_2 + \frac{\left(24 + 73N + 73N^2\right)}{18N\,(N+1)}\right\}, \tag{9.18}
$$

3 The small x limit of the LO anomalous dimension was calculated in [35].

4 For calculations in the non-forward case, see [34, 38–40].

$$\overline{a}_{qq,Q}^{(2),\mathrm{TR}} = C_F T_F \Biggl\{ -\left[\frac{656}{81} + \frac{20}{9}\zeta_2 + \frac{8}{9}\zeta_3\right] S_1 + \left[\frac{112}{27} + \frac{4}{3}\zeta_2\right] S_2 - \frac{20}{9} S_3 \quad + \frac{4}{3} S_4 + \frac{1}{6}\zeta_2 + \frac{2}{3}\zeta_3 + \frac{\left(-144 - 48N + 757N^2 + 1034N^3 + 517N^4\right)}{216N^2 (N+1)^2} \Biggr\}. \tag{9.19}$$

The renormalized 2-loop massive OME (9.14) reads

$$\Delta_T A_{qq,Q}^{(2),\mathrm{NS},\overline{\mathrm{MS}}} = C_F T_F \Biggl\{ \left[-\frac{8}{3} S_1 + 2\right] \ln^2\left(\frac{m^2}{\mu^2}\right) + \left[-\frac{80}{9} S_1 + \frac{2}{3} + \frac{16}{3} S_2\right] \ln\left(\frac{m^2}{\mu^2}\right) - \frac{8}{3} S_3 + \frac{40}{9} S_2 - \frac{224}{27} S_1 + \frac{24 + 73N + 73N^2}{18N(N+1)} \Biggr\}. \tag{9.20}$$

Using MATAD, we calculated the moments $N = 1, \ldots, 13$ at $O(a_s^2)$ and $O(a_s^3)$. At the 2-loop level, we find complete agreement with the results presented in Eqs. 9.16–9.19. At $O(a_s^3)$, we also obtain $\hat{\gamma}_{qq}^{(2),\mathrm{TR}}$, which can be compared to the T_F-terms in the calculation [41–44] for $N = 1, \ldots, 8$. This is the first re-calculation of these terms and we find agreement. For the moments $N = 9, \ldots, 13$ this contribution to the transversity anomalous dimension is calculated for the first time. We list the anomalous dimensions in Appendix G. There, also the constant contributions $a_{qq,Q}^{(3),\mathrm{TR}}$ are given for $N = 1, \ldots, 13$, which is a new result. Furthermore, we obtain in the 3-loop calculation the moments $N = 1, \ldots, 13$ of the complete 2-loop anomalous dimensions. These are in accordance with Refs. [28, 36, 37].

Finally, we show as examples the first moments of the $\overline{\mathrm{MS}}$–renormalized $O(a_s^3)$ massive transversity OME. Unlike the case for the vector current, the first moment does not vanish, since there is no conservation law to enforce this.

$$\Delta_T A_{qq,Q}^{(3),\mathrm{NS},\overline{\mathrm{MS}}}(1) = C_F T_F \Biggl\{ \left(\frac{44}{27} C_A - \frac{16}{27} T_F (n_f + 2)\right) \ln^3\left(\frac{m^2}{\mu^2}\right) + \left(\frac{32}{3} C_F - \frac{106}{9} C_A - \frac{104}{27} T_F\right) \ln^2\left(\frac{m^2}{\mu^2}\right) + \left[\left(\frac{233}{9} + 16\zeta_3\right) C_F + \left(-\frac{2233}{81} - 16\zeta_3\right) C_A - \frac{604}{81} n_f T_F - \frac{496}{81} T_F\right] \ln\left(\frac{m^2}{\mu^2}\right) + \left(-\frac{16}{3} B_4 + 24\zeta_4 - \frac{278}{9}\zeta_3 + \frac{7511}{81}\right) \times C_F + \left(\frac{8}{3} B_4 - 24\zeta_4 + \frac{437}{27}\zeta_3 - \frac{34135}{729}\right) C_A + \left(-\frac{6556}{729} + \frac{128}{27}\zeta_3\right) \times T_F n_f + \left(\frac{2746}{729} - \frac{224}{27}\zeta_3\right) T_F \Biggr\}, \tag{9.21}$$

$$\Delta_T A_{qq,Q}^{(3),\mathsf{NS},\overline{\mathsf{MS}}}(2) = C_F T_F \left\{ \left(\frac{44}{9} C_A - \frac{16}{9} T_F (n_f + 2) \right) \ln^3 \left(\frac{m^2}{\mu^2} \right) + \left(-\frac{34}{3} C_A 8 T_F \right) \right.$$
$$-\ln^2 \left(\frac{m^2}{\mu^2} \right) + \left[(15 + 48\zeta_3) C_F + \left(-\frac{73}{9} - 48\zeta_3 \right) C_A - \frac{196}{9} n_f T_F - \frac{496}{27} T_F \right]$$
$$\times \ln \left(\frac{m^2}{\mu^2} \right) + \left(-16 B_4 + 72\zeta_4 - \frac{310}{3}\zeta_3 + \frac{4133}{27} \right) C_F + \left(8 B_4 - 72\zeta_4 \right.$$
$$\left. \left. + \frac{533}{9}\zeta_3 - 56 \right) C_A + \left(-\frac{1988}{81} + \frac{128}{9}\zeta_3 \right) T_F n_f + \left(\frac{338}{27} - \frac{224}{9}\zeta_3 \right) T_F \right\}. \tag{9.22}$$

The structure of the result and the contributing numbers are the same as in the unpolarized case, cf. Eq. 7.41. We checked the moments $N = 1, \ldots, 4$ keeping the complete dependence on the gauge-parameter ξ and find that it cancels in the final result. Again, we observe that the massive OMEs do not depend on ζ_2, cf. Sect. 7.3.

Since the light flavor Wilson coefficients for the processes from which the transversity distribution can be extracted are not known to 2- and 3-loop order, phenomenological studies on the effect of the heavy flavor contributions cannot yet be performed. However, the results of this Section can be used in comparisons with upcoming lattice simulations with (2+1+1)-dynamical fermions including the charm quark. More details on this calculation are given in [45].

References

1. J.P. Ralston, D.E. Soper, Nucl. Phys. **B152**, 109 (1979)
2. R.L. Jaffe, X.-D. Ji, Phys. Rev. Lett. **67**, 552 (1991)
3. R.L. Jaffe, X.-D. Ji, Nucl. Phys. **B375**, 527 (1992)
4. J.L. Cortes, B. Pire, J.P. Ralston, Z. Phys. **C55**, 409 (1992)
5. X. Artru, M. Mekhfi, Z. Phys. **C45**, 669 (1990)
6. J.C. Collins, Nucl. Phys. **B396**, 161 (1993) hep-ph/9208213
7. J.C. Collins, Phys. Rev. Lett. **71**, 2547 (1993) hep-ph/9307329
8. R.D. Tangerman, P.J. Mulders, Polarized twist-three distributions g_T and h_L and the role of intrinsic transverse momentum, (1994) hep-ph/9408305.
9. D. Boer, P.J. Mulders, Phys. Rev. **D57**, 5780 (1998) hep-ph/9711485
10. V. Barone, A. Drago, P.G. Ratcliffe, Phys. Rept. **359**, 1 (2002) hep-ph/0104283
11. HERMES Collaboration, A. Airapetian et al., Phys. Rev. Lett. **94**, 012002 (2005) hep-ex/0408013
12. HERMES Collaboration, A. Airapetian et al., JHEP **06**, 017 (2008), hep-ex/0803.2367
13. COMPASS Collaboration, V.Y. Alexakhin et al., Phys. Rev. Lett. **94**, 202002 (2005) hep-ex/0503002
14. A. Afanasev et al., Transversity and transverse spin in nucleon structure through SIDIS at Jefferson Lab (2007) hep-ph/0703288
15. COMPASS collaboration, M. Alekseev et al., Phys. Lett. **B673**, 127 (2009) hep-ex/0802.2160
16. The PANDA collaboration, M.F.M. Lutz, B. Pire, O. Scholten, R. Timmermans, (2009). hep-ex/0903.3905
17. M. Anselmino et al., Phys. Rev. D **75**, 054032 (2007) hep-ph/0701006

18. M. Anselmino et al., Nucl. Phys. Proc. Suppl. **191**, 98 (2009) hep-ph/0812.4366
19. LHPC Collaboration, D. Dolgov et al., Phys. Rev.**D66**, 034506 (2002) hep-lat/0201021
20. S. Aoki, M. Doui, T. Hatsuda, Y. Kuramashi, Phys. Rev. D **56**, 433 (1997) hep-lat/9608115
21. M. Göckeler et al., Nucl. Phys. Proc. Suppl. **53**, 315 (1997) hep-lat/9609039
22. A.A. Khan et al., Nucl. Phys. Proc. Suppl. **140**, 408 (2005) hep-lat/0409161
23. QCDSF Collaboration, M. Diehl et al., (2005) hep-ph/0511032
24. QCDSF Collaboration, M. Göckeler et al., Phys. Rev. Lett. **98**, 222001 (2007) hep-lat/0612032
25. X.-D. Ji, Phys. Rev. **D49**, 114 (1994) hep-ph/9307235
26. A.. Bacchetta, P.J. Mulders, Phys. Rev. **D62**, 114004 (2000) hep-ph/0007120
27. W. Vogelsang, A. Weber, Phys. Rev. D **48**, 2073 (1993)
28. W. Vogelsang, Phys. Rev. **D57**, 1886 (1998) hep-ph/9706511
29. H. Shimizu, G. Shimizu, W. Vogelsang, H. Yokoya, Phys. Rev. **D71**, 114007 (2005) hep-ph/0503270
30. F. Baldracchini, N.S. Craigie, V. Roberto, M. Socolovsky, Fortschr. Phys. **30**, 505 (1981)
31. M.A. Shifman, M.I. Vysotsky, Nucl. Phys. **B186**, 475 (1981)
32. A.P. Bukhvostov, G.V. Frolov, L.N. Lipatov, E.A. Kuraev, Nucl. Phys. **B258**, 601 (1985)
33. A. Mukherjee, D. Chakrabarti, Phys. Lett. **B506**, 283 (2001) hep-ph/0102003
34. J. Blümlein, Eur. Phys. J. **C20**, 683 (2001) hep-ph/0104099
35. R. Kirschner, L. Mankiewicz, A. Schafer, L. Szymanowski, Z. Phys. **C74**, 501 (1997) hep-ph/9606267
36. A. Hayashigaki, Y. Kanazawa, Y. Koike, Phys. Rev. **D56**, 7350 (1997) hep-ph/9707208
37. S. Kumano, M. Miyama, Phys. Rev. **D56**, 2504 (1997) hep-ph/9706420
38. A.V. Belitsky, D. Müller, Phys. Lett. **B417**, 129 (1998) hep-ph/9709379
39. P. Hoodbhoy, X.-D. Ji, Phys. Rev. **D58**, 054006 (1998) hep-ph/9709379
40. A.V. Belitsky, A. Freund, D. Müller, Phys. Lett. **B493**, 341 (2000) hep-ph/0008005
41. J.A. Gracey, Nucl. Phys. **B662**, 247 (2003) hep-ph/0304113
42. J.A. Gracey, Nucl. Phys. **B667**, 242 (2003) hep-ph/0306163
43. J.A. Gracey, JHEP **10**, 040 (2006) hep-ph/0609231
44. J.A. Gracey, Phys. Lett. **B643**, 374 (2006) hep-ph/0611071
45. J. Blümlein, S. Klein, B. Tödtli, Phys. Lett. **D80**, 094010 (2009) hep-ph/0909.1547

Chapter 10
First Steps Towards a Calculation of $A_{ij}^{(3)}$ for all Moments

In Chap. 7, we described how the various massive OMEs are calculated for fixed integer values of the Mellin variable N at 3-loop order using MATAD. The ultimate goal is to calculate these quantities for general values of N. So far no massive single scale calculation at $O(a_s^3)$ has been performed. In the following we would like to present some first results and a general method, which may be of use in later work calculating the general N-dependence of the massive OMEs $A_{ij}^{(3)}$.

In Sect. 10.1, we solve, as an example, a 3-loop ladder graph contributing to $A_{Qg}^{(3)}$ for general values of N by direct integration, avoiding the integration-by-parts method. In Sect. 10.2, Refs. [1, 2], we discuss a general algorithm which allows to determine from a sufficiently large but finite number of moments for a recurrent quantity its general N-dependence. This algorithm has been successfully applied in [2] to reconstruct the 3-loop anomalous dimensions, [3, 4], and massless 3-loop Wilson coefficients, [5], from their moments. These are the largest single scale quantities known at the moment and are well suited to demonstrate the power of this formalism. Similarly, one may apply this method to new problems of smaller size which emerge in course of the calculation of the OMEs $A_{ij}^{(3)}$ for general values of N.

10.1 Results for all-N Using Generalized Hypergeometric Functions

In Sect. 6.1, we showed that there is only one basic 2-loop massive tadpole which needs to be considered. From it, all diagrams contributing to the massive 2-loop OMEs can be derived by attaching external quark-, gluon- and ghost-lines, respectively, and including one operator insertion according to the Feynman rules given in Appendix B. The corresponding parameter-integrals are then all of the same structure, Eq. 6.5. If one knows a method to calculate the basic topology for arbitrary integer powers of the propagators, the calculation of the 2-loop OMEs is straightforward for fixed values of N. In the general case, we arrived at infinite sums containing the

DOI: 10.1007/978-3-642-23286-2_10,

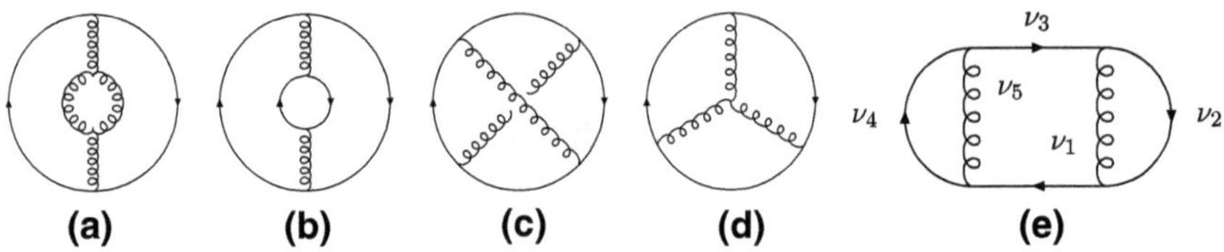

Fig. 10.1 Basic 3-loop topologies. Straight lines: quarks, curly lines: gluons. The gluon loop in **a** can also be replaced by a ghost loop; **e** 3-loop ladder graph

parameter N. To calculate these sums, additional tools are needed, e.g. the program Sigma, cf. Sect. 6.2.

We would like to follow the same approach in the 3-loop case. Here, five basic topologies need to considered, which are shown in Fig. 10.1. Diagram (a) and (b)—if one of the quark loops corresponds to a massless quark—can be reduced to 2-loop integrals, because the massless loop can be integrated trivially. For the remaining terms, this is not the case. Diagrams (c) and (d) are the most complex topologies, the former giving rise to the number B_4, Eq. 4.89, whereas the latter yields single ζ-values up to weight 4, cf. e.g. [6]. Diagram (b)—if both quarks are massive—and (e) are ladder topologies and of less complexity. Let us, as an example, consider diagram (e).

Our notation is the same as in Sect. 6.1. The scalar D-dimensional integral corresponding to diagram (e) reads for arbitrary exponents of the propagators

$$T_e = \iiint \frac{d^Dq d^Dk d^Dl}{(2\pi)^{3D}} \frac{i(-1)^{\nu_{12345}}(m^2)^{\nu_{12345}-3D/2}(4\pi)^{3D/2}}{(k^2)^{\nu_1}((k-l)^2-m^2)^{\nu_2}(l^2-m^2)^{\nu_3}((q-l)^2-m^2)^{\nu_4}(q^2)^{\nu_5}}. \tag{10.1}$$

Again, we have attached suitable normalization factors for convenience. After loop-by-loop integration of the momenta $k,\ q,\ l$ (in this order) using Feynman-parameters, one obtains after a few steps the following parameter integral

$$\begin{aligned} T_e &= \Gamma\left[\begin{matrix}\nu_{12345}-6-3\varepsilon/2\\ \nu_1,\nu_2,\nu_3,\nu_4,\nu_5\end{matrix}\right] \int_0^1 dw_1 \ldots \int_0^1 dw_4 \theta(1-w_1-w_2) \\ &\times \frac{w_1^{-3-\varepsilon/2+\nu_{12}} w_2^{-3-\varepsilon/2+\nu_{45}}(1-w_1-w_2)^{\nu_3-1}}{(1+w_1\frac{w_3}{1-w_3}+w_2\frac{w_4}{1-w_4})^{\nu_{12345}-6-3\varepsilon/2}} \\ &\times w_3^{1+\varepsilon/2-\nu_1}(1-w_3)^{1+\varepsilon/2-\nu_2} w_4^{1+\varepsilon/2-\nu_5}(1-w_4)^{1+\varepsilon/2-\nu_4}. \end{aligned} \tag{10.2}$$

The θ-function enforces $w_1+w_2 \le 1$. In order to perform the $\{w_1, w_2\}$ integration, one considers

$$I = \int_0^1 dw_1 \int_0^1 dw_2 \theta(1-w_1-w_2) w_1^{b-1} w_2^{b'-1}(1-w_1-w_2)^{c-b-b'-1}(1-w_1x-w_2y)^{-a}. \tag{10.3}$$

The parameters a, b, b', c shall be such that this integral is convergent. It can be expressed in terms of the Appell function F_1 via, [7] [1]

$$I = \Gamma\left[\begin{matrix} b, b', c-b-b' \\ c \end{matrix}\right] \sum_{m,n=0}^{\infty} \frac{(a)_{m+n}(b)_n(b')_m}{(1)_m(1)_n(c)_m+n} x^n y^m \tag{10.4}$$

$$= \Gamma\left[\begin{matrix} b, b', c-b-b' \\ c \end{matrix}\right] F_1\left[a; b, b'; c; x, y\right]. \tag{10.5}$$

The parameters x, y correspond to $w_3/(1-w_3)$ and $w_4/(1-w_4)$ in Eq. 10.2, respectively. Hence the integral over these variables would yield a divergent sum. Therefore one uses the following analytic continuation relation for F_1, [7],

$$F_1\left[a; b, b'; c; \frac{x}{x-1}, \frac{y}{y-1}\right] = (1-x)^b(1-y)^{b'} F_1[c-a; b, b'; c; x, y]. \tag{10.6}$$

Finally one arrives at an infinite double sum

$$\begin{aligned} T_e = \Gamma&\left[\begin{matrix} -2-\varepsilon/2+\nu_{12}, -2-\varepsilon/2+\nu_{45}, -6-3\varepsilon/2+\nu_{12345} \\ \nu_2, \nu_4, -4-\varepsilon+\nu_{12345} \end{matrix}\right] \\ &\times \sum_{m,n=0}^{\infty} \Gamma\left[\begin{matrix} 2+m+\varepsilon/2-\nu_1, 2+n+\varepsilon/2-\nu_5 \\ 1+m, 1+n, 2+m+\varepsilon/2, 2+n+\varepsilon/2 \end{matrix}\right] \\ &\times \frac{(2+\varepsilon/2)_{n+m}(-2-\varepsilon/2+\nu_{12})_m(-2-\varepsilon/2+\nu_{45})_n}{(-4-\varepsilon+\nu_{12345})_{n+m}}. \end{aligned} \tag{10.7}$$

Here, we have adopted the notation for the Γ-function defined in (C.8) and $(a)_b$ is Pochhammer's symbol, Eq. C.14. As one expects, Eq. 10.7 is symmetric w.r.t. exchanges of the indices $\{\nu_1, \nu_2\} \leftrightarrow \{\nu_4, \nu_5\}$. For any values of ν_i of the type $\nu_i = a_i + b_i\varepsilon$, with $a_i \in \mathbb{N}, b_i \in \mathbb{C}$, the Laurent-series in ε of Eq. 10.7 can be calculated using e.g. Summer, [8]. We have checked Eq. 10.7 for various values of the ν_i using MATAD, cf. Sect. 7.2.

Next, let us consider the diagram shown in Fig. 10.2, which contributes to $A_{Qg}^{(3)}$ and derives from diagram (e). We treat the case where all exponents of the propagators are equal to one.

Including the factor $i(m^2)^{2-3\varepsilon/2}(4\pi)^{3D/2}$ and integrating q, k, l (in this order), we obtain

$$\begin{aligned} I_{ex} = &\Gamma(2-3\varepsilon/2) \int_0^1 dw_i \theta(1-w_1-w_2) \frac{w_1^{-\varepsilon/2} w_2^{-\varepsilon/2}(1-w_1-w_2)}{(1+w_1\frac{1-w_3}{w_3}+w_2\frac{1-w_4}{w_4})^{2-3\varepsilon/2}} \\ &\times w_3^{-1+\varepsilon/2}(1-w_3)^{\varepsilon/2} w_4^{-1+\varepsilon/2}(1-w_4)^{\varepsilon/2} \\ &\times (1-w_5w_1-w_6w_2-(1-w_1-w_2)w_7)^N, \end{aligned} \tag{10.8}$$

[1] Note that Eq. 8.2.2 of Ref. [7] contains typos.,

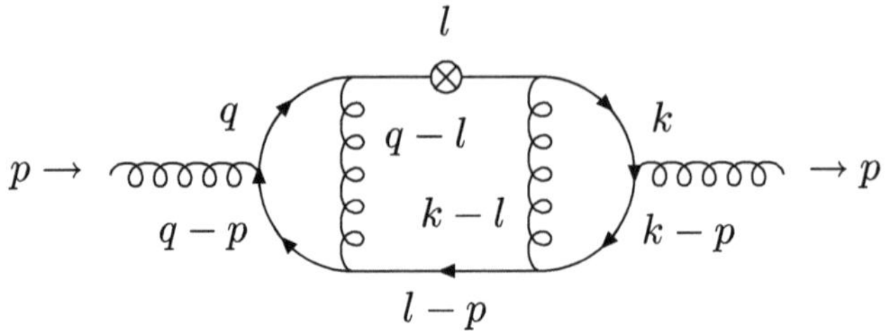

Fig. 10.2 Example 3-loop graph

where all parameters $w_1 \dots w_7$ have to be integrated from $0 \dots 1$. As in the 2-loop case, (6.5), one observes that the integral-kernel given by the corresponding massive tadpole integral (10.2) is multiplied with a polynomial containing various integration parameters to the power N. The same holds true for the remaining 3-loop diagrams. Hence, if a general sum representation for the corresponding tadpoles integrals is known and one knows how to evaluate the corresponding sums, at least fixed moments of the 3-loop massive OMEs can be calculated right away. The presence of the polynomial to the power N (which may also involve a finite sum, cf. the Feynman-rules in Appendix B,) complicates the calculation further. One has to find a suitable way to deal with this situation, which depends on the integral considered. For I_{ex}, we split it up into several finite sums, rendering the integrals calculable in the same way as for T_e. We obtain

$$\begin{aligned}
I_{ex} &= \frac{-\Gamma(2-3\varepsilon/2)}{(N+1)(N+2)(N+3)} \\
&\times \sum_{m,n=0}^{\infty} \Bigg\{ \sum_{t=1}^{N+2} \binom{3+N}{t} \frac{(t-\varepsilon/2)_m (2+N+\varepsilon/2)_{n+m} (3-t+N-\varepsilon/2)_n}{(4+N-\varepsilon)_{n+m}} \\
&\qquad \times \Gamma\left[\begin{matrix} t, t-\varepsilon/2, 1+m+\varepsilon/2, 1+n+\varepsilon/2, 3-t+N, 3-t+N-\varepsilon/2 \\ 4+N-\varepsilon, 1+m, 1+n, 1+t+m+\varepsilon/2, 4-t+n+N+\varepsilon/2 \end{matrix}\right] \\
&\qquad - \sum_{s=1}^{N+3} \sum_{r=1}^{s-1} \binom{s}{r} \binom{3+N}{s} (-1)^s \frac{(r-\varepsilon/2)_m (-1+s+\varepsilon/2)_{n+m} (s-r-\varepsilon/2)_n}{(1+s-\varepsilon)_n + m} \\
&\qquad \times \Gamma\left[\begin{matrix} r, r-\varepsilon/2, s-r, 1+m+\varepsilon/2, 1+n+\varepsilon/2, s-r-\varepsilon/2 \\ 1+m, 1+n, 1+r+m+\varepsilon/2, 1+s-r+n+\varepsilon/2, 1+s-\varepsilon \end{matrix}\right] \Bigg\}.
\end{aligned} \tag{10.9}$$

After expanding in ε, the summation can be performed using Sigma and the summation techniques explained in Sect. 6.2. The result reads

$$\begin{aligned}
I_{ex} &= -\frac{4(N+1)S_1+4}{(N+1)^2(N+2)}\zeta_3 + 2S_{2,1,1}(N+2)(N+3) + \frac{1}{(N+1)(N+2)(N+3)} \\
&\times \Bigg\{ -2(3N+5)S_{3,1} - \frac{{S_1}^4}{4} + \frac{4(N+1)S_1 - 4N}{N+1} S_{2,1} + 2\left((2N+3)S_1 + \frac{5N+6}{N+1}\right) S_3
\end{aligned}$$

$$
+\frac{9+4N}{4}S_2{}^2+\left(2\frac{7N+11}{(N+1)(N+2)}+\frac{5N}{N+1}S_1-\frac{5}{2}S_1{}^2\right)S_2+\frac{2(3N+5)S_1{}^2}{(N+1)(N+2)}
+\frac{N}{N+1}S_1{}^3+\frac{4(2N+3)S_1}{(N+1)^2(N+2)}-\frac{(2N+3)S_4}{2}+8\frac{2N+3}{(N+1)^3(N+2)}\Bigg\}+O(\varepsilon), \quad (10.10)
$$

which agrees with the fixed moments $N = 1, \ldots, 10$ obtained using MATAD, cf. Sect. 7.2.

We have shown that in principle one can be apply similar techniques as on the 2-loop level, Sect. 6.1, to calculate the massive 3-loop OMEs considering only the five basic topologies. In this approach the integration-by-parts method is not used. We have given the necessary formulas for one non-trivial topology (e) and showed for one of the cases there how the calculation proceeds keeping the all-N dependence. In order to obtain complete results for the massive OMEs, suitable integral representations for diagrams (b), (c) and (d) of Fig. 10.1 have to be derived first. This will allow for a calculation of fixed moments not relying on MATAD. Next, an automatization of the step from (10.8) to (10.9) has to be found in order to obtain sums which can be handled e.g. by Sigma. The latter step is not trivial, since it depends on the respective diagram and the flow of the outer momentum p through it.

10.2 Reconstructing General-N Relations from a Finite Number of Mellin-Moments

Higher order calculations in Quantum Field Theories easily become tedious due to the large number of terms emerging and the sophisticated form of the contributing Feynman parameter integrals. This applies already to zero scale and single scale quantities. Even more this is the case for problems containing at least two scales. While in the latter case the mathematical structure of the solution of the Feynman integrals is widely unknown, it is explored to a certain extent for zero- and single scale quantities. Zero scale quantities emerge as the expansion coefficients of the running couplings and masses, as fixed moments of splitting functions, etc.. They can be expressed by rational numbers and certain special numbers as multiple zeta-values (MZVs), [9, 10] and related quantities.

Single scale quantities depend on a scale z which may be given as a ratio of Lorentz invariants s'/s in the respective physical problem. One may perform a Mellin transform over z, Eq. 2.65. All subsequent calculations are then carried out in Mellin space and one assumes $N \in \mathbb{N}$, $N > 0$. By this transformation, the problem at hand becomes discrete and one may seek a description in terms of difference equations, [11, 12]. Zero scale problems are obtained from single scale problems treating N as a fixed integer or considering the limit $N \to \infty$.

A main question concerning zero scale quantities is: do the corresponding Feynman integrals always lead to MZVs? In the lower orders this is the case.

However, starting at some order, even for single-mass problems, other special numbers will occur, [13–15]. Since one has to known the respective basis completely, this makes it difficult to use methods like PSLQ, [16], to determine the analytic structure of the corresponding terms even if one may calculate them numerically at high enough precision. Zero scale problems are much easier to calculate than single scale problems. In some analogy to the determination of the analytic structure in zero scale problems through integer relations over a known basis (PSLQ) one may think of an automated reconstruction of the all-N relation out of a finite number of Mellin moments given in analytic form. This is possible for recurrent quantities. At least up to 3-loop order, presumably even to higher orders, single scale quantities belong to this class. Here we report on a general algorithm for this purpose, which we applied to the problem being currently the most sophisticated one: the anomalous dimensions and massless Wilson coefficients to 3-loop order for unpolarized DIS, [3, 4, 5]. Details of our calculation are given in Refs. [1, 2].

10.2.1 Single Scale Feynman Integrals as Recurrent Quantities

For a large variety of massless problems single scale Feynman integrals can be represented as polynomials in the ring formed by the nested harmonic sums, cf. Appendix C.4, and the MZVs $\zeta_{a_1,\ldots,a_l}$, which we set equal to the σ-values defined in Eq. C.35. Rational functions in N and harmonic sums obey recurrence relations. Thus, due to closure properties, [17, 18], also any polynomial expression in such terms is a solution of a recurrence. Consider as an example the recursion

$$F(N+1) - F(N) = \frac{\text{sign}(a)^{N+1}}{(N+1)^{|a|}}. \tag{10.11}$$

It is solved by the harmonic sum $S_a(N)$. Corresponding difference equations hold for harmonic sums of deeper nestedness. Feynman integrals can often be decomposed into a combination containing terms of the form

$$\int_0^1 dz \frac{z^{N-1} - 1}{1 - z} H_{\vec{a}}(z), \int_0^1 dz \frac{(-z)^{N-1} - 1}{1 + z} H_{\vec{a}}(z), \tag{10.12}$$

with $H_{\vec{a}}(z)$ being a harmonic polylogarithm, [19]. This structure also leads to recurrences, [20]. Therefore, it is very likely that single scale Feynman diagrams do always obey difference equations.

10.2.2 Establishing and Solving Recurrences

Suppose we are given a finite array of rational numbers,

$$q_1, q_2, \ldots, q_K,$$

which are the first terms of an infinite sequence *F(N)*, i.e., $F(1) = q_1$ $F(2) = q_2$, etc. Let us assume that *F(N)* represents a physical quantity and satisfies a recurrence of type

$$\sum_{k=0}^{l}\left(\sum_{i=0}^{d} c_{i,k} N^i\right) F(N+k) = 0, \tag{10.13}$$

which we would like to deduce from the given numbers q_m. In a strict sense, this is not possible without knowing how the sequence continues for $N > K$. One thing we can do is to determine the recurrence equations satisfied by the data we are given. Any recurrence for *F(N)* must certainly be among those.

To find the recurrence equations of *F(N)* valid for the first terms, the simplest way to proceed is by making an ansatz with undetermined coefficients. Let us fix an order $l \in \mathbb{N}$ and a degree $d \in \mathbb{N}$ and consider the generic recurrence (10.13), where the $c_{i,k}$ are unknown. For each specific choice $N = 1, 2, \ldots, K - l$, we can evaluate the ansatz, because we know all the values of $F(N+k)$ in this range, and we obtain a system of $K - l$ homogeneous linear equations for $(l+1)(d+1)$ unknowns $c_{i,j}$.

If $K - l > (l+1)(d+1)$, this system is under-determined and is thus guaranteed to have nontrivial solutions. All these solutions will be valid recurrences for *F(N)* for $N = 1, \ldots, K - l$, but they will most typically fail to hold beyond. If, on the other hand, $K - l \leq (l+1)(d+1)$, then the system is overdetermined and nontrivial solutions are not to be expected. But at least recurrence equations valid for all *N*, if there are any, must appear among the solutions. We therefore expect in this case that the solution set will precisely consist of the recurrences of *F(N)* of order *l* and degree *d* valid for all *N*.

As an example, let us consider the contribution to the gluon splitting function $\propto C_A$ at leading order, $P_{gg}^{(0)}(N)$. The first 20 terms, starting with $N=3$, of the sequence $F(N)$ are

$$\frac{14}{5}, \frac{21}{5}, \frac{181}{35}, \frac{83}{14}, \frac{4129}{630}, \frac{319}{45}, \frac{26186}{3465}, \frac{18421}{2310}, \frac{752327}{90090}, \frac{71203}{8190}, \frac{811637}{90090}, \frac{128911}{13860}, \frac{29321129}{3063060},$$
$$\frac{2508266}{255255}, \frac{292886261}{29099070}, \frac{7045513}{684684}, \frac{611259269}{58198140}, \frac{1561447}{145860}, \frac{4862237357}{446185740}, \frac{988808455}{89237148}.$$

Making an ansatz for a recurrence of order 3 with polynomial coefficients of degree 3 leads to an overdetermined homogeneous linear system with 16 unknowns and 17 equations. Despite of being overdetermined and dense, this system has two linearly independent solutions. Using bounds for the absolute value of determinants depending on the size of a matrix and the bit size of its coefficients, one can very roughly estimate the probability for this to happen "by coincidence" to about 10^{-65}. And in fact, it did not happen by coincidence. The solutions to the system correspond to the two recurrence equations

$$(7N^3 + 113N^2 + 494N + 592)F(N) - (12N^3 + 233N^2 + 1289N + 2156)F(N+1) + (3N^3 + 118N^2 + 1021N + 2476)F(N+2) + (2N^3 + 2N^2 - 226N - 912)F(N+3) = 0 \tag{10.14}$$

and

$$(4N^3 + 64N^2 + 278N + 332)F(N) - (7N^3 + 134N^2 + 735N + 1222)F(N+1) + (2N^3 + 71N^2 + 595N + 1418)F(N+2) + (N^3 - N^2 - 138N - 528)F(N+3) = 0, \tag{10.15}$$

which both are valid for all $N \geq 1$. If we had found that the linear system did not have a nontrivial solution, then we could have concluded that the sequence $F(N)$ would *definitely* (i.e. without any uncertainty) not satisfy a recurrence of order 3 and degree 3. It might then still have satisfied recurrences with larger order or degree, but more terms of the sequence had to be known for detecting those.

The method of determining (potential) recurrence equations for sequences as just described is not new. It is known to the experimental mathematics community as automated guessing and is frequently applied in the study of combinatorial sequences. Standard software packages for generating functions such as gfun [17] for MAPLE or GeneratingFunctions.m [18] for MATHEMATICA provide functions which take as input a finite array of numbers, thought of as the first terms of some infinite sequence, and produce as output recurrence equations that are, with high probability, satisfied by the infinite sequence.

These packages apply the method described above more or less literally, and this is perfectly sufficient for small examples. But if thousands of terms of a sequence are needed, there is no way to solve the linear systems using rational number arithmetic. Even worse, already for medium sized problems from our collection, the size of the linear system exceeds by far typical memory capacities of 16–64 Gb. Let us consider as an example the difference equation associated to the contribution of the color factor C_F^3 for the 3-loop Wilson coefficient $C_{2,q}^{(3)}$ in unpolarized deeply inelastic scattering. 11 Tb of memory would be required to establish (10.13) in a naive way. Therefore refined methods have to be applied. We use arithmetic in finite fields together with Chinese remaindering, [21–23], which reduces the storage requirements to a few Gb of memory. The linear system approximately minimizes for $l \approx d$. If one finds more than one recurrence the different recurrences are joined to reduce l to a minimal value. It seems to be a general phenomenon that the recurrence of minimal order is that with the smallest integer coefficients, cf. also [24]. For even larger problems than those dealt with in the present analysis, a series of further technical improvements may be carried out, [25, 26].

For the solution of the recurrence low orders are clearly preferred. It is solved in depth-optimal $\Pi\Sigma$ fields, [27–34]; here we apply advanced symbolic summation methods as: efficient recurrence solvers and refined telescoping algorithms. They are available in the summation package Sigma, [28, 35].

The solutions are found as linear combinations of rational terms in N combined with functions, which cannot be further reduced in the $\Pi\Sigma$ fields. In the present application they turn out to be nested harmonic sums, cf. Appendix C.4. Other or higher order applications may lead to sums of different type as well, which are uniquely found by the present algorithm.

10.2.3 Determination of the 3-Loop Anomalous Dimensions and Wilson Coefficients

We apply the method to determine the unpolarized anomalous dimensions and massless Wilson coefficients to 3-loop order. Here we apply the above method to the contributions stemming from a single color/ζ_i-factor. These are 186 terms. As input we use the respective Mellin moments, which were calculated by a MAPLE-code based on the harmonic sum representation calculated in Refs. [3–5]. We need very high moments and calculate the input recursively. As an example, let us illustrate the size of the moments for the C_F^3-contribution to the Wilson coefficient $C_{2,q}^{(3)}$. The highest moment required is $N = 5114$. It cannot be calculated simply using Summer, [8], and we used a recursive algorithm in MAPLE for it.

The corresponding difference equations (10.13) are determined by a recurrence finder. Furthermore, the order of the difference equation is reduced to the smallest value possible. The difference equations are then solved order by order using the summation package . For the C_F^3-term in $C_{q,2}^{(3)}$, the recurrence was established after 20.7 days of CPU time. Here 4 h were required for the modular prediction of the dimension of the system, 5.8 days were spent on solving modular linear systems, and 11 days for the modular operator GCDs. The Chinese remainder method and rational reconstruction took 3.8 days. 140 word size primes were needed. As output one obtains a recurrence of 31 Mb, which is of order 35 and degree 938, with a largest integer of 1,227 digits. The recurrence was solved by Sigma after 5.9 days. We reached a compactification from 289 harmonic sums needed in [3–5] to 58 harmonic sums. The determination of the 3-loop anomalous dimensions is a much smaller problem. Here the computation takes only about 18 h for the complete result.

For the three most complicated cases, establishing and solving of the difference equations took 3 + 1 weeks each, requiring $\leq$10 Gb on a 2 GHz processor. This led to an overall computation time of about sixteen weeks.

In the final representation, we account for algebraic reduction, [36]. For this task we used the package HarmonicSums, [37], which complements the functionality of Sigma. One observes that different color factor contributions lead to the same, or nearly the same, amount of sums for a given quantity. This points to the fact that the amount of sums contributing, after the algebraic reduction has been carried out, is governed by topology rather than the field- and color structures involved. The linear harmonic sum representations used in [3–5] require many more sums than in the representation reached by the present analysis. A further reduction can be obtained using the structural relations, which leads to maximally 35 different sums up to the level of the 3-loop Wilson coefficients, [20, 38, 39].It is not unlikely that the present method can be applied to single scale problems in even higher orders. As has been found before in [40–50], representing a large number of 2- and 3-loop processes in terms of harmonic sums, the basis elements emerging are always the same.

In practice no method does yet exist to calculate such a high number of moments ab initio as required for the determination of the all-N formulas in the 3-loop case. On the other hand, a proof of existence has been delivered of a quite general and powerful automatic difference-equation solver, standing rather demanding tests. It opens up good prospects for the development of even more powerful methods, which can be applied in establishing and solving difference equations for single scale quantities such as the classes of Feynman-parameter integrals contributing to the massive operator matrix elements for general values of N.

References

1. J. Blümlein, M. Kauers, S. Klein, C. Schneider, PoS ACAT 2008, 106 (2008), hep-ph/0902.4095
2. J. Blümlein, M. Kauers, S. Klein, C. Schneider, Comput. Phys. Commun. **180**, 2143 (2009), hep-ph/0902.4091
3. S. Moch, J.A.M. Vermaseren, A. Vogt, Nucl. Phys. **B688**, 101 (2004), hepph/ 0403192
4. A. Vogt, S. Moch, J.A.M. Vermaseren, Nucl. Phys. **B691**, 129 (2004), hepph/ 0404111
5. J.A.M. Vermaseren, A. Vogt, S. Moch, Nucl. Phys. **B724**, 3 (2005), hepph/ 0504242
6. D.J. Broadhurst, Z. Phys. **C54**, 599 (1992)
7. L. Slater, *Generalized Hypergeometric Functions*, (Cambridge University Press, Cambridge, 1966), pp. 273.
8. J.A.M. Vermaseren, Int. J. Mod. Phys. **A14**, 2037 (1999), hep-ph/9806280
9. J.M. Borwein, D.M. Bradley, D.J. Broadhurst, P. Lisonek, Trans. Am. Math. Soc. **353**, 907 (2001), math/9910045
10. J. Blümlein, D. Broadhurst, J. Vermaseren, Comput. Phys. Commun. **181**, 582 (2010), math-ph/0907.2557
11. N. Nörlund, *Vorlesungen über Differenzenrechnung*, (Springer, Berlin, 1924), pp. 551.
12. L. Milne-Thomson, *The Calculus of Finite Differences*, (MacMillan, London, 1951), pp. 273.
13. D.J. Broadhurst, Eur. Phys. J. **C8**, 311 (1999), hep-th/9803091
14. Y. André, in *Proceedings of the International Conference*, Motives, Quantum Field Theory, an Pseudo Differential Operators, Clay Mathematical Institute, Boston, June 2008
15. F. Brown, Commun. Math. Phys. **287**, 925 (2009), math.AG/0804.1660
16. H. Ferguson, D.H. Bailey, A polynomial time, numerically stable integer relation algorithm, RNR Technical Report RNR-91-032, 199 (1991)
17. B. Salvy, P. Zimmermann, ACM Trans. Math. Softw. **2**(20), 163 (1994)
18. C. Mallinger, Algorithmic manipulations and transformations of univariate holonomic functions and sequences, Master Thesis, J. Kepler University, Linz, 1996
19. E. Remiddi, J.A.M. Vermaseren, Int. J. Mod. Phys. **A15**, 725 (2000), hep-ph/9905237
20. J. Blümlein, Nucl. Phys. Proc. Suppl. **183**, 232 (2008), math-ph/0807.0700
21. K. Geddes, S. Czapor, G. Labahn, *Algorithms for Computer Algebra*, (Kluwer Academic Publishers, Boston, 1992), pp. 585.
22. J. vonzur Gathen, J. Gerhard, *Modern Computer Algebra* , (Cambridge University Press, Cambridge, 1999), pp. 754.
23. M. Kauers, Nucl. Phys. Proc. Suppl. **183**, 245 (2008)
24. A. Bostan, M. Kauers, Proc. AMS **138**, 3063 (2010)
25. B. Beckermann, G. Labahn, Numer. Algorithms **3**, 45 (1992)
26. B. Beckermann, G. Labahn, SIAM J. Matrix Anal. Appl. **22**, 114 (2000)
27. C. Schneider, J. Algebra Appl. **6**(3), 415 (2007)

28. C. Schneider, J. Diffr. Equ. Appl. **11**(9), 799 (2005)
29. M. Karr, J. ACM **28**, 305 (1981)
30. M. Karr, J. Symb. Comput. **1**, 303 (1985)
31. C. Schneider, Small symbolic summation in difference fields, Ph.D. Thesis, RISC–Linz, J. Kepler University, Linz, 2001
32. C. Schneider, Symbolic summation finds optimal nested sum representations, SFB-Report 2007-26, SFB F013, J. Kepler University, Linz, 2007
33. C. Schneider, J. Symb. Comput. **43**, 611 (2008)
34. C. Schneider, Ann. Comb. **14**, 533 (2010)
35. C. Schneider, Sém. Lothar. Combin. **56** Article B56b and Habilitation Thesis, JKU Linz (2007)
36. J. Blümlein, Comput. Phys. Commun. **159**, 19 (2004), hep-ph/0311046
37. J. Ablinger, A computer algebra toolbox for harmonic sums related to particle physics, Diploma Thesis, J. Kepler University, Linz (2009)
38. J. Blümlein, Comput. Phys. Commun. **180**, 2218 (2009), hep-ph/0901.3106
39. J. Blümlein, Structural relations of harmonic sums and mellin transforms at weight w=6, in *Proceedings of the Workshop "Motives, Quantum Field Theory, and Pseudodifferential Operators, June (2008)"*, (Clay Institute, Boston University, 2009), math-ph/0901.0837
40. M. Buza, Y. Matiounine, J. Smith, R. Migneron, W.L. van Neerven, Nucl. Phys. **B472**, 611 (1996), hep-ph/9601302
41. J. Blümlein, A. De Freitas, W. L. van Neerven, and S. Klein, Nucl. Phys. **B755**, 272 (2006), hep-ph/0608024
42. I. Bierenbaum, J. Blümlein, S. Klein, Nucl. Phys. **B780**, 40 (2007), hep-ph/0703285
43. I. Bierenbaum, J. Blümlein, S. Klein, C. Schneider, Nucl. Phys. **B803**, 1 (2008), hep-ph/0803.0273
44. J. Blümlein, S. Klein, PoS ACAT 2007, 084 (2007), hep-ph/0706.2426
45. J. Blümlein, V. Ravindran, Nucl. Phys. **B716**, 128 (2005), hep-ph/0501178
46. J. Blümlein, V. Ravindran, Nucl. Phys. **B749**, 1 (2006), hep-ph/0604019
47. F.A. Berends, W.L. van Neerven, G.J.H. Burgers, Nucl. Phys. **B297**, 429 (1988); [Erratum-ibid.] **B304**, 921 (1988)
48. W.L. van Neerven, Acta Phys. Polon. **B28**, 2715 (1997), hep-ph/9708452
49. J. Blümlein, S. Moch, in preparation
50. J. Blümlein, A. de Freitas, W. van Neerven, PoS RADCOR 2007, 005 (2007), hep-ph/0812.1588

Chapter 11
Conclusions

In this thesis, we extended the description of the contributions of a single heavy quark to the unpolarized Wilson coefficients $C^{S,PS,NS}_{(q,g),2}$ to $O(a_s^3)$. In upcoming precision analyzes of deep-inelastic data, this will allow more precise determinations of parton distribution functions and of the strong coupling constant. We applied a factorization relation for the complete inclusive heavy flavor Wilson coefficients, which holds in the limit $Q^2 \gg 10\,\mathrm{m}^2$ in case of $F_2(x, Q^2)$, [1], at the level of twist-2. It relates the asymptotic heavy flavor Wilson coefficients to a convolution of the corresponding light flavor Wilson coefficients, which are known up to $O(a_s^3)$, [2], and describe all process dependence, with the massive operator matrix elements. The latter are process independent quantities and describe all mass-dependent contributions but the power-suppressed terms $((m^2/Q^2)^k,\ k \geq 1)$. They are obtained from the unpolarized twist-2 local composite operators stemming from the light-cone expansion of the electromagnetic current between on-shell partonic states, including virtual heavy quark lines. The first calculation of fixed moments of all three-loop massive OMEs is the main result of this thesis.

In Sect. 3.2, we applied the factorization formula at the $O(a_s^3)$-level. It holds for the inclusive heavy flavor Wilson coefficients, including radiative corrections due to heavy quark loops. In order to describe the production of heavy quarks in the final states only, further assumptions have to be made. This description succeeded at the two-loop level in Ref. [1] because of the possible comparison with the exact calculation in Refs. [3–5] and since the contributing virtual heavy flavor corrections are easily identified, cf. Sect. 5.1. At $O(a_s^3)$ this is not possible anymore and only the inclusive description should be used, as has been done in Ref. [6] in order to derive heavy flavor parton densities. These are obtained as convolutions of the light flavor densities with the massive OMEs, cf. Sect. 3.3.

In Chap. 4, we derived and presented in detail the renormalization of the massive operator matrix elements up to $O(a_s^3)$. This led to an intermediary representation in a defined MOM-scheme to maintain the partonic description required for the factorization of the heavy flavor Wilson coefficients into OMEs and the light flavor Wilson coefficients. Finally, we applied the $\overline{\mathrm{MS}}$-scheme for coupling constant

S. Klein, *Charm Production in Deep Inelastic Scattering*, Springer Theses,
DOI: 10.1007/978-3-642-23286-2_11,

renormalization in order to refer to the inclusive heavy flavor Wilson coefficients and to be able to combine our results with the light flavor Wilson coefficients, which have been calculated in the same scheme. For mass renormalization we chose the on-mass-shell-scheme and provided in Chap. 5 all necessary formulas to transform between the MOM and the on-shell-scheme, respectively, and the $\overline{\text{MS}}$-scheme.

For renormalization at $O(a_s^3)$, all $O(a_s^2)$ massive OMEs A_{Qg}, A_{Qq}^{PS}, $A_{qq,Q}^{NS}$, $A_{gg,Q}$, $A_{gq,Q}$ are needed up to $O(\varepsilon)$ in dimensional regularization. In Chap. 6, we newly calculated all the corresponding $O(\varepsilon)$ contributions in Mellin space for general values of N. This involved a first re-calculation of the complete terms $A_{gg,Q}^{(2)}$ and $A_{gq,Q}^{(2)}$, in which we agree with the literature [6]. We made use of the representation of the Feynman-parameter integrals in terms of generalized hypergeometric functions. The $O(\varepsilon)$-expansion led to new infinite sums which had to be solved by analytic and algebraic methods. The results can be expressed in terms of polynomials of the basic nested harmonic sums up to weight $w = 4$ and derivatives thereof. They belong to the complexity-class of the general two-loop Wilson coefficients or hard scattering cross sections in massless QED and QCD and are described by six basic functions and their derivatives in Mellin space. The package Sigma, [7–12], proved to be a useful tool to solve the sums occurring in the present problem, leading to extensions of this code by the author.

The main part of the thesis was the calculation of fixed moments of all three-loop massive operator matrix elements A_{Qg}, $A_{qg,Q}$, A_{Qq}^{PS}, $A_{qq,Q}^{PS}$, $A_{qq,Q}^{NS}$, $A_{gq,Q}$, $A_{gg,Q}$, cf. Chap. 7. These are needed to describe the asymptotic heavy flavor Wilson coefficients at $O(a_s^3)$ and to derive massive quark-distributions at the same level [6]. We developed computer algebra codes which allow based on QGRAF, [13], the automatic generation of three-loop Feynman diagrams with local operator insertions. These were then projected onto massive tadpole diagrams for fixed values of the Mellin variable N. For the final calculation of the diagrams, use was made of the FORM-code MATAD [14]. The representation of the massive OMEs is available for general values of N in analytic form, apart from the constant terms $a_{ij}^{(3)}$ of the unrenormalized three-loop OMEs. This is achieved by combining our general expressions for the renormalized results, the all-N results up to $O(a_s^2\varepsilon)$ and results given in the literature. A number of fixed Mellin moments of the terms $a_{ij}^{(3)}$ were calculated, reaching up to $N = 10, 12, 14$, depending on the complexity of the corresponding operator matrix element. The computation required about 250 CPU days on 32/64 Gb-machines.

Through the renormalization of the massive OMEs, the corresponding moments of the complete two-loop anomalous dimensions and the T_F-terms of the three-loop anomalous dimensions were obtained, as were the moments of the complete anomalous dimensions $\gamma_{qq}^{(2),PS}$ and $\gamma_{qg}^{(2)}$, in which we agree with the literature. This provides a first independent check of the moments of the fermionic contributions to the three-loop anomalous dimensions, which have been obtained in Refs. [15, 16].

In Chap. 8, we presented results on the effects of heavy quarks in polarized deep-inelastic scattering, using essentially the same description as in the unpolarized case. We worked in the scheme for γ_5 in dimensional regularization used in Ref. [17] and could confirm the results given there for the two-loop massive OMEs ΔA_{Qq}^{PS} and ΔA_{Qg}. Additionally, we newly presented the $O(\varepsilon)$ contributions of these terms.

We calculated the two-loop massive OMEs of transversity for all-N and the three-loop terms for the moments $N = 1, \ldots, 13$ in Chap. 9. This calculation is not yet of phenomenological use, since the corresponding light flavor Wilson coefficients have not been calculated so far. However, these results could be obtained by making only minor changes to the computer programs written for the unpolarized case. We confirmed for the first time the moments $N = 1, \ldots, 8$ of the fermionic contributions to the three-loop transversity anomalous dimension obtained in Refs. [18–21]. Our results can, however, be used in comparison with lattice calculations.

Several steps were undertaken towards an all-N calculation of the massive OMEs. Four non-trivial three-loop massive topologies contribute. We presented in an example a first all-N result for a ladder-topology in Sect. 10.1.

In Sect. 10.2, we described a general algorithm to calculate the exact expression for single scale quantities from a finite (suitably large) number of moments, which are zero scale quantities. The latter are much more easily calculable than single scale quantities. We applied the method to the anomalous dimensions and massless Wilson coefficients up to three-loop order [22–24]. Solving three-loop problems in this way directly is not possible at present, since the number of required moments is too large for the methods available. Yet this method constitutes a proof of principle and may find application in medium-sized problems in the future.

References

1. M. Buza, Y. Matiounine, J. Smith, R. Migneron, W.L. van Neerven, Nucl. Phys. **B472**, 611 (1996), hep-ph/9601302
2. J.A.M. Vermaseren, A. Vogt, S. Moch, Nucl. Phys. **B724**, 3 (2005), hep-ph/0504242
3. E. Laenen, S. Riemersma, J. Smith, van W.L. Neerven, Nucl. Phys. **B392**, 162 (1993a)
4. E. Laenen, S. Riemersma, J. Smith, van W.L. Neerven, Nucl. Phys. **B392**, (1993b)
5. S. Riemersma, J. Smith, W.L. van Neerven, Phys. Lett. **B347**, 143 (1995), hep-ph/9411431
6. M. Buza, Y. Matiounine, J. Smith, W.L. van Neerven, Eur. Phys. J. **C1**, 301 (1998), hep-ph/9612398
7. C. Schneider, Ann. Comb. **9**(1) 75 (2005)
8. C. Schneider, in *Proceedings of ISSAC'05*, (ACM Press, New York, 2005), p. 285
9. C. Schneider, in *Proceedings of FPSAC'07* (2007), p. 1
10. C. Schneider, J. Algebra Appl. **6**(3), 415 (2007)
11. C. Schneider, J. Diff. Equ. Appl. **11**(9), 799 (2005)
12. C. Schneider, Sém. Lothar. Combin. **56** (2007) Article B56b and Habilitation Thesis, JKU Linz, (2007)
13. P. Nogueira, J. Comput. Phys. **105**, 279 (1993)
14. M. Steinhauser, Comput. Phys. Commun. **134**, 335 (2001), hep-ph/0009029
15. S.A. Larin, P. Nogueira, T. van Ritbergen, J.A.M. Vermaseren, Nucl. Phys. **B492**, 338 (1997), hep-ph/9605317
16. A. Retey, J.A.M. Vermaseren, Nucl. Phys. **B604**, 281 (2001), hep-ph/0007294
17. M. Buza, Y. Matiounine, J. Smith, W.L. van Neerven, Nucl. Phys. **B485**, 420 (1997), hep-ph/9608342
18. J.A. Gracey, Nucl. Phys. **B662**, 247 (2003a), hep-ph/0304113
19. J.A. Gracey, Nucl. Phys. **B667**, 242 (2003b), hep-ph/0306163

20. J.A. Gracey, JHEP **10**, 040 (2006a), hep-ph/0609231
21. J.A. Gracey, Phys. Lett. **B643**, 374 (2006b), hep-ph/0611071
22. G. Zweig, An SU (3) model for the strong interaction symmetry and its breaking, CERN-TH-401, 412 (1964)
23. S. Moch, J.A.M. Vermaseren, A. Vogt, Nucl. Phys. **B688**, 101 (2004), hep-ph/0403192
24. A. Vogt, S. Moch, J.A.M. Vermaseren, Nucl. Phys. **B691**, 129 (2004), hep-ph/0404111

Appendix A
Conventions

We use natural units

$$\hbar = 1, \quad c = 1, \quad \epsilon_0 = 1, \tag{A.1}$$

where $\hbar$ denotes Planck's constant, c the vacuum speed of light and ϵ_0 the permittivity of vacuum. The electromagnetic fine-structure constant α is given by

$$\alpha = \alpha'(\mu^2 = 0) = \frac{e^2}{4\pi\epsilon_0\hbar c} = \frac{e^2}{4\pi} \approx \frac{1}{137.03599911(46)}. \tag{A.2}$$

In this convention, energies and momenta are given in the same units, electron volt (eV).

The space–time dimension is taken to be $D = 4 + \epsilon$ and the metric tensor $g_{\mu\nu}$ in Minkowski-space is defined as

$$g_{00} = 1, \quad g_{ii} = -1, \;\; i = 1 \ldots D-1, \quad g_{ij} = 0, \;\; i \neq j. \tag{A.3}$$

Einstein's summation convention is used, i.e.

$$x_\mu y^\mu := \sum_{\mu=0}^{D-1} x_\mu y^\mu. \tag{A.4}$$

Bold-faced symbols represent $(D-1)$-dimensional spatial vectors:

$$x = (x_0, \mathbf{x}). \tag{A.5}$$

If not stated otherwise, Greek indices refer to the D-component space–time vector and Latin ones to the $D-1$ spatial components only. The dot product of two vectors is defined by

$$p \cdot q = p_0 q_0 - \sum_{i=1}^{D-1} p_i q_i. \tag{A.6}$$

S. Klein, *Charm Production in Deep Inelastic Scattering*, Springer Theses,
DOI: 10.1007/978-3-642-23286-2, © Springer-Verlag Berlin Heidelberg 2012

The γ-matrices γ_μ are taken to be of dimension D and fulfill the anti-commutation relation

$$\{\gamma_\mu, \gamma_\nu\} = 2g_{\mu\nu}. \tag{A.7}$$

It follows that

$$\gamma_\mu \gamma^\mu = D \tag{A.8}$$

$$Tr(\gamma_\mu \gamma_\nu) = 4g_{\mu\nu} \tag{A.9}$$

$$Tr(\gamma_\mu \gamma_\nu \gamma_\alpha \gamma_\beta) = 4[g_{\mu\nu}g_{\alpha\beta} + g_{\mu\beta}g_{\nu\alpha} - g_{\mu\alpha}g_{\nu\beta}]. \tag{A.10}$$

The slash-symbol for a D-momentum p is defined by

$$\not{p} := \gamma_\mu p^\mu. \tag{A.11}$$

The conjugate of a bi-spinor u of a particle is given by

$$\overline{u} = u^\dagger \gamma_0, \tag{A.12}$$

where $\dagger$ denotes Hermitian and $*$ complex conjugation, respectively. The bi-spinors u and v fulfill the free Dirac-equation,

$$(\not{p} - m)u(p) = 0, \quad \overline{u}(p)(\not{p} - m) = 0 \tag{A.13}$$

$$(\not{p} + m)v(p) = 0, \quad \overline{v}(p)(\not{p} + m) = 0. \tag{A.14}$$

Bi-spinors and polarization vectors are normalized to

$$\sum_\sigma u(p,\sigma)\overline{u}(p,\sigma) = \not{p} + m \tag{A.15}$$

$$\sum_\sigma v(p,\sigma)\overline{v}(p,\sigma) = \not{p} - m \tag{A.16}$$

$$\sum_\lambda \epsilon^\mu(k,\lambda)\epsilon^\nu(k,\lambda) = -g^{\mu\nu}, \tag{A.17}$$

where λ and σ represent the spin.

The commonly used caret "^" to signify an operator, e.g. $\hat{O}$, is omitted if confusion is not to be expected.

The gauge symmetry group of QCD is the Lie-Group $SU(3)_c$. We consider the general case of $SU(N_c)$. The non-commutative generators are denoted by t^a, where a runs from 1 to $N_c^2 - 1$. The generators can be represented by Hermitian, traceless matrices [1]. The structure constants f^{abc} and d^{abc} of $SU(N_c)$ are defined via the commutation and anti-commutation relations of its generators [2],

$$[t^a, t^b] = i f^{abc} t^c \tag{A.18}$$

$$\{t^a, t^b\} = d^{abc} t^c + \frac{1}{N_c}\delta_{ab}. \tag{A.19}$$

The indices of the color matrices, in a certain representation, are denoted by $i, j, k, l, \ldots$ The color invariants most commonly encountered are

$$\delta_{ab} C_A = f^{acd} f^{bcd} \tag{A.20}$$

$$\delta_{ij} C_F = t^a_{il} t^a_{lj} \tag{A.21}$$

$$\delta_{ab} T_F = t^a_{ik} t^b_{ki}. \tag{A.22}$$

These constants evaluate to

$$C_A = N_c, \quad C_F = \frac{N_c^2 - 1}{2N_c}, \quad T_F = \frac{1}{2}. \tag{A.23}$$

At higher loops, more color-invariants emerge. At 3-loop order, one additionally obtains

$$d^{abd} d_{abc} = (N_c^2 - 1)(N_c^2 - 4)/N_c. \tag{A.24}$$

In case of $SU(3)_c, C_A = 3, C_F = 4/3, d^{abc} d_{abc} = 40/3$ holds.

Appendix B
Feynman Rules

For the QCD Feynman rules, Fig. B.1, we follow Ref. [2], cf. also Refs. [3, 4]. D-dimensional momenta are denoted by p_i and Lorentz-indices by Greek letters. Color indices are $a, b, \ldots$ and i, j are indices of the color matrices. Solid lines represent fermions, wavy lines gluons and dashed lines ghosts. Arrows denote the direction of the momenta. A factor (-1) has to be included for each closed fermion—or ghost loop.

The Feynman rules for the quarkonic composite operators are given in Fig. B.2. Up to $O(g^2)$ they can be found in Ref. [5] and also in [6]. Note that the $O(g)$ term in the former reference contains a typographical error. We have checked these terms and agree up to normalization factors, which may be due to other conventions being applied there. We newly derived the rule with three external gluons. The terms $\gamma_\pm$ refer to the unpolarized $(+)$ and polarized $(-)$ case, respectively. Gluon momenta are taken to be incoming.

The Feynman rules for the unpolarized gluonic composite operators are given in Fig. B.3. Up to $O(g^2)$, they can be found in Refs. [7, 8]. We have checked these terms and agree up to $O(g^0)$. At $O(g)$, we agree with [7], but not with [8]. At $O(g^2)$, we do not agree with either of these results, which even differ from each other.[1]

[1] We would like to thank J. Smith for the possibility to compare with their FORM-code used in Refs. [9–12], to which we agree.

S. Klein, *Charm Production in Deep Inelastic Scattering*, Springer Theses,
DOI: 10.1007/978-3-642-23286-2, © Springer-Verlag Berlin Heidelberg 2012

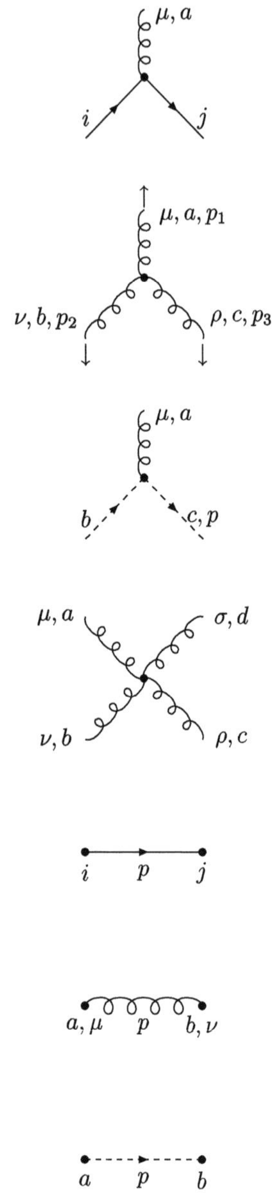

$$ig_s\gamma_\mu t^a_{ji}$$

$$-g_s f^{abc}[(p_1-p_2)_\rho g_{\mu\nu} + (p_2-p_3)_\mu g_{\nu\rho} + (p_3-p_1)_\nu g_{\mu\rho}]$$

$$-g_s f^{abc} p_\mu$$

$$-ig_s^2 \sum_e \Big\{ f^{abe} f^{cde}[g_{\mu\rho}g_{\nu\sigma} - g_{\mu\sigma}g_{\nu\rho}] + f^{ace} f^{bde}[g_{\mu\nu}g_{\rho\sigma} - g_{\mu\sigma}g_{\nu\rho}] + f^{ade} f^{cbe}[g_{\mu\rho}g_{\nu\sigma} - g_{\mu\nu}g_{\rho\sigma}] \Big\}$$

$$\frac{i}{\not p - m + i0}\delta_{ij}$$

$$\frac{i}{p^2+i0}(-g_{\mu\nu} + \xi p_\mu p_\nu/(p^2+i0))\delta_{ab}$$

$$\frac{i}{p^2+i0}\delta_{ab}$$

Fig. B.1 Feynman rules of QCD

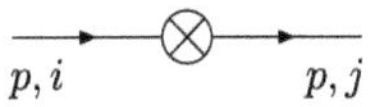

$$\delta^{ij}\not\Delta\gamma_{\pm}(\Delta\cdot p)^{N-1}\,,\quad N\geq 1$$

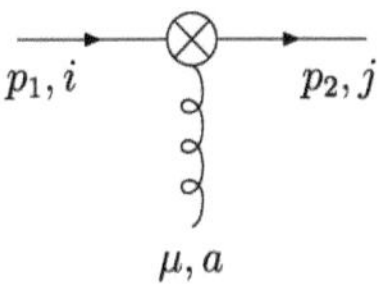

$$g t^{a}_{ji}\Delta^{\mu}\not\Delta\gamma_{\pm}\sum_{j=0}^{N-2}(\Delta\cdot p_1)^{j}(\Delta\cdot p_2)^{N-j-2}\,,\quad N\geq 2$$

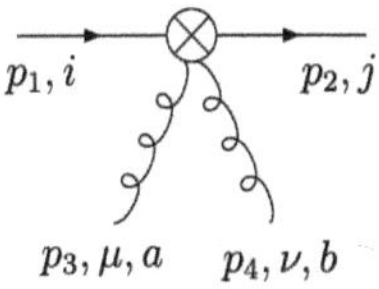

$$g^{2}\Delta^{\mu}\Delta^{\nu}\not\Delta\gamma_{\pm}\sum_{j=0}^{N-3}\sum_{l=j+1}^{N-2}(\Delta p_2)^{j}(\Delta p_1)^{N-l-2}$$
$$\left[(t^{a}t^{b})_{ji}(\Delta p_1+\Delta p_4)^{l-j-1}+(t^{b}t^{a})_{ji}(\Delta p_1+\Delta p_3)^{l-j-1}\right],$$
$$N\geq 3$$

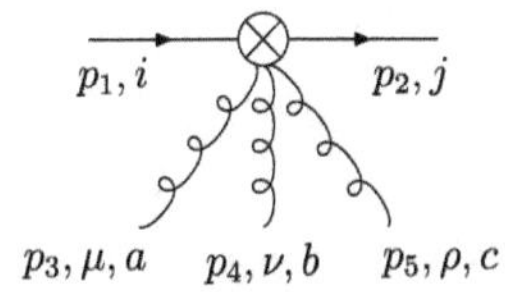

$$g^{3}\Delta^{\mu}\Delta^{\nu}\Delta^{\rho}\not\Delta\gamma_{\pm}\sum_{j=0}^{N-4}\sum_{l=j+1}^{N-3}\sum_{m=l+1}^{N-2}(\Delta.p_2)^{j}(\Delta.p_1)^{N-m-2}$$
$$\Big[\ (t^{a}t^{b}t^{c})_{ji}(\Delta.p_4+\Delta.p_5+\Delta.p_1)^{l-j-1}(\Delta.p_5+\Delta.p_1)^{m-l-1}$$
$$+(t^{a}t^{c}t^{b})_{ji}(\Delta.p_4+\Delta.p_5+\Delta.p_1)^{l-j-1}(\Delta.p_4+\Delta.p_1)^{m-l-1}$$
$$+(t^{b}t^{a}t^{c})_{ji}(\Delta.p_3+\Delta.p_5+\Delta.p_1)^{l-j-1}(\Delta.p_5+\Delta.p_1)^{m-l-1}$$
$$+(t^{b}t^{c}t^{a})_{ji}(\Delta.p_3+\Delta.p_5+\Delta.p_1)^{l-j-1}(\Delta.p_3+\Delta.p_1)^{m-l-1}$$
$$+(t^{c}t^{a}t^{b})_{ji}(\Delta.p_3+\Delta.p_4+\Delta.p_1)^{l-j-1}(\Delta.p_4+\Delta.p_1)^{m-l-1}$$
$$+(t^{c}t^{b}t^{a})_{ji}(\Delta.p_3+\Delta.p_4+\Delta.p_1)^{l-j-1}(\Delta.p_3+\Delta.p_1)^{m-l-1}\Big],$$
$$N\geq 4$$

$\gamma_{+}=1\,,\quad \gamma_{-}=\gamma_5\,.$ For transversity, one has to replace: $\not\Delta\gamma_{\pm}\rightarrow\ \sigma^{\mu\nu}\Delta_{\nu}$.

Fig. B.2 Feynman rules for quarkonic composite operators. Δ denotes a light-like four-vector, $\Delta^2=0$; N is a suitably large positive integer

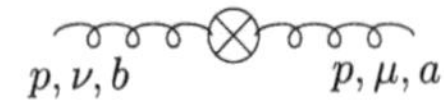

$$\frac{1+(-1)^N}{2}\delta^{ab}(\Delta\cdot p)^{N-2}\Big[g_{\mu\nu}(\Delta\cdot p)^2-(\Delta_\mu p_\nu+\Delta_\nu p_\mu)\Delta\cdot p+p^2\Delta_\mu\Delta_\nu\Big]\,,\quad N\geq 2$$

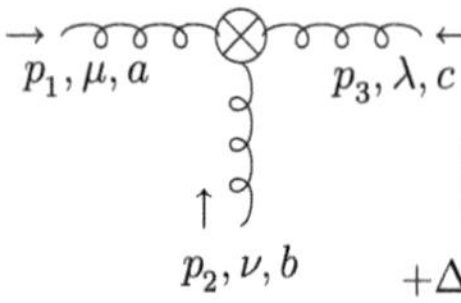

$$-ig\frac{1+(-1)^N}{2}f^{abc}\Bigg(\Big[(\Delta_\nu g_{\lambda\mu}-\Delta_\lambda g_{\mu\nu})\Delta\cdot p_1+\Delta_\mu(p_{1,\nu}\Delta_\lambda-p_{1,\lambda}\Delta_\nu)\Big](\Delta\cdot p_1)^{N-2}$$
$$+\Delta_\lambda\Big[\Delta\cdot p_1\,p_{2,\mu}\Delta_\nu+\Delta\cdot p_2\,p_{1,\nu}\Delta_\mu-\Delta\cdot p_1\,\Delta\cdot p_2\,g_{\mu\nu}-p_1\cdot p_2\Delta_\mu\Delta_\nu\Big]$$
$$\times\sum_{j=0}^{N-3}(-\Delta\cdot p_1)^j(\Delta\cdot p_2)^{N-3-j}$$
$$+\left\{\begin{smallmatrix}p_1\to p_2\to p_3\to p_1\\ \mu\to\nu\to\lambda\to\mu\end{smallmatrix}\right\}+\left\{\begin{smallmatrix}p_1\to p_3\to p_2\to p_1\\ \mu\to\lambda\to\nu\to\mu\end{smallmatrix}\right\}\Bigg)\,,\quad N\geq 2$$

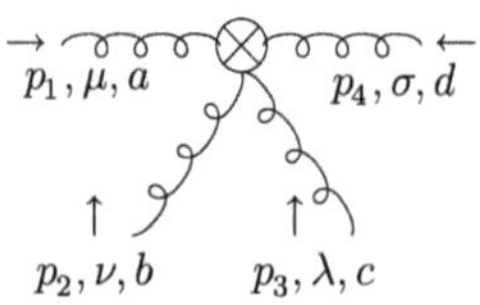

$$g^2\frac{1+(-1)^N}{2}\Bigg(f^{abe}f^{cde}O_{\mu\nu\lambda\sigma}(p_1,p_2,p_3,p_4)$$
$$+f^{ace}f^{bde}O_{\mu\lambda\nu\sigma}(p_1,p_3,p_2,p_4)+f^{ade}f^{bce}O_{\mu\sigma\nu\lambda}(p_1,p_4,p_2,p_3)\Bigg)\,,$$
$$O_{\mu\nu\lambda\sigma}(p_1,p_2,p_3,p_4)=\Delta_\nu\Delta_\lambda\Bigg\{-g_{\mu\sigma}(\Delta\cdot p_3+\Delta\cdot p_4)^{N-2}$$
$$+[p_{4,\mu}\Delta_\sigma-\Delta\cdot p_4\,g_{\mu\sigma}]\sum_{i=0}^{N-3}(\Delta\cdot p_3+\Delta\cdot p_4)^i(\Delta\cdot p_4)^{N-3-i}$$
$$-[p_{1,\sigma}\Delta_\mu-\Delta\cdot p_1\,g_{\mu\sigma}]\sum_{i=0}^{N-3}(-\Delta\cdot p_1)^i(\Delta\cdot p_3+\Delta\cdot p_4)^{N-3-i}$$
$$+[\Delta\cdot p_1\,\Delta\cdot p_4\,g_{\mu\sigma}+p_1\cdot p_4\,\Delta_\mu\Delta_\sigma-\Delta\cdot p_4\,p_{1,\sigma}\Delta_\mu-\Delta\cdot p_1\,p_{4,\mu}\Delta_\sigma]$$
$$\times\sum_{i=0}^{N-4}\sum_{j=0}^{i}(-\Delta\cdot p_1)^{N-4-i}(\Delta\cdot p_3+\Delta\cdot p_4)^{i-j}(\Delta\cdot p_4)^j\Bigg\}$$
$$-\left\{\begin{smallmatrix}p_1\leftrightarrow p_2\\ \mu\leftrightarrow\nu\end{smallmatrix}\right\}-\left\{\begin{smallmatrix}p_3\leftrightarrow p_4\\ \lambda\leftrightarrow\sigma\end{smallmatrix}\right\}+\left\{\begin{smallmatrix}p_1\leftrightarrow p_2,\ p_3\leftrightarrow p_4\\ \mu\leftrightarrow\nu,\ \lambda\leftrightarrow\sigma\end{smallmatrix}\right\}\,,\quad N\geq 2$$

Fig. B.3 Feynman rules for gluonic composite operators. Δ denotes a light-like four-vector, $\Delta^2=0$; N is an integer

Appendix C
Special Functions

In the following we summarize for convenience some relations for special functions which occur repeatedly in quantum field theory and are used in this thesis.

C.1 The Γ-Function

The Γ-function, cf. [13, 14], is analytic in the whole complex plane except at single poles at the non-positive integers. Its inverse is given by Euler's infinite product

$$\frac{1}{\Gamma(z)} = z\exp(\gamma_E z)\prod_{i=1}^{\infty}\left[\left(1+\frac{z}{i}\right)\exp(-z/i)\right]. \tag{C.1}$$

The residues of the Γ-function at its poles are given by

$$\mathrm{Res}[\Gamma(z)]_{z=-N} = \frac{(-1)^N}{N!}, \quad N \in \mathbb{N} \cup 0. \tag{C.2}$$

In case of $\mathrm{Re}(z) > 0$, it can be expressed by Euler's integral

$$\Gamma(z) = \int_0^{\infty} \exp(-t)t^{z-1}dt, \tag{C.3}$$

from which one infers the well known functional equation of the Γ-function

$$\Gamma(z+1) = z\Gamma(z), \tag{C.4}$$

S. Klein, *Charm Production in Deep Inelastic Scattering*, Springer Theses,
DOI: 10.1007/978-3-642-23286-2,

which is used for analytic continuation. Around $z = 1$, the following series expansion is obtained

$$\Gamma(1-\varepsilon) = \exp(\varepsilon\gamma_E)\exp\left\{\sum_{i=2}^{\infty}\zeta_i\frac{\varepsilon^i}{i}\right\}, \tag{C.5}$$

$$|\varepsilon|<1. \tag{C.6}$$

Here and in (C.1), γ_E denotes the Euler–Mascheroni constant, see Eq. 4.7. In C.5 Riemann's ζ-function is given by

$$\zeta_k = \sum_{i=1}^{\infty}\frac{1}{i^k}, \quad 2 \leq k \in \mathbb{N}. \tag{C.7}$$

A shorthand notation for rational functions of Γ-functions is

$$\Gamma\left[\begin{matrix} a_1,\ldots,a_i \\ b_1,\ldots,b_j \end{matrix}\right] := \frac{\Gamma(a_1)\ldots\Gamma(a_i)}{\Gamma(b_1)\ldots\Gamma(b_j)}. \tag{C.8}$$

Functions closely related to the Γ-function are the function $\psi(x)$, the Beta-function $B(A,C)$ and the function $\beta(x)$.

The Beta-function can be defined by Eq. C.8

$$B(A,C) = \Gamma\left[\begin{matrix} A,C \\ A+C \end{matrix}\right]. \tag{C.9}$$

If $\mathrm{Re}(A),\mathrm{Re}(C) > 0$, the following integral representation is valid

$$B(A,C) = \int_0^1 dx\, x^{A-1}(1-x)^{C-1}. \tag{C.10}$$

For arbitrary values of A and C, (C.10) can be continued analytically using Eqs. C.1, C.9. Its expansion around singularities can be performed via Eqs. C.2, C.5. The ψ-function and $\beta(x)$ are defined as derivatives of the Γ-function via

$$\psi(x) = \frac{1}{\Gamma(x)}\frac{d}{dx}\Gamma(x). \tag{C.11}$$

$$\beta(x) = \frac{1}{2}\left[\psi\left(\frac{x+1}{2}\right) - \psi\left(\frac{x}{2}\right)\right]. \tag{C.12}$$

C.2 The Generalized Hypergeometric Function

The generalized hypergeometric function ${}_PF_Q$ is defined by, cf. [15–17],

$${}_PF_Q\left[\begin{matrix}a_1,\ldots,a_P\\ b_1,\ldots,b_Q\end{matrix};z\right]=\sum_{i=0}^{\infty}\frac{(a_1)_i\ldots(a_P)_i}{(b_1)_i\ldots(b_Q)_i}\frac{z^i}{\Gamma(i+1)}. \tag{C.13}$$

Here $(c)_n$ is Pochhammer's symbol

$$(c)_n=\frac{\Gamma(c+n)}{\Gamma(c)}, \tag{C.14}$$

for which the following relation holds

$$(N+1)_{-i}=\frac{(-1)^i}{(-N)_i}, \quad N\in\mathbb{N}. \tag{C.15}$$

In (C.13), there are P numerator parameters $a_1\ldots a_P$, Q denominator parameters $b_1\ldots b_Q$ and one variable z, all of which may be real or complex. Additionally, the denominator parameters must not be negative integers, since in that case (C.13) is not defined. The generalized hypergeometric series ${}_PF_Q$ are evaluated at a certain value of z, which in this thesis is always $z=1$ for the final integrals.

Gauss was the first to study this kind of functions, introducing the Gauss function ${}_2F_1$, and proved the theorem, cf. [15],

$${}_2F_1[a,b;c;1]=\Gamma\left[\begin{matrix}c,c-a-b\\ c-a,c-b\end{matrix}\right], \quad \mathrm{Re}(c-a-b)>0 \tag{C.16}$$

which is called Gauss' theorem. An integral representation for the Gauss function is given by the integral, cf. [15],

$${}_2F_1\left[\begin{matrix}a,b+1\\ c+b+2\end{matrix};z\right]=\Gamma\left[\begin{matrix}c+b+2\\ c+1,b+1\end{matrix}\right]\int_0^1 dx\, x^b(1-x)^c(1-zx)^{-a}, \tag{C.17}$$

provided that the conditions

$$|z|<1, \quad \mathrm{Re}(c+1),\mathrm{Re}(b+1)>0, \tag{C.18}$$

are obeyed. Applying Eq. C.17 recursively, one obtains the following integral representation for a general ${}_{P+1}F_P$-function

$${}_{P+1}F_P\left[\begin{matrix}a_0,a_1,\ldots,a_P\\ b_1,\ldots,b_P\end{matrix};z\right]=\Gamma\left[\begin{matrix}b_1,\ldots,b_P\\ a_1,\ldots,a_P,b_1-a_1,\ldots,b_P-a_P\end{matrix}\right]$$
$$\times\int_0^1 dx_1\ldots\int_0^1 dx_P x_1^{a_1-1}(1-x_1)^{b_1-a_1-1}\ldots x_P^{a_P-1}(1-x_P)^{b_P-a_P-1}(1-zx_1\ldots x_P)^{-a_0}, \tag{C.19}$$

under similar conditions as in Eq. C.18.

C.3 Mellin–Barnes Integrals

For the Gauss function, there exists a representation in terms of a complex contour integral over Γ-functions. It is given by, cf. [15],

$$ {}_2F_1\left[\begin{matrix} a,b \\ c \end{matrix} ;z\right] = \frac{\Gamma(c)}{2\pi i\Gamma(a)\Gamma(b)} \int_{-i\infty+\alpha}^{i\infty+\alpha} \frac{\Gamma(a+s)\Gamma(b+s)\Gamma(-s)}{\Gamma(c+s)}(-z)^s ds, \quad \text{(C.20)} $$

under the conditions

$$ |z|<1, \quad |\arg(-z)|<\pi. \quad \text{(C.21)} $$

(C.20) only holds if one chooses the integration contour in the complex plane and the positive constant α in such a way that the poles of the Γ-functions containing $(+s)$ are separated from those arising from the Γ-functions containing $(-s)$ and closes the contour to the right.

Setting $b=1, c=1$ in (C.20) one obtains

$$ {}_1F_0[a;z] = \frac{1}{(1-z)^a}, \quad \text{(C.22)} $$

which yields the Mellin–Barnes transformation, cf. [18–22],

$$ \frac{1}{(X+Y)^\lambda} = \frac{1}{2\pi i\Gamma(\lambda)} \int_{-i\infty+\alpha}^{+i\infty+\alpha} ds\Gamma(\lambda+s)\Gamma(-s)\frac{Y^s}{X^{\lambda+s}}. \quad \text{(C.23)} $$

The contour has to be chosen as in (C.20) and the conditions $0<\alpha<\mathrm{Re}(\lambda), |\arg(Y/X)|<\pi$ have to be fulfilled.

C.4 Harmonic Sums and Nielsen-Integrals

Expanding the Γ-function in ϵ, its logarithmic derivatives, the $\psi^{(k)}$-functions, emerge. In many applications of perturbative QCD and QED, harmonic sums occur, cf. [23, 24], which can be considered as generalization of the ψ-function and the β-function. These are defined by

$$ S_{a_1,\ldots,a_m}(N) = \sum_{n_1=1}^{N}\sum_{n_2=1}^{n_1}\cdots\sum_{n_m=1}^{n_{m-1}} \frac{(\mathrm{sign}(a_1))^{n_1}}{n_1^{|a_1|}}\frac{(\mathrm{sign}(a_2))^{n_2}}{n_2^{|a_2|}}\cdots\frac{(\mathrm{sign}(a_m))^{n_m}}{n_m^{|a_m|}}, $$

$$ N\in\mathbb{N},\ \forall l a_l\in\mathbb{Z}\setminus 0, \quad \text{(C.24)} $$

$$ S_\emptyset = 1. \quad \text{(C.25)} $$

We adopt the convention

$$S_{a_1,\ldots,a_m} \equiv S_{a_1,\ldots,a_m}(N), \tag{C.26}$$

i.e. harmonic sums are taken at argument (N), if no argument is indicated. Related quantities are the Z-sums defined by

$$Z_{m_1,\ldots,m_k}(N) = \sum_{N \geq i_1 > i_2 \cdots > i_k > 0} \frac{\prod_{l=1}^{k} [\text{sign}(m_l)]^{i_l}}{i_l^{|m_l|}}. \tag{C.27}$$

The depth d and the weight w of a harmonic sum are given by

$$d := m, \tag{C.28}$$

$$w := \sum_{i=1}^{m} |a_i|. \tag{C.29}$$

Harmonic sums of depth $d = 1$ are referred to as single harmonic sums. The complete set of algebraic relations connecting harmonic sums to other harmonic sums of the same or lower weight is known [25], see also [24] for an implementation in $FORM$. Thus the number of independent harmonic sums can be reduced significantly, e.g., for $w = 3$ the 18 possible harmonic sums can be expressed algebraically in terms of 8 basic harmonic sums only. One introduces a product for the harmonic sums, the shuffle product ⧢, cf. [25]. For the product of a single and a general finite harmonic sum it is given by

$$S_{a_1}(N) ⧢ S_{b_1,\ldots,b_m}(N) = S_{a_1,b_1,\ldots,b_m}(N) + S_{b_1,a_1,b_2,\ldots,b_m}(N) + \cdots + S_{b_1,b_2,\ldots,b_m,a_1}(N). \tag{C.30}$$

For sums $S_{a_1,\ldots,a_n}(N)$ and $S_{b_1,\ldots,b_m}(N)$ of arbitrary depth, the shuffle product is then the sum of all harmonic sums of depth $m + n$ in the index set of which a_i occurs left of a_j for $i < j$, likewise for b_k and b_l for $k < l$. Note that the shuffle product is symmetric. One can show that the following relation holds, cf. [25],

$$\begin{aligned} S_{a_1}(N) \cdot S_{b_1,\ldots,b_m}(N) = {} & S_{a_1}(N) ⧢ S_{b_1,\ldots,b_m}(N) - S_{a_1 \wedge b_1,b_2,\ldots,b_m}(N) \\ & - \cdots - S_{b_1,b_2,\ldots,a_1 \wedge b_m}(N), \end{aligned} \tag{C.31}$$

where the $\wedge$ symbol is defined as

$$a \wedge b = \text{sign}(a)\text{sign}(b)(|a| + |b|). \tag{C.32}$$

Due to the additional terms containing wedges ($\wedge$) between indices, harmonic sums form a quasi-shuffle algebra [26, 27]. By summing (C.31) over permutations, one obtains the symmetric algebraic relations between harmonic sums. At depth 2 and 3 these read, [23],

$$S_{m,n} + S_{n,m} = S_m S_n + S_{m \wedge n}, \tag{C.33}$$

$$\sum_{\text{perm}\{l,m,n\}} S_{l,m,n} = S_l S_m S_n + \sum_{\text{inv perm}\{l,m,n\}} S_l S_{m\wedge n} + 2S_{l\wedge m\wedge n}, \tag{C.34}$$

which we used extensively to simplify our expressions. In (C.33, C.34), "perm" denotes all permutations and "inv perm" invariant ones.

The limit $N \to \infty$ of finite harmonic sums exists only if $a_1 \neq 1$ in (C.24). Additionally, one defines all σ-values symbolically as

$$\sigma_{k_l,\ldots,k_1} = \lim_{N\to\infty} S_{a_1,\ldots,a_l}(N). \tag{C.35}$$

The finite σ-values are related to multiple ζ-values, [23, 24, 28–30], Eq. C.7. Further we define the symbol

$$\sigma_0 := \sum_{i=1}^{\infty} 1. \tag{C.36}$$

It is useful to include these σ-values into the algebra, since they allow to treat parts of sums individually, accounting for the respective divergences, cf. also [24]. These divergent pieces cancel in the end if the overall sum is finite.

The relation of single harmonic sums with positive or negative indices to the $\psi^{(k)}$-functions is then given by

$$S_1(N) = \psi(N+1) + \gamma_E, \tag{C.37}$$

$$S_a(N) = \frac{(-1)^{a-1}}{\Gamma(a)}\psi^{(a-1)}(N+1) + \zeta_a, \quad a \geq 2, \tag{C.38}$$

$$S_{-1}(N) = (-1)^N \beta(N+1) - \ln(2), \tag{C.39}$$

$$S_{-a}(N) = -\frac{(-1)^{N+a}}{\Gamma(a)}\beta^{(a-1)}(N+1) - \left(1 - 2^{1-a}\right)\zeta_a, \quad a \geq 2. \tag{C.40}$$

Thus single harmonic sums can be analytically continued to complex values of N by these relations. At higher depths, harmonic sums can be expressed in terms of Mellin-transforms of polylogarithms and the more general Nielsen-integrals [31–34]. The latter are defined by

$$\mathrm{S}_{n,p}(z) = \frac{(-1)^{n+p-1}}{(n-1)!p!}\int_0^1 \frac{dx}{x}\log^{n-1}(x)\log^p(1-zx) \tag{C.41}$$

and fulfill the relation

$$\frac{d\mathrm{S}_{n,p}(x)}{d\log(x)} = \mathrm{S}_{n-1,p}(x). \tag{C.42}$$

If $p = 1$, one obtains the polylogarithms

$$\mathrm{Li}_n(x) = \mathrm{S}_{n-1,1}(x), \tag{C.43}$$

where

$$\mathrm{Li}_0(x) = \frac{x}{1-x}. \tag{C.44}$$

These functions do not suffice for arbitrary harmonic sums, in which case the harmonic polylogarithms have to be considered [35]. The latter functions obey a direct shuffle algebra, cf. [25, 28]. The representation in terms of Mellin-transforms then allows an analytic continuation of arbitrary harmonic sums to complex N, cf. [36–39]. Equivalently, one may express harmonic sums by factorial series, [14, 40, 41], up to polynomials of $S_1(N)$ and harmonic sums of lower degree, and use this representation for the analytic continuation to $N \in \mathbb{C}$, cf. [42, 43].

Appendix D
Finite and Infinite Sums

In this appendix, we list some examples for infinite sums which were needed in the present analysis and are newly calculated. The calculation was done using the *Sigma*-package as explained in Sect. 6.2. A complete set of sums contributing to the calculation of the 2-loop massive OMEs can be found in Appendix B of Refs. [44, 45].

$$\sum_{i=1}^{\infty}\frac{B(N-2,i)}{(i+N)^3} = (-1)^N\frac{4S_{1,-2}+2S_{-3}+2\zeta_2S_1+2\zeta_3-6S_{-2}-3\zeta_2}{(N-2)(N-1)N} + \frac{1}{(N-2)(N-1)N^2}. \tag{D.1}$$

$$\sum_{i=1}^{\infty}\frac{B(N-2,i)}{(i+N)^2}S_1(i+N-2) = \frac{(-1)^{N+1}}{(N-2)(N-1)N} \times\left(8S_{1,-2}-4S_{-3}-4S_1S_{-2}-2\zeta_3+2\zeta_2S_1-10S_{-2}-5\zeta_2\right) + \frac{N^2-3N+3}{(N-2)(N-1)^2N^2}S_1-\frac{N^3-5N+3}{(N-2)(N-1)^3N^3}. \tag{D.2}$$

S. Klein, *Charm Production in Deep Inelastic Scattering*, Springer Theses,
DOI: 10.1007/978-3-642-23286-2,

$$\sum_{i=1}^{\infty} \frac{B(N,i)}{i+N+2} S_1(i) S_1(N+i) = \frac{(-1)^N}{N(N+1)(N+2)} \times \left(4S_{-2,1} - 6S_{-3} - 4S_{-2}S_1 - 2\zeta_3 \right.$$
$$\left. - 2\zeta_2 S_1 - 2\frac{\zeta_2}{(N+1)} - 4\frac{S_{-2}}{(N+1)}\right)$$
$$+ \frac{-2S_3 - S_1 S_2 + \zeta_2 S_1 + 2\zeta_3}{N+2}$$
$$+ \frac{2+7N+7N^2+5N^3+N^4}{N^3(N+1)^3(N+2)} S_1$$
$$+ 2\frac{2+7N+9N^2+4N^3+N^4}{N^4(N+1)^3(N+2)}. \quad \text{(D.3)}$$

$$\sum_{i=1}^{\infty} \frac{S_1(i+N) S_1^2(i)}{i+N} = \frac{\sigma_1^4}{4} - \frac{3\zeta_2^2}{4} + \left(\frac{2}{N} - 2S_1\right)\zeta_3 + \left(\frac{S_1}{N} - \frac{S_1^2}{2} - \frac{S_2}{2}\right)\zeta_2 + \frac{S_1^3}{N}$$
$$- \frac{S_1^4}{4} + S_1^2\left(-\frac{1}{N^2} - \frac{3S_2}{2}\right) - \frac{S_2}{N^2} - \frac{S_2^2}{4} - \frac{S_{2,1}}{N}$$
$$+ S_1\left(3\frac{S_2}{N} + S_{2,1} - 2S_3\right) + 2\frac{S_3}{N} + S_{3,1} - S_4. \quad \text{(D.4)}$$

$$\sum_{i=1}^{\infty}\left(S_1(i+N) - S_1(i)\right)^3 = -\frac{3}{2}S_1^2 - S_1^3 - \frac{1}{2}S_2 + 3NS_{2,1} - NS_3 + N\zeta_3. \quad \text{(D.5)}$$

$$\sum_{k=1}^{\infty} \frac{B(k+\varepsilon/2, N+1)}{N+k} = (-1)^N\left[2S_{-2} + \zeta_2\right]$$
$$+ \frac{\varepsilon}{2}(-1)^N\left[-\zeta_3 + \zeta_2 S_1 + 2S_{1,-2} - 2S_{-2,1}\right]$$
$$+ \frac{\varepsilon^2}{4}(-1)^N\left[\frac{2}{5}\zeta_2^2 - \zeta_3 S_1 + \zeta_2 S_{1,1} + 2\left\{S_{1,1,-2} + S_{-2,1,1} - S_{1,-2,1}\right\}\right]$$
$$+ \varepsilon^3(-1)^N\left[-\frac{\zeta_5}{8} + \frac{S_1}{20}\zeta_2^2 - \frac{S_{1,1}}{8}\zeta_3 + \frac{S_{1,1,1}}{8}\zeta_2 + \frac{S_{1,-2,1,1} + S_{1,1,1,-2} - S_{-2,1,1,1} - S_{1,1,-2,1}}{4}\right]$$
$$+ O(\varepsilon^4). \quad \text{(D.6)}$$

An example for a double infinite sum we encountered is given by

$$N\sum_{i,j=1}^{\infty}\frac{S_1(i)S_1(i+j+N)}{i(i+j)(j+N)} = 4S_{2,1,1} - 2S_{3,1} + S_1\left(-3S_{2,1} + \frac{4S_3}{3}\right) - \frac{S_4}{2}$$
$$- S_2^2 + S_1^2 S_2 + \frac{S_1^4}{6} + 6S_1\zeta_3 + \zeta_2\left(2S_1^2 + S_2\right). \tag{D.7}$$

A detailed description of the method to calculate this sum can be found in Appendix B of Ref. [45].

Appendix E
Moments of the Fermionic Contributions to the 3-Loop Anomalous Dimensions

The pole terms of the unrenormalized OMEs in our calculation agree with the general structure we presented in Eqs. 4.94, 4.103, 4.104, 4.116, 4.117, 4.124, 4.134. Using the lower order renormalization coefficients and the constant terms of the 2-loop results, Eqs. 6.34, 6.53, 6.60, 6.68, 6.80, allows to determine the T_F-terms of the 3-loop anomalous dimensions for fixed values of N. All our results agree with the results of Refs. [46–51]. Note that in this way we obtain the complete expressions for the terms $\gamma_{qg}^{(2)}$ and $\gamma_{qq}^{(2),PS}$, since they always involve an overall factor T_F. For them we obtain

(i) $\hat{\gamma}_{qg}^{(2)}$:

$$\hat{\gamma}_{qg}^{(2)}(2)=T_F\Biggl[(1+2n_f)T_F\left(\frac{8464}{243}C_A-\frac{1384}{243}C_F\right) +\frac{\zeta_3}{3}\left(-416C_AC_F+288C_A^2+128C_F^2\right)-\frac{7178}{81}C_A^2+\frac{556}{9}C_AC_F-\frac{8620}{243}C_F^2\Biggr], \tag{E.1}$$

$$\hat{\gamma}_{qg}^{(2)}(4) = T_F\Biggl[(1+2n_f)T_F\left(\frac{4481539}{303750}C_A+\frac{9613841}{3037500}C_F\right) +\frac{\zeta_3}{25}\left(2832C_A^2-3876C_AC_F+1044C_F^2\right) -\frac{295110931}{3037500}C_A^2+\frac{278546497}{2025000}C_AC_F-\frac{757117001}{12150000}C_F^2\Biggr], \tag{E.2}$$

DOI: 10.1007/978-3-642-23286-2,

$$\hat{\gamma}_{qg}^{(2)}(6) = T_F\Bigg[(1+2n_f)T_F\left(\frac{86617163}{11668860}C_A + \frac{1539874183}{340341750}C_F\right) + \frac{\zeta_3}{735}\left(69864C_A^2 - 94664C_AC_F + 24800C_F^2\right) - \frac{58595443051}{653456160}C_A^2 + \frac{1199181909343}{8168202000}C_AC_F - \frac{2933980223981}{40841010000}C_F^2\Bigg], \quad \text{(E.3)}$$

$$\hat{\gamma}_{qg}^{(2)}(8) = T_F\Bigg[(1+2n_f)T_F\left(\frac{10379424541}{2755620000}C_A + \frac{7903297846481}{1620304560000}C_F\right) + \zeta_3\left(\frac{128042}{1575}C_A^2 - \frac{515201}{4725}C_AC_F + \frac{749}{27}C_F^2\right) - \frac{24648658224523}{289340100000}C_A^2 + \frac{4896295442015177}{32406091200000}C_AC_F - \frac{4374484944665803}{56710659600000}C_F^2\Bigg], \quad \text{(E.4)}$$

$$\hat{\gamma}_{qg}^{(2)}(10) = T_F\Bigg[(1+2n_f)T_F\left(\frac{1669885489}{988267500}C_A + \frac{1584713325754369}{323600780868750}C_F\right) + \zeta_3\left(\frac{1935952}{27225}C_A^2 - \frac{2573584}{27225}C_AC_F + \frac{70848}{3025}C_F^2\right) - \frac{21025430857658971}{255684567600000}C_A^2 + \frac{926990216580622991}{6040547909550000}C_AC_F - \frac{1091980048536213833}{13591232796487500}C_F^2\Bigg]. \quad \text{(E.5)}$$

(ii) $\hat{\gamma}_{qq}^{(2),PS}$:

$$\hat{\gamma}_{qq}^{(2),PS}(2) = T_FC_F\Bigg[-(1+2n_f)T_F\frac{5024}{243} + \frac{256}{3}\left(C_F - C_A\right)\zeta_3 + \frac{10136}{243}C_A - \frac{14728}{243}C_F\Bigg], \quad \text{(E.6)}$$

$$\hat{\gamma}_{qq}^{(2),PS}(4) = T_FC_F\Bigg[-(1+2n_f)T_F\frac{618673}{151875} + \frac{968}{75}\left(C_F - C_A\right)\zeta_3 + \frac{2485097}{506250}C_A - \frac{2217031}{675000}C_F\Bigg], \quad \text{(E.7)}$$

$$\hat{\gamma}_{qq}^{(2),PS}(6) = T_F C_F \Bigg[-(1+2n_f) T_F \frac{126223052}{72930375} + \frac{3872}{735}\Bigg(C_F - C_A\Bigg)\zeta_3 + \frac{1988624681}{4084101000} C_A + \frac{11602048711}{10210252500} C_F \Bigg], \tag{E.8}$$

$$\hat{\gamma}_{qq}^{(2),PS}(8) = T_F C_F \Bigg[-(1+2n_f) T_F \frac{13131081443}{13502538000} + \frac{2738}{945}\Bigg(C_F - C_A\Bigg)\zeta_3 - \frac{343248329803}{648121824000} C_A + \frac{39929737384469}{22684263840000} C_F \Bigg], \tag{E.9}$$

$$\hat{\gamma}_{qq}^{(2),PS}(10) = T_F C_F \Bigg[-(1+2n_f) T_F \frac{265847305072}{420260754375} + \frac{50176}{27225}\Bigg(C_F - C_A\Bigg)\zeta_3 - \frac{1028766412107043}{1294403123475000} C_A + \frac{839864254987192}{485401171303125} C_F \Bigg], \tag{E.10}$$

$$\hat{\gamma}_{qq}^{(2),PS}(12) = T_F C_F \Bigg[-(1+2n_f) T_F \frac{2566080055386457}{5703275664286200} + \frac{49928}{39039}\Bigg(C_F - C_A\Bigg)\zeta_3 - \frac{69697489543846494691}{83039693672007072000} C_A + \frac{86033255402443256197}{54806197823524667520} C_F \Bigg]. \tag{E.11}$$

For the remaining terms, only the projection onto the color factor T_F can be obtained : (iii) $\hat{\gamma}_{qq}^{(2),NS,+}$:

$$\hat{\gamma}_{qq}^{(2),NS,+}(2) = T_F C_F \Bigg[-(1+2n_f) T_F \frac{1792}{243} + \frac{256}{3}\Bigg(C_F - C_A\Bigg)\zeta_3 - \frac{12512}{243} C_A - \frac{13648}{243} C_F \Bigg], \tag{E.12}$$

$$\hat{\gamma}_{qq}^{(2),NS,+}(4) = T_F C_F \Bigg[-(1+2n_f) T_F \frac{384277}{30375} + \frac{2512}{15}\Bigg(C_F - C_A\Bigg)\zeta_3 - \frac{8802581}{121500} C_A - \frac{165237563}{1215000} C_F \Bigg], \tag{E.13}$$

$$\hat{\gamma}_{qq}^{(2),NS,+}(6) = T_F C_F \Bigg[-(1+2n_f) T_F \frac{160695142}{10418625} + \frac{22688}{105}\Big(C_F - C_A\Big)\zeta_3 - \frac{13978373}{171500} C_A - \frac{44644018231}{243101250} C_F \Bigg], \tag{E.14}$$

$$\hat{\gamma}_{qq}^{(2),NS,+}(8) = T_F C_F \Bigg[-(1+2n_f) T_F \frac{38920977797}{2250423000} + \frac{79064}{315}\Big(C_F - C_A\Big)\zeta_3 - \frac{1578915745223}{18003384000} C_A - \frac{91675209372043}{420078960000} C_F \Bigg], \tag{E.15}$$

$$\hat{\gamma}_{qq}^{(2),NS,+}(10) = T_F C_F \Bigg[-(1+2n_f) T_F \frac{27995901056887}{1497656506500} + \frac{192880}{693}\Big(C_F - C_A\Big)\zeta_3 - \frac{9007773127403}{97250422500} C_A - \frac{75522073210471127}{307518802668000} C_F \Bigg], \tag{E.16}$$

$$\hat{\gamma}_{qq}^{(2),NS,+}(12) = T_F C_F \Bigg[-(1+2n_f) T_F \frac{65155853387858071}{3290351344780500} + \frac{13549568}{45045}\Big(C_F - C_A\Big)\zeta_3 - \frac{25478252190337435009}{263228107582440000} C_A - \frac{35346062280941906036867}{131745667845011220000} C_F \Bigg], \tag{E.17}$$

$$\hat{\gamma}_{qq}^{(2),NS,+}(14) = T_F C_F \Bigg[-(1+2n_f) T_F \frac{68167166257767019}{3290351344780500} + \frac{2881936}{9009}\Big(C_F - C_A\Big)\zeta_3 - \frac{9253131636331924154\,9}{921298376538540000} C_A - \frac{37908544797975614512733}{131745667845011220000} C_F \Bigg]. \tag{E.18}$$

(iv) $\hat{\gamma}_{qq}^{(2),NS,-}$:

$$\hat{\gamma}_{qq}^{(2),NS,-}(1) = 0, \tag{E.19}$$

$$\hat{\gamma}_{qq}^{(2),NS,-}(3) = T_F C_F \Bigg[-(1+2n_f) T_F \frac{2569}{243} + \frac{400}{3}\Big(C_F - C_A\Big)\zeta_3 - \frac{62249}{972} C_A - \frac{203627}{1944} C_F \Bigg], \tag{E.20}$$

$$\hat{\gamma}_{qq}^{(2),NS,-}(5) = T_F C_F \left[-(1+2n_f) T_F \frac{431242}{30375} + \frac{2912}{15} \left(C_F - C_A \right) \zeta_3 \right.$$
$$\left. - \frac{38587}{500} C_A - \frac{5494973}{33750} C_F \right], \tag{E.21}$$

$$\hat{\gamma}_{qq}^{(2),NS,-}(7) = T_F C_F \left[-(1+2n_f) T_F \frac{1369936511}{83349000} + \frac{8216}{35} \left(C_F - C_A \right) \zeta_3 \right.$$
$$\left. - \frac{2257057261}{26671680} C_A - \frac{3150205788689}{15558480000} C_F \right], \tag{E.22}$$

$$\hat{\gamma}_{qq}^{(2),NS,-}(9) = T_F C_F \left[-(1+2n_f) T_F \frac{20297329837}{1125211500} + \frac{16720}{63} \left(C_F - C_A \right) \zeta_3 \right.$$
$$\left. - \frac{126810403414}{1406514375} C_A - \frac{1630263834317}{7001316000} C_F \right], \tag{E.23}$$

$$\hat{\gamma}_{qq}^{(2),NS,-}(11) = T_F C_F \left[-(1+2n_f) T_F \frac{28869611542843}{1497656506500} + \frac{1005056}{3465} \left(C_F - C_A \right) \zeta_3 \right.$$
$$\left. - \frac{1031510572686647}{10892047320000} C_A - \frac{1188145134622636787}{4612782040020000} C_F \right], \tag{E.24}$$

$$\hat{\gamma}_{qq}^{(2),NS,-}(13) = T_F C_F \left[-(1+2n_f) T_F \frac{66727681292862571}{3290351344780500} \right.$$
$$+ \frac{13995728}{45045} \left(C_F - C_A \right) \zeta_3 - \frac{90849626920977361109}{921298376538540000} C_A$$
$$\left. - \frac{36688336888519925613757}{131745667845011220000} C_F \right]. \tag{E.25}$$

(v) $\hat{\gamma}_{gg}^{(2)}$:

$$\hat{\gamma}_{gg}^{(2)}(2) = T_F \left[(1+2n_f) T_F \left(-\frac{8464}{243} C_A + \frac{1384}{243} C_F \right) \right.$$
$$+ \frac{\zeta_3}{3} \left(-288 C_A^2 + 416 C_A C_F - 128 C_F^2 \right) + \frac{7178}{81} C_A^2$$
$$\left. - \frac{556}{9} C_A C_F + \frac{8620}{243} C_F^2 \right], \tag{E.26}$$

$$\hat{\gamma}_{gg}^{(2)}(4) = T_F\Bigg[(1+2n_f)T_F\left(-\frac{757861}{30375}C_A - \frac{979774}{151875}C_F\right) + \frac{\zeta_3}{25}\left(-6264C_A^2 + 6528C_AC_F - 264C_F^2\right) + \frac{53797499}{607500}C_A^2 - \frac{235535117}{1012500}C_AC_F + \frac{2557151}{759375}C_F^2\Bigg], \quad \text{(E.27)}$$

$$\hat{\gamma}_{gg}^{(2)}(6) = T_F\Bigg[(1+2n_f)T_F\left(-\frac{52781896}{2083725}C_A - \frac{560828662}{72930375}C_F\right) + \zeta_3\left(-\frac{75168}{245}C_A^2 + \frac{229024}{735}C_AC_F - \frac{704}{147}C_F^2\right) + \frac{9763460989}{116688600}C_A^2 - \frac{9691228129}{32672808}C_AC_F - \frac{11024749151}{10210252500}C_F^2\Bigg], \quad \text{(E.28)}$$

$$\hat{\gamma}_{gg}^{(2)}(8) = T_F\Bigg[(1+2n_f)T_F\left(-\frac{420970849}{16074450}C_A - \frac{6990254812}{843908625}C_F\right) + \zeta_3\left(-\frac{325174}{945}C_A^2 + \frac{327764}{945}C_AC_F - \frac{74}{27}C_F^2\right) + \frac{2080130771161}{25719120000}C_A^2 - \frac{220111823810087}{648121824000}C_AC_F - \frac{14058417959723}{5671065960000}C_F^2\Bigg], \quad \text{(E.29)}$$

$$\hat{\gamma}_{gg}^{(2)}(10) = T_F\Bigg[(1+2n_f)T_F\left(-\frac{2752314359}{101881395}C_A - \frac{3631303571944}{420260754375}C_F\right) + \zeta_3\left(-\frac{70985968}{190575}C_A^2 + \frac{71324656}{190575}C_AC_F - \frac{5376}{3025}C_F^2\right) + \frac{43228502203851731}{549140719050000}C_A^2 - \frac{337408133551 7123191}{9060821864325000}C_FC_A - \frac{3009386129483453}{970802342606250}C_F^2\Bigg]. \quad \text{(E.30)}$$

(vi) $\hat{\gamma}_{gq}^{(2)}$:

$$\hat{\gamma}_{gq}^{(2)}(2) = T_F C_F \left[(1+2n_f) T_F \frac{2272}{81} + \frac{512}{3}(C_A - C_F)\zeta_3 + \frac{88}{9} C_A + \frac{28376}{243} C_F \right], \tag{E.31}$$

$$\hat{\gamma}_{gq}^{(2)}(4) = T_F C_F \left[(1+2n_f) T_F \frac{109462}{10125} + \frac{704}{15}(C_A - C_F)\zeta_3 - \frac{799}{12150} C_A + \frac{14606684}{759375} C_F \right], \tag{E.32}$$

$$\hat{\gamma}_{gq}^{(2)}(6) = T_F C_F \left[(1+2n_f) T_F \frac{22667672}{3472875} + \frac{2816}{105}(C_A - C_F)\zeta_3 - \frac{253841107}{145860750} C_A + \frac{20157323311}{2552563125} C_F \right], \tag{E.33}$$

$$\hat{\gamma}_{gq}^{(2)}(8) = T_F C_F \left[(1+2n_f) T_F \frac{339184373}{75014100} + \frac{1184}{63}(C_A - C_F)\zeta_3 - \frac{3105820553}{1687817250} C_A + \frac{8498139408671}{2268426384000} C_F \right], \tag{E.34}$$

$$\hat{\gamma}_{gq}^{(2)}(10) = T_F C_F \left[(1+2n_f) T_F \frac{1218139408}{363862125} + \frac{7168}{495}(C_A - C_F)\zeta_3 - \frac{18846629176433}{11767301122500} C_A + \frac{529979902254031}{323600780868750} C_F \right], \tag{E.35}$$

$$\hat{\gamma}_{gq}^{(2)}(12) = T_F C_F \left[(1+2n_f) T_F \frac{13454024393417}{5222779912350} + \frac{5056}{429}(C_A - C_F)\zeta_3 - \frac{64190493078139789}{48885219979596000} C_A + \frac{1401404001326440151}{3495293228541114000} C_F \right], \tag{E.36}$$

$$\hat{\gamma}_{gq}^{(2)}(14) = T_F C_F \left[(1+2n_f) T_F \frac{19285002274}{9495963477} + \frac{13568}{1365}(C_A - C_F)\zeta_3 - \frac{37115284124613269}{35434552943790000} C_A - \frac{40163401444446690479}{104797690331258925000} C_F \right]. \tag{E.37}$$

Appendix F
The $O(\varepsilon^0)$ Contributions to $\hat{\hat{A}}_{ij}^{(3)}$

Finally, we present all moments we calculated. We only give the constant term in ε of the unrenormalized result, cf. Eqs. 4.94, 4.103, 4.104, 4.116, 4.117, 4.124, 4.134. These terms have to be inserted into the general results on the renormalized level, cf. Eqs. 4.96, 4.105, 4.106, 4.118, 4.119, 4.126, 4.137. We obtain

(i) $a_{Qq}^{(3),PS}$:

$$\begin{aligned} a_{Qq}^{(3),PS}(2) = {} & T_F C_F C_A \left(\frac{117290}{2187} + \frac{64}{9} B_4 - 64\zeta_4 + \frac{1456}{27}\zeta_3 + \frac{224}{81}\zeta_2 \right) \\ & + T_F C_F^2 \left(\frac{42458}{243} - \frac{128}{9} B_4 + 64\zeta_4 - \frac{9664}{81}\zeta_3 + \frac{704}{27}\zeta_2 \right) \\ & + T_F^2 C_F \left(-\frac{36880}{2187} - \frac{4096}{81}\zeta_3 - \frac{736}{81}\zeta_2 \right) \\ & + n_f T_F^2 C_F \left(-\frac{76408}{2187} + \frac{896}{81}\zeta_3 - \frac{112}{81}\zeta_2 \right), \end{aligned} \tag{F.1}$$

$$\begin{aligned} a_{Qq}^{(3),PS}(4) = {} & T_F C_F C_A \left(\frac{23115644813}{1458000000} + \frac{242}{225} B_4 - \frac{242}{25}\zeta_4 + \frac{1403}{180}\zeta_3 + \frac{283481}{270000}\zeta_2 \right) \\ & + T_F C_F^2 \left(-\frac{181635821459}{8748000000} - \frac{484}{225} B_4 + \frac{242}{25}\zeta_4 + \frac{577729}{40500}\zeta_3 + \frac{4587077}{1620000}\zeta_2 \right) \\ & + T_F^2 C_F \left(-\frac{2879939}{5467500} - \frac{15488}{2025}\zeta_3 - \frac{1118}{2025}\zeta_2 \right) \\ & + n_f T_F^2 C_F \left(-\frac{474827503}{109350000} + \frac{3388}{2025}\zeta_3 - \frac{851}{20250}\zeta_2 \right), \end{aligned} \tag{F.2}$$

S. Klein, *Charm Production in Deep Inelastic Scattering*, Springer Theses,
DOI: 10.1007/978-3-642-23286-2,

$$a_{Qq}^{(3),PS}(6)$$
$$=T_FC_FC_A\left(\frac{111932846538053}{10291934520000}+\frac{968}{2205}B_4-\frac{968}{245}\zeta_4+\frac{2451517}{1852200}\zeta_3+\frac{5638039}{7779240}\zeta_2\right)$$
$$+T_FC_F^2\left(-\frac{238736626635539}{5145967260000}-\frac{1936}{2205}B_4+\frac{968}{245}\zeta_4+\frac{19628197}{555660}\zeta_3+\frac{8325229}{10804500}\zeta_2\right)$$
$$+T_F^2C_F\left(\frac{146092097}{1093955625}-\frac{61952}{19845}\zeta_3-\frac{7592}{99225}\zeta_2\right)$$
$$+n_fT_F^2C_F\left(-\frac{82616977}{45378900}+\frac{1936}{2835}\zeta_3-\frac{16778}{694575}\zeta_2\right), \tag{F.3}$$

$$a_{Qq}^{(3),PS}(8)=T_FC_FC_A\left(\frac{314805694173451777}{32665339929600000}+\frac{1369}{5670}B_4-\frac{1369}{630}\zeta_4-\frac{202221853}{137168640}\zeta_3\right.$$
$$\left.+\frac{1888099001}{3429216000}\zeta_2\right)+T_FC_F^2\left(-\frac{25652839216168097959}{457314759014400000}-\frac{1369}{2835}B_4\right.$$
$$\left.+\frac{1369}{630}\zeta_4+\frac{2154827491}{48988800}\zeta_3+\frac{12144008761}{48009024000}\zeta_2\right)$$
$$+T_F^2C_F\left(\frac{48402207241}{272211166080}-\frac{43808}{25515}\zeta_3+\frac{1229}{142884}\zeta_2\right)$$
$$+n_fT_F^2C_F\left(-\frac{16194572439593}{15122842560000}+\frac{1369}{3645}\zeta_3-\frac{343781}{14288400}\zeta_2\right), \tag{F.4}$$

$$a_{Qq}^{(3),PS}(10)=T_FC_FC_A\left(\frac{989015303211567766373}{107642563748181000000}+\frac{12544}{81675}B_4\right.$$
$$\left.-\frac{12544}{9075}\zeta_4-\frac{1305489421}{431244000}\zeta_3+\frac{2903694979}{6670805625}\zeta_2\right)$$
$$+T_FC_F^2\left(-\frac{493601383301409762635563}{8073192281113575000 0}-\frac{25088}{81675}B_4+\frac{12544}{9075}\zeta_4\right.$$
$$\left.+\frac{94499430133}{1940598000}\zeta_3+\frac{282148432}{4002483375}\zeta_2\right)$$
$$+T_F^2C_F\left(\frac{430570223624411}{2780024890190625}-\frac{802816}{735075}\zeta_3+\frac{319072}{11026125}\zeta_2\right)$$
$$+n_fT_F^2C_F\left(-\frac{454721266324013}{624087220246875}+\frac{175616}{735075}\zeta_3-\frac{547424}{24257475}\zeta_2\right), \tag{F.5}$$

$$a_{Qq}^{(3),PS}(12) = T_F C_F C_A \left(\frac{968307050156826905398206547}{107727062441920086477312000} + \frac{12482}{117117} B_4 - \frac{12482}{13013} \zeta_4 - \frac{64839185833913}{16206444334080} \zeta_3 + \frac{489403711559293}{1382612282251200} \zeta_2 \right)$$
$$+ T_F C_F^2 \left(-\frac{190211298439834685159055148289}{29624942171528023781260800000} - \frac{24964}{117117} B_4 + \frac{12482}{13013} \zeta_4 + \frac{418408135384633}{8103222167040} \zeta_3 - \frac{72904483229177}{15208735104763200} \zeta_2 \right)$$
$$+ T_F^2 C_F \left(\frac{1727596215111011341}{1355098297834401 1200} - \frac{798848}{1054053} \zeta_3 + \frac{11471393}{347837490} \zeta_2 \right)$$
$$+ n_f T_F^2 C_F \left(-\frac{66215577092930561601 77}{12331394510293050192000} + \frac{24964}{150579} \zeta_3 - \frac{1291174013}{63306423180} \zeta_2 \right). \tag{F.6}$$

(ii) $a_{qq,Q}^{(3),PS}$:

$$a_{qq,Q}^{(3),PS}(2) = n_f T_F^2 C_F \left(-\frac{100096}{2187} + \frac{896}{81} \zeta_3 - \frac{256}{81} \zeta_2 \right), \tag{F.7}$$

$$a_{qq,Q}^{(3),PS}(4) = n_f T_F^2 C_F \left(-\frac{118992563}{21870000} + \frac{3388}{2025} \zeta_3 - \frac{4739}{20250} \zeta_2 \right), \tag{F.8}$$

$$a_{qq,Q}^{(3),PS}(6) = n_f T_F^2 C_F \left(-\frac{17732294117}{10210252500} + \frac{1936}{2835} \zeta_3 - \frac{9794}{694575} \zeta_2 \right), \tag{F.9}$$

$$a_{qq,Q}^{(3),PS}(8) = n_f T_F^2 C_F \left(-\frac{20110404913057}{27221116608000} + \frac{1369}{3645} \zeta_3 + \frac{135077}{4762800} \zeta_2 \right), \tag{F.10}$$

$$a_{qq,Q}^{(3),PS}(10) = n_f T_F^2 C_F \left(-\frac{308802524517334}{873722108345625} + \frac{175616}{735075} \zeta_3 + \frac{4492016}{121287375} \zeta_2 \right), \tag{F.11}$$

$$a_{qq,Q}^{(3),PS}(12) = n_f T_F^2 C_F \left(-\frac{6724380801633998071}{38535607844665781850} + \frac{24964}{150579} \zeta_3 + \frac{583767694}{15826605795} \zeta_2 \right), \tag{F.12}$$

$$a_{qq,Q}^{(3),PS}(14) = n_f T_F^2 C_F \Biggl(-\frac{6161646154432563473333}{754543370385064260000} + \frac{22472}{184275}\zeta_3 + \frac{189601441}{5533778250}\zeta_2 \Biggr). \tag{F.13}$$

(iii) $a_{Qg}^{(3)}$:

$$\begin{aligned} a_{Qg}^{(3)}(2) = {} & T_F C_A^2 \Biggl(\frac{170227}{4374} - \frac{88}{9}B_4 + 72\zeta_4 - \frac{31367}{324}\zeta_3 + \frac{1076}{81}\zeta_2 \Biggr) \\ & + T_F C_F C_A \Biggl(-\frac{154643}{729} + \frac{208}{9}B_4 - 104\zeta_4 + \frac{7166}{27}\zeta_3 - 54\zeta_2 \Biggr) \\ & + T_F C_F^2 \Biggl(-\frac{15574}{243} - \frac{64}{9}B_4 + 32\zeta_4 - \frac{3421}{81}\zeta_3 + \frac{704}{27}\zeta_2 \Biggr) \\ & + T_F^2 C_A \Biggl(-\frac{20542}{2187} + \frac{4837}{162}\zeta_3 - \frac{670}{81}\zeta_2 \Biggr) + T_F^2 C_F \Biggl(\frac{11696}{729} + \frac{569}{81}\zeta_3 + \frac{256}{9}\zeta_2 \Biggr) \\ & - \frac{64}{27}T_F^3\zeta_3 + n_f T_F^2 C_A \Biggl(-\frac{6706}{2187} - \frac{616}{81}\zeta_3 - \frac{250}{81}\zeta_2 \Biggr) \\ & + n_f T_F^2 C_F \Biggl(\frac{158}{243} + \frac{896}{81}\zeta_3 + \frac{40}{9}\zeta_2 \Biggr), \end{aligned} \tag{F.14}$$

$$\begin{aligned} a_{Qg}^{(3)}(4) = {} & T_F C_A^2 \Biggl(-\frac{425013969083}{2916000000} - \frac{559}{50}B_4 + \frac{2124}{25}\zeta_4 - \frac{352717109}{5184000}\zeta_3 - \frac{4403923}{270000}\zeta_2 \Biggr) \\ & + T_F C_F C_A \Biggl(-\frac{95898493099}{874800000} + \frac{646}{25}B_4 - \frac{2907}{25}\zeta_4 + \frac{172472027}{864000}\zeta_3 - \frac{923197}{40500}\zeta_2 \Biggr) \\ & + T_F C_F^2 \Biggl(-\frac{87901205453}{699840000} - \frac{174}{25}B_4 + \frac{783}{25}\zeta_4 + \frac{937829}{12960}\zeta_3 + \frac{62019319}{3240000}\zeta_2 \Biggr) \\ & + T_F^2 C_A \Biggl(\frac{960227179}{29160000} + \frac{1873781}{51840}\zeta_3 + \frac{120721}{13500}\zeta_2 \Biggr) \\ & + T_F^2 C_F \Biggl(-\frac{1337115617}{874800000} + \frac{73861}{324000}\zeta_3 + \frac{8879111}{810000}\zeta_2 \Biggr) \\ & - \frac{176}{135}T_F^3\zeta_3 + n_f T_F^2 C_A \Biggl(\frac{947836283}{72900000} - \frac{18172}{2025}\zeta_3 - \frac{11369}{13500}\zeta_2 \Biggr) \\ & + n_f T_F^2 C_F \Biggl(\frac{8164734347}{4374000000} + \frac{130207}{20250}\zeta_3 + \frac{1694939}{810000}\zeta_2 \Biggr), \end{aligned} \tag{F.15}$$

$$
\begin{aligned}
a_{Qg}^{(3)}(6) = T_F C_A^2 \Biggl(& -\frac{48989733311629681}{263473523712000} - \frac{2938}{315} B_4 + \frac{17466}{245} \zeta_4 - \frac{748603616077}{11379916800} \zeta_3 \\
& - \frac{93013721}{3457440} \zeta_2 \Biggr) \\
& + T_F C_F C_A \Biggl(\frac{712876107019}{55319040000} + \frac{47332}{2205} B_4 - \frac{23666}{245} \zeta_4 + \frac{276158927731}{1896652800} \zeta_3 \\
& + \frac{4846249}{11113200} \zeta_2 \Biggr) \\
& + T_F C_F^2 \Biggl(-\frac{3873986781136413}{137225793600000} - \frac{2480}{441} B_4 + \frac{1240}{49} \zeta_4 + \frac{148514798653}{711244800} \zeta_3 \\
& + \frac{4298936309}{388962000} \zeta_2 \Biggr) \\
& + T_F^2 C_A \Biggl(\frac{706058069789557}{18819537408000} + \frac{3393002903}{116121600} \zeta_3 + \frac{6117389}{555660} \zeta_2 \Biggr) \\
& + T_F^2 C_F \Biggl(-\frac{447496496568703}{54890317440000} - \frac{666922481}{284497920} \zeta_3 + \frac{49571129}{9724050} \zeta_2 \Biggr) \\
& - \frac{176}{189} T_F^3 \zeta_3 + n_f T_F^2 C_A \Biggl(\frac{12648331693}{735138180} - \frac{4433}{567} \zeta_3 + \frac{23311}{111132} \zeta_2 \Biggr) \\
& + n_f T_F^2 C_F \Biggl(-\frac{8963002169173}{1715322420000} + \frac{111848}{19845} \zeta_3 + \frac{11873563}{19448100} \zeta_2 \Biggr),
\end{aligned}
\tag{F.16}
$$

$$
\begin{aligned}
a_{Qg}^{(3)}(8) = T_F C_A^2 \Biggl(& -\frac{358497428780844484961}{2389236291993600000} - \frac{899327}{113400} B_4 + \frac{64021}{1050} \zeta_4 \\
& - \frac{123211748184444641}{112368549888000} \zeta_3 - \frac{19581298057}{612360000} \zeta_2 \Biggr) \\
& + T_F C_F C_A \Biggl(\frac{941315502886297276939}{8362327021977600000} + \frac{515201}{28350} B_4 - \frac{515201}{6300} \zeta_4 \\
& + \frac{5580970944338269}{56184274944000} \zeta_3 + \frac{495290785657}{34292160000} \zeta_2 \Biggr) \\
& + T_F C_F^2 \Biggl(-\frac{23928053971795796451443}{365851807211520000000} - \frac{749}{162} B_4 + \frac{749}{36} \zeta_4 \\
& + \frac{719875828314061}{1404606873600} \zeta_3 + \frac{2484799653079}{480090240000} \zeta_2 \Biggr)
\end{aligned}
$$

$$
\begin{aligned}
&+T_F^2 C_A\left(\frac{156313300657148129}{4147979673600000}+\frac{58802880439}{2388787200}\zeta_3+\frac{46224083}{4082400}\zeta_2\right)\\
&+T_F^2 C_F\left(-\frac{98650562736291304 7}{87107573145600000}-\frac{185046016777}{50164531200}\zeta_3+\frac{7527074663}{3429216000}\zeta_2\right)\\
&-\frac{296}{405}T_F^3\zeta_3+n_f T_F^2 C_A\left(\frac{24718362393463}{1322697600000}-\frac{125356}{18225}\zeta_3+\frac{2118187}{2916000}\zeta_2\right)\\
&+n_f T_F^2 C_F\left(-\frac{291376419801571603}{32665339929600000}+\frac{887741}{174960}\zeta_3-\frac{139731073}{1143072000}\zeta_2\right),
\end{aligned}
\tag{F.17}
$$

$$
\begin{aligned}
a_{Qg}^{(3)}(10) = {}& T_F C_A^2\left(\frac{68303634635669246922 53659}{68585057506396569600 0000}-\frac{563692}{81675}B_4+\frac{483988}{9075}\zeta_4\right.\\
&\left.-\frac{103652031822049723}{415451499724800}\zeta_3-\frac{20114890664357}{581101290000}\zeta_2\right)\\
&+T_F C_F C_A\left(\frac{872201479486471797889957487}{29928025093700321280 00000}+\frac{1286792}{81675}B_4\right.\\
&\left.-\frac{643396}{9075}\zeta_4-\frac{761897167477437907}{33236119977984000}\zeta_3+\frac{15455008277}{660342375}\zeta_2\right)\\
&+T_F C_F^2\left(-\frac{247930147349635960148869654541}{148143724213816590336000000}-\frac{11808}{3025}B_4\right.\\
&\left.+\frac{53136}{3025}\zeta_4+\frac{9636017147214304991}{7122025709568000}\zeta_3+\frac{14699237127551}{15689734830000}\zeta_2\right)\\
&+T_F^2 C_A\left(\frac{23231189758106199645229}{633397356480430080000}+\frac{123553074914173}{5755172290560}\zeta_3\right.\\
&\left.+\frac{4206955789}{377338500}\zeta_2\right)\\
&+T_F^2 C_F\left(-\frac{18319931182630444611912149}{1410892611560158003200000}-\frac{502987059528463}{113048027136000}\zeta_3\right.\\
&\left.+\frac{24683221051}{46695639375}\zeta_2\right)-\frac{896}{1485}T_F^3\zeta_3\\
&+n_f T_F^2 C_A\left(\frac{297277185134077151}{15532837481700000}-\frac{1505896}{245025}\zeta_3+\frac{189965849}{188669250}\zeta_2\right)\\
&+n_f T_F^2 C_F\left(-\frac{1178560772273339822317}{107642563748181000000}+\frac{62292104}{13476375}\zeta_3\right.\\
&\left.-\frac{49652772817}{93391278750}\zeta_2\right).
\end{aligned}
\tag{F.18}
$$

(iv) $a_{qg,Q}^{(3)}$:

$$a_{qg,Q}^{(3)}(2) = n_f T_F^2 C_A \left(\frac{83204}{2187} - \frac{616}{81}\zeta_3 + \frac{290}{81}\zeta_2 \right) + n_f T_F^2 C_F \left(-\frac{5000}{243} + \frac{896}{81}\zeta_3 - \frac{4}{3}\zeta_2 \right), \tag{F.19}$$

$$a_{qg,Q}^{(3)}(4) = n_f T_F^2 C_A \left(\frac{835586311}{14580000} - \frac{18172}{2025}\zeta_3 + \frac{71899}{13500}\zeta_2 \right) + n_f T_F^2 C_F \left(-\frac{21270478523}{874800000} + \frac{130207}{20250}\zeta_3 - \frac{1401259}{810000}\zeta_2 \right), \tag{F.20}$$

$$a_{qg,Q}^{(3)}(6) = n_f T_F^2 C_A \left(\frac{277835781053}{5881105440} - \frac{4433}{567}\zeta_3 + \frac{2368823}{555660}\zeta_2 \right) + n_f T_F^2 C_F \left(-\frac{36123762156197}{1715322420000} + \frac{111848}{19845}\zeta_3 - \frac{26095211}{19448100}\zeta_2 \right), \tag{F.21}$$

$$a_{qg,Q}^{(3)}(8) = n_f T_F^2 C_A \left(\frac{157327027056457}{3968092800000} - \frac{125356}{18225}\zeta_3 + \frac{7917377}{2268000}\zeta_2 \right) + n_f T_F^2 C_F \left(-\frac{201046808090490443}{10888446643200000} + \frac{887741}{174960}\zeta_3 - \frac{3712611349}{3429216000}\zeta_2 \right), \tag{F.22}$$

$$a_{qg,Q}^{(3)}(10) = n_f T_F^2 C_A \left(\frac{6542127929072987}{191763425700000} - \frac{1505896}{245025}\zeta_3 + \frac{1109186999}{377338500}\zeta_2 \right) + n_f T_F^2 C_F \left(-\frac{353813854966442889041}{21528512749636200000} + \frac{62292104}{13476375}\zeta_3 - \frac{83961181063}{93391278750}\zeta_2 \right). \tag{F.23}$$

(v) $a_{gq,Q}^{(3)}$:

$$a_{gq,Q}^{(3)}(2) = T_F C_F C_A \left(-\frac{126034}{2187} - \frac{128}{9} B_4 + 128\zeta_4 - \frac{9176}{81}\zeta_3 - \frac{160}{81}\zeta_2 \right)$$
$$+ T_F C_F^2 \left(-\frac{741578}{2187} + \frac{256}{9} B_4 - 128\zeta_4 + \frac{17296}{81}\zeta_3 - \frac{4496}{81}\zeta_2 \right)$$
$$+ T_F^2 C_F \left(\frac{21872}{729} + \frac{2048}{27}\zeta_3 + \frac{416}{27}\zeta_2 \right)$$
$$+ n_f T_F^2 C_F \left(\frac{92200}{729} - \frac{896}{27}\zeta_3 + \frac{208}{27}\zeta_2 \right), \qquad \text{(F.24)}$$

$$a_{gq,Q}^{(3)}(4) = T_F C_F C_A \left(-\frac{5501493631}{218700000} - \frac{176}{45} B_4 + \frac{176}{5}\zeta_4 - \frac{8258}{405}\zeta_3 + \frac{13229}{8100}\zeta_2 \right)$$
$$+ T_F C_F^2 \left(-\frac{12907539571}{145800000} + \frac{352}{45} B_4 - \frac{176}{5}\zeta_4 + \frac{132232}{2025}\zeta_3 - \frac{398243}{27000}\zeta_2 \right)$$
$$+ T_F^2 C_F \left(\frac{1914197}{911250} + \frac{2816}{135}\zeta_3 + \frac{1252}{675}\zeta_2 \right)$$
$$+ n_f T_F^2 C_F \left(\frac{50305997}{1822500} - \frac{1232}{135}\zeta_3 + \frac{626}{675}\zeta_2 \right), \qquad \text{(F.25)}$$

$$a_{gq,Q}^{(3)}(6) = T_F C_F C_A \left(-\frac{384762916141}{24504606000} - \frac{704}{315} B_4 + \frac{704}{35}\zeta_4 - \frac{240092}{19845}\zeta_3 + \frac{403931}{463050}\zeta_2 \right)$$
$$+ T_F C_F^2 \left(-\frac{40601579774533}{918922725000} + \frac{1408}{315} B_4 - \frac{704}{35}\zeta_4 + \frac{27512264}{694575}\zeta_3 \right.$$
$$\left. - \frac{24558841}{3472875}\zeta_2 \right)$$
$$+ T_F^2 C_F \left(-\frac{279734446}{364651875} + \frac{11264}{945}\zeta_3 + \frac{8816}{33075}\zeta_2 \right)$$
$$+ n_f T_F^2 C_F \left(\frac{4894696577}{364651875} - \frac{704}{135}\zeta_3 + \frac{4408}{33075}\zeta_2 \right), \qquad \text{(F.26)}$$

$$a_{gq,Q}^{(3)}(8) = T_F C_F C_A \Biggl(-\frac{10318865954633473}{816633498240000} - \frac{296}{189} B_4 + \frac{296}{21}\zeta_4 - \frac{1561762}{178605}\zeta_3 + \frac{30677543}{85730400}\zeta_2 \Biggr)$$
$$+ T_F C_F^2 \Biggl(-\frac{305405135103422947}{11432868975360000} + \frac{592}{189} B_4 - \frac{296}{21}\zeta_4 + \frac{124296743}{4286520}\zeta_3 - \frac{4826251837}{1200225600}\zeta_2 \Biggr)$$
$$+ T_F^2 C_F \Biggl(-\frac{864658160833}{567106596000} + \frac{4736}{567}\zeta_3 - \frac{12613}{59535}\zeta_2 \Biggr)$$
$$+ n_f T_F^2 C_F \Biggl(\frac{9330164983967}{1134213192000} - \frac{296}{81}\zeta_3 - \frac{12613}{119070}\zeta_2 \Biggr), \qquad \text{F.27}$$

$$a_{gq,Q}^{(3)}(10) = T_F C_F C_A \Biggl(-\frac{1453920909405842897}{130475834846280000} - \frac{1792}{1485} B_4 + \frac{1792}{165}\zeta_4 - \frac{1016096}{147015}\zeta_3 + \frac{871711}{26952750}\zeta_2 \Biggr)$$
$$+ T_F C_F^2 \Biggl(-\frac{117033382372448370173}{6672059736457500000} + \frac{3584}{1485} B_4 - \frac{1792}{165}\zeta_4 + \frac{62282416}{2695275}\zeta_3 - \frac{6202346032}{2547034875}\zeta_2 \Biggr)$$
$$+ T_F^2 C_F \Biggl(-\frac{1346754066466}{756469357875} + \frac{28672}{4455}\zeta_3 - \frac{297472}{735075}\zeta_2 \Biggr)$$
$$+ n_f T_F^2 C_F \Biggl(\frac{4251185859247}{756469357875} - \frac{12544}{4455}\zeta_3 - \frac{148736}{735075}\zeta_2 \Biggr), \qquad \text{(F.28)}$$

$$a_{gq,Q}^{(3)}(12) = T_F C_F C_A \Biggl(-\frac{1515875996003174876943331}{147976734123516602304000} - \frac{1264}{1287} B_4 + \frac{1264}{143}\zeta_4 - \frac{999900989}{173918745}\zeta_3 - \frac{693594486209}{3798385390800}\zeta_2 \Biggr)$$
$$+ T_F C_F^2 \Biggl(-\frac{48679935129017185612582919}{4069360188396706563360000} + \frac{2528}{1287} B_4 - \frac{1264}{143}\zeta_4 + \frac{43693776149}{2260943685}\zeta_3 - \frac{2486481253717}{1671289571952}\zeta_2 \Biggr)$$

$$+T_F^2C_F\left(-\frac{2105210836073143063}{1129248581528667600}+\frac{20224}{3861}\zeta_3-\frac{28514494}{57972915}\zeta_2\right)$$
$$+n_fT_F^2C_F\left(\frac{9228836319135394697}{2258497163057335200}-\frac{8848}{3861}\zeta_3-\frac{14257247}{57972915}\zeta_2\right), \qquad (F.29)$$

$$a_{gq,Q}^{(3)}(14)=T_FC_FC_A\left(-\frac{191825356953814257271820 9}{19919944978165696464000 0}-\frac{3392}{4095}B_4\right.$$
$$\left.+\frac{3392}{455}\zeta_4-\frac{2735193382}{553377825}\zeta_3-\frac{1689839813797}{5113211103000}\zeta_2\right)$$
$$+T_FC_F^2\left(-\frac{1437971805100351708026209 17}{1742995185589498440600000 0}+\frac{6784}{4095}B_4\right.$$
$$\left.-\frac{3392}{455}\zeta_4+\frac{12917466836}{774728955}\zeta_3-\frac{4139063104013}{4747981738500}\zeta_2\right)$$
$$+T_F^2C_F\left(-\frac{337392441268078561}{179653183425015300}+\frac{54272}{12285}\zeta_3-\frac{98112488}{184459275}\zeta_2\right)$$
$$+n_fT_F^2C_F\left(\frac{222188365726202803}{71861273370006120}-\frac{3392}{1755}\zeta_3-\frac{49056244}{184459275}\zeta_2\right). \qquad (F.30)$$

(vi) $a_{gg,Q}^{(3)}$:

$$a_{gg,Q}^{(3)}(2)=T_FC_A^2\left(-\frac{170227}{4374}+\frac{88}{9}B_4-72\zeta_4+\frac{31367}{324}\zeta_3-\frac{1076}{81}\zeta_2\right)$$
$$+T_FC_FC_A\left(\frac{154643}{729}-\frac{208}{9}B_4+104\zeta_4-\frac{7166}{27}\zeta_3+54\zeta_2\right)$$
$$+T_FC_F^2\left(\frac{15574}{243}+\frac{64}{9}B_4-32\zeta_4+\frac{3421}{81}\zeta_3-\frac{704}{27}\zeta_2\right)$$
$$+T_F^2C_A\left(\frac{20542}{2187}-\frac{4837}{162}\zeta_3+\frac{670}{81}\zeta_2\right)$$
$$+T_F^2C_F\left(-\frac{11696}{729}-\frac{569}{81}\zeta_3-\frac{256}{9}\zeta_2\right)+\frac{64}{27}T_F^3\zeta_3$$
$$+n_fT_F^2C_A\left(-\frac{76498}{2187}+\frac{1232}{81}\zeta_3-\frac{40}{81}\zeta_2\right)$$
$$+n_fT_F^2C_F\left(\frac{538}{27}-\frac{1792}{81}\zeta_3-\frac{28}{9}\zeta_2\right), \qquad (F.31)$$

$$a_{gg,Q}^{(3)}(4) = T_F C_A^2 \left(\frac{29043652079}{291600000} + \frac{533}{25} B_4 - \frac{4698}{25} \zeta_4 + \frac{610035727}{2592000} \zeta_3 + \frac{92341}{6750} \zeta_2 \right)$$
$$+ T_F C_F C_A \left(\frac{272542528639}{874800000} - \frac{1088}{25} B_4 + \frac{4896}{25} \zeta_4 - \frac{3642403}{17280} \zeta_3 + \frac{73274237}{810000} \zeta_2 \right)$$
$$+ T_F C_F^2 \left(\frac{41753961371}{1749600000} + \frac{44}{25} B_4 - \frac{198}{25} \zeta_4 + \frac{2676077}{64800} \zeta_3 - \frac{4587077}{1620000} \zeta_2 \right)$$
$$+ T_F^2 C_A \left(-\frac{1192238291}{14580000} - \frac{2134741}{25920} \zeta_3 - \frac{16091}{675} \zeta_2 \right)$$
$$+ T_F^2 C_F \left(-\frac{785934527}{43740000} - \frac{32071}{8100} \zeta_3 - \frac{226583}{8100} \zeta_2 \right) + \frac{64}{27} T_F^3 \zeta_3$$
$$+ n_f T_F^2 C_A \left(-\frac{271955197}{1822500} + \frac{13216}{405} \zeta_3 - \frac{6526}{675} \zeta_2 \right)$$
$$+ n_f T_F^2 C_F \left(-\frac{465904519}{27337500} - \frac{6776}{2025} \zeta_3 - \frac{61352}{10125} \zeta_2 \right), \qquad \text{(F.32)}$$

$$a_{gg,Q}^{(3)}(6) = T_F C_A^2 \left(\frac{37541473421359}{448084224000} + \frac{56816}{2205} B_4 - \frac{56376}{245} \zeta_4 + \frac{926445489353}{2844979200} \zeta_3 + \frac{11108521}{555660} \zeta_2 \right)$$
$$+ T_F C_F C_A \left(\frac{18181142251969309}{54890317440000} - \frac{114512}{2205} B_4 + \frac{57256}{245} \zeta_4 - \frac{12335744909}{67737600} \zeta_3 + \frac{94031857}{864360} \zeta_2 \right)$$
$$+ T_F C_F^2 \left(\frac{16053159907363}{635304600000} + \frac{352}{441} B_4 - \frac{176}{49} \zeta_4 + \frac{3378458681}{88905600} \zeta_3 - \frac{8325229}{10804500} \zeta_2 \right)$$

$$+T_F^2C_A\left(-\frac{670098465769}{6001128000}-\frac{25725061}{259200}\zeta_3-\frac{96697}{2835}\zeta_2\right)$$
$$+T_F^2C_F\left(-\frac{8892517283287}{490092120000}-\frac{12688649}{2540160}\zeta_3-\frac{2205188}{77175}\zeta_2\right)+\frac{64}{27}T_F^3\zeta_3$$
$$+n_fT_F^2C_A\left(-\frac{245918019913}{1312746750}+\frac{3224}{81}\zeta_3-\frac{250094}{19845}\zeta_2\right)$$
$$+n_fT_F^2C_F\left(-\frac{71886272797}{3403417500}-\frac{3872}{2835}\zeta_3-\frac{496022}{77175}\zeta_2\right), \tag{F.33}$$

$$a_{gg,Q}^{(3)}(8)=T_FC_A^2\left(\frac{5129033047123476076}{18665908531200000}+\frac{108823}{3780}B_4-\frac{162587}{630}\zeta_4\right.$$
$$\left.+\frac{2735007975361}{6502809600}\zeta_3+\frac{180224911}{7654500}\zeta_2\right)$$
$$+T_FC_FC_A\left(\frac{134895840434433199991}{43553786572800000}-\frac{163882}{2835}B_4+\frac{81941}{315}\zeta_4\right.$$
$$\left.-\frac{3504113623243}{25082265600}\zeta_3+\frac{414844703639}{3429216000}\zeta_2\right)$$
$$+T_FC_F^2\left(\frac{599012727207322546 7}{228657379507200000}+\frac{37}{81}B_4-\frac{37}{18}\zeta_4+\frac{3222019505879}{87787929600}\zeta_3\right.$$
$$\left.-\frac{12144008761}{48009024000}\zeta_2\right)$$
$$+T_F^2C_A\left(-\frac{16278325750483243}{124439390208000}-\frac{871607413}{7962624}\zeta_3-\frac{591287}{14580}\zeta_2\right)$$
$$+T_F^2C_F\left(-\frac{7458367007740639}{408316749120000}-\frac{291343229}{52254720}\zeta_3-\frac{2473768763}{85730400}\zeta_2\right)$$
$$+\frac{64}{27}T_F^3\zeta_3+n_fT_F^2C_A\left(-\frac{102747532985051}{486091368000}+\frac{54208}{1215}\zeta_3-\frac{737087}{51030}\zeta_2\right)$$
$$+n_fT_F^2C_F\left(-\frac{1145917332616927}{51039593640000}-\frac{2738}{3645}\zeta_3-\frac{70128089}{10716300}\zeta_2\right), \tag{F.34}$$

$$a_{gg,Q}^{(3)}(10)=T_F C_A^2\Biggl(-\frac{1543448346233166100527575 9}{327337774462347264000000}+\frac{17788828}{571725}B_4 -\frac{17746492}{63525}\zeta_4+\frac{269094476549521109}{519314374656000}\zeta_3+\frac{1444408720649}{55468759500}\zeta_2\Biggr)$$
$$+T_F C_F C_A\Biggl(\frac{207095356146239371087405921}{77158189694696140800000 0}-\frac{35662328}{571725}B_4 +\frac{17831164}{63525}\zeta_4-\frac{3288460968359099}{37093883904000}\zeta_3+\frac{6078270984602}{46695639375}\zeta_2\Biggr)$$
$$+T_F C_F^2\Biggl(\frac{553777925867720521493231}{2066737223965075200000 0}+\frac{896}{3025}B_4-\frac{4032}{3025}\zeta_4 +\frac{7140954579599}{198717235200}\zeta_3-\frac{282148432}{4002483375}\zeta_2\Biggr)$$
$$+T_F^2 C_A\Biggl(-\frac{630598434818955028 07}{433789788579840000}-\frac{85188238297}{729907200}\zeta_3-\frac{33330316}{735075}\zeta_2\Biggr)$$
$$+T_F^2 C_F\Biggl(-\frac{6556905805599587741 57}{35787657557836800000}-\frac{71350574183}{12043468800}\zeta_3-\frac{3517889264}{121287375}\zeta_2\Biggr)$$
$$+\frac{64}{27}T_F^3\zeta_3+n_f T_F^2 C_A\Biggl(-\frac{6069333056458984}{26476427525625}+\frac{215128}{4455}\zeta_3-\frac{81362132}{5145525}\zeta_2\Biggr)$$
$$+n_f T_F^2 C_F\Biggl(-\frac{100698363899844296}{4368610541728125}-\frac{351232}{735075}\zeta_3-\frac{799867252}{121287375}\zeta_2\Biggr). \tag{F.35}$$

(vii) $a_{qq,Q}^{(3),NS}$:

$$a_{qq,Q}^{(3),NS}(1)=0, \tag{F.36}$$

$$a_{qq,Q}^{(3),NS}(2)=T_F C_F C_A\Biggl(\frac{8744}{2187}+\frac{64}{9}B_4-64\zeta_4+\frac{4808}{81}\zeta_3-\frac{64}{81}\zeta_2\Biggr)$$
$$+T_F C_F^2\Biggl(\frac{359456}{2187}-\frac{128}{9}B_4+64\zeta_4-\frac{848}{9}\zeta_3+\frac{2384}{81}\zeta_2\Biggr)$$
$$+T_F^2 C_F\Biggl(-\frac{28736}{2187}-\frac{2048}{81}\zeta_3-\frac{512}{81}\zeta_2\Biggr)$$
$$+n_f T_F^2 C_F\Biggl(-\frac{100096}{2187}+\frac{896}{81}\zeta_3-\frac{256}{81}\zeta_2\Biggr), \tag{F.37}$$

$$a_{qq,Q}^{(3),NS}(3) = T_F C_F C_A \left(\frac{522443}{34992} + \frac{100}{9} B_4 - 100\zeta_4 + \frac{15637}{162}\zeta_3 + \frac{175}{162}\zeta_2 \right)$$
$$+ T_F C_F^2 \left(\frac{35091701}{139968} - \frac{200}{9} B_4 + 100\zeta_4 - \frac{1315}{9}\zeta_3 + \frac{29035}{648}\zeta_2 \right)$$
$$+ T_F^2 C_F \left(-\frac{188747}{8748} - \frac{3200}{81}\zeta_3 - \frac{830}{81}\zeta_2 \right)$$
$$+ n_f T_F^2 C_F \left(-\frac{1271507}{17496} + \frac{1400}{81}\zeta_3 - \frac{415}{81}\zeta_2 \right), \qquad \text{(F.38)}$$

$$a_{qq,Q}^{(3),NS}(4) = T_F C_F C_A \left(\frac{419369407}{21870000} + \frac{628}{45} B_4 - \frac{628}{5}\zeta_4 + \frac{515597}{4050}\zeta_3 + \frac{10703}{4050}\zeta_2 \right)$$
$$+ T_F C_F^2 \left(\frac{137067007129}{437400000} - \frac{1256}{45} B_4 + \frac{628}{5}\zeta_4 - \frac{41131}{225}\zeta_3 + \frac{4526303}{81000}\zeta_2 \right)$$
$$+ T_F^2 C_F \left(-\frac{151928299}{5467500} - \frac{20096}{405}\zeta_3 - \frac{26542}{2025}\zeta_2 \right)$$
$$+ n_f T_F^2 C_F \left(-\frac{1006358899}{10935000} + \frac{8792}{405}\zeta_3 - \frac{13271}{2025}\zeta_2 \right), \qquad \text{(F.39)}$$

$$a_{qq,Q}^{(3),NS}(5) = T_F C_F C_A \left(\frac{816716669}{43740000} + \frac{728}{45} B_4 - \frac{728}{5}\zeta_4 + \frac{12569}{81}\zeta_3 + \frac{16103}{4050}\zeta_2 \right)$$
$$+ T_F C_F^2 \left(\frac{13213297537}{36450000} - \frac{1456}{45} B_4 + \frac{728}{5}\zeta_4 - \frac{142678}{675}\zeta_3 + \frac{48391}{750}\zeta_2 \right)$$
$$+ T_F^2 C_F \left(-\frac{9943403}{303750} - \frac{23296}{405}\zeta_3 - \frac{31132}{2025}\zeta_2 \right) \qquad \text{(F.40)}$$
$$+ n_f T_F^2 C_F \left(-\frac{195474809}{1822500} + \frac{10192}{405}\zeta_3 - \frac{15566}{2025}\zeta_2 \right), \qquad \text{(F.41)}$$

$$a_{qq,Q}^{(3),NS}(6)=T_FC_FC_A\Biggl(\frac{1541550898907}{105019740000}+\frac{5672}{315}B_4-\frac{5672}{35}\zeta_4+\frac{720065}{3969}\zeta_3+\frac{1016543}{198450}\zeta_2\Biggr)+T_FC_F^2\Biggl(\frac{186569400917}{463050000}-\frac{11344}{315}B_4+\frac{5672}{35}\zeta_4-\frac{7766854}{33075}\zeta_3+\frac{55284811}{771750}\zeta_2\Biggr)+T_F^2C_F\Biggl(-\frac{26884517771}{729303750}-\frac{181504}{2835}\zeta_3-\frac{1712476}{99225}\zeta_2\Biggr)+n_fT_F^2C_F\Biggl(-\frac{524427335513}{4375822500}+\frac{11344}{405}\zeta_3-\frac{856238}{99225}\zeta_2\Biggr),\tag{F.42}$$

$$a_{qq,Q}^{(3),NS}(7)=T_FC_FC_A\Biggl(\frac{5307760084631}{672126336000}+\frac{2054}{105}B_4-\frac{6162}{35}\zeta_4+\frac{781237}{3780}\zeta_3+\frac{19460531}{3175200}\zeta_2\Biggr)+T_FC_F^2\Biggl(\frac{4900454072126579}{11202105600000}-\frac{4108}{105}B_4+\frac{6162}{35}\zeta_4-\frac{8425379}{33075}\zeta_3+\frac{1918429937}{24696000}\zeta_2\Biggr)+T_F^2C_F\Biggl(-\frac{8488157192423}{210039480000}-\frac{65728}{945}\zeta_3-\frac{3745727}{198450}\zeta_2\Biggr)+n_fT_F^2C_F\Biggl(-\frac{54861581223623}{420078960000}+\frac{4108}{135}\zeta_3-\frac{3745727}{396900}\zeta_2\Biggr),\tag{F.43}$$

$$a_{qq,Q}^{(3),NS}(8)=T_FC_FC_A\Biggl(-\frac{37259291367883}{38887309440000}+\frac{19766}{945}B_4-\frac{19766}{105}\zeta_4+\frac{1573589}{6804}\zeta_3+\frac{200739467}{28576800}\zeta_2\Biggr)+T_FC_F^2\Biggl(\frac{3817101976847353531}{8166334982400000}-\frac{39532}{945}B_4+\frac{19766}{105}\zeta_4-\frac{80980811}{297675}\zeta_3+\frac{497748102211}{6001128000}\zeta_2\Biggr)+T_F^2C_F\Biggl(-\frac{740566685766263}{17013197880000}-\frac{632512}{8505}\zeta_3-\frac{36241943}{1786050}\zeta_2\Biggr)+n_fT_F^2C_F\Biggl(-\frac{4763338626853463}{34026395760000}+\frac{39532}{1215}\zeta_3-\frac{36241943}{3572100}\zeta_2\Biggr),\tag{F.44}$$

$$a_{qq,Q}^{(3),NS}(9) = T_F C_F C_A \Biggl(-\frac{3952556872585211}{340263957600000} + \frac{4180}{189} B_4 - \frac{4180}{21}\zeta_4 + \frac{21723277}{85050}\zeta_3 + \frac{559512437}{71442000}\zeta_2 \Biggr) + T_F C_F^2 \Biggl(\frac{1008729211999128667}{2041583745600000} - \frac{8360}{189} B_4 + \frac{4180}{21}\zeta_4 - \frac{85539428}{297675}\zeta_3 + \frac{131421660271}{1500282000}\zeta_2 \Biggr) + T_F^2 C_F \Biggl(-\frac{393938732805271}{8506598940000} - \frac{133760}{1701}\zeta_3 - \frac{19247947}{893025}\zeta_2 \Biggr) + n_f T_F^2 C_F \Biggl(-\frac{2523586499054071}{17013197880000} + \frac{8360}{243}\zeta_3 - \frac{19247947}{1786050}\zeta_2 \Biggr), \quad \text{(F.45)}$$

$$a_{qq,Q}^{(3),NS}(10) = T_F C_F C_A \Biggl(-\frac{10710275715721975271}{452891327565600000} + \frac{48220}{2079} B_4 - \frac{48220}{231}\zeta_4 + \frac{2873636069}{10291050}\zeta_3 + \frac{961673201}{112266000}\zeta_2 \Biggr) + T_F C_F^2 \Biggl(\frac{170291990048723954490137}{328799103812625600000} - \frac{96440}{2079} B_4 + \frac{48220}{231}\zeta_4 - \frac{10844970868}{36018675}\zeta_3 + \frac{183261101886701}{1996875342000}\zeta_2 \Biggr) + T_F^2 C_F \Biggl(-\frac{6080478350275977191}{124545115080540000} - \frac{1543040}{18711}\zeta_3 - \frac{2451995507}{108056025}\zeta_2 \Biggr) + n_f T_F^2 C_F \Biggl(-\frac{38817494524177585991}{249090230161080000} + \frac{96440}{2673}\zeta_3 - \frac{2451995507}{216112050}\zeta_2 \Biggr), \quad \text{(F.46)}$$

$$a_{qq,Q}^{(3),NS}(11)=T_FC_FC_A\Biggl(-\frac{22309979286641292041}{603855103420800000}+\frac{251264}{10395}B_4$$
$$-\frac{251264}{1155}\zeta_4+\frac{283300123}{935550}\zeta_3+\frac{1210188619}{130977000}\zeta_2\Biggr)$$
$$+T_FC_F^2\Biggl(\frac{177435748292579058982241}{328799103812625600000}-\frac{502528}{10395}B_4$$
$$+\frac{251264}{1155}\zeta_4-\frac{451739191}{1440747}\zeta_3+\frac{47705202493793}{499218835500}\zeta_2\Biggr)$$
$$+T_F^2C_F\Biggl(-\frac{6365809346912279423}{1245451150805400000}-\frac{8040448}{93555}\zeta_3-\frac{512808781}{21611205}\zeta_2\Biggr)$$
$$+n_fT_F^2C_F\Biggl(-\frac{40517373495580091423}{2490902301610800000}+\frac{502528}{13365}\zeta_3-\frac{512808781}{43222410}\zeta_2\Biggr), \tag{F.47}$$

$$a_{qq,Q}^{(3),NS}(12)=T_FC_FC_A\Biggl(-\frac{126207343604156227942043}{24638150869716384000000}+\frac{3387392}{135135}B_4$$
$$-\frac{3387392}{15015}\zeta_4+\frac{51577729507}{158107950}\zeta_3+\frac{2401246832561}{243486243000}\zeta_2\Biggr)$$
$$+T_FC_F^2\Biggl(\frac{68296027149155250557867961293}{122080805651901196900800000}-\frac{6774784}{135135}B_4$$
$$+\frac{3387392}{15015}\zeta_4-\frac{79117185295}{243486243}\zeta_3+\frac{108605787257580461}{1096783781593500}\zeta_2\Biggr)$$
$$+T_F^2C_F\Biggl(-\frac{189306988923316881320303}{35571330318153029400000}-\frac{108396544}{1216215}\zeta_3$$
$$-\frac{90143221429}{3652293645}\zeta_2\Biggr)$$
$$+n_fT_F^2C_F\Biggl(-\frac{12017333911777204697723303}{711426606363060588000000}$$
$$+\frac{6774784}{173745}\zeta_3-\frac{90143221429}{7304587290}\zeta_2\Biggr), \tag{F.48}$$

$$a_{qq,Q}^{(3),NS}(13) = T_F C_F C_A \Bigg(-\frac{1203212324638987356550337 3}{1810904088924154224000 00} + \frac{3498932}{135135} B_4 - \frac{3498932}{15015}\zeta_4 + \frac{2288723461}{6548850}\zeta_3 + \frac{106764723181157}{10226422206000}\zeta_2 \Bigg) + T_F C_F^2 \Bigg(\frac{10076195142551036234891679659}{17440115093128742414400000} - \frac{6997864}{135135} B_4 + \frac{3498932}{15015}\zeta_4 - \frac{81672622894}{243486243}\zeta_3 + \frac{448416864235277759}{4387135126374000}\zeta_2 \Bigg) + T_F^2 C_F \Bigg(-\frac{1962430666520403825353 03}{3557133031815302940000} - \frac{111965824}{1216215}\zeta_3 - \frac{93360116539}{3652293645}\zeta_2 \Bigg) + n_f T_F^2 C_F \Bigg(-\frac{124284081287434258846730 3}{71142660636306058800 00} + \frac{6997864}{173745}\zeta_3 - \frac{93360116539}{7304587290}\zeta_2 \Bigg), \tag{F.49}$$

$$a_{qq,Q}^{(3),NS}(14) = T_F C_F C_A \Bigg(-\frac{9947745876145368730238 63}{120726939261610281600 00} + \frac{720484}{27027} B_4 - \frac{720484}{3003}\zeta_4 + \frac{6345068237}{17027010}\zeta_3 + \frac{37428569944327}{3408807402000}\zeta_2 \Bigg) + T_F C_F^2 \Bigg(\frac{72598193631729215117875463981}{122080805651901196900800000} - \frac{1440968}{27027} B_4 + \frac{720484}{3003}\zeta_4 - \frac{2101051892878}{6087156075}\zeta_3 + \frac{461388998135343407}{4387135126374000}\zeta_2 \Bigg) + T_F^2 C_F \Bigg(-\frac{405400320636508947082 51}{711426606363060588000} - \frac{23055488}{243243}\zeta_3 - \frac{481761665447}{18261468225}\zeta_2 \Bigg) + n_f T_F^2 C_F \Bigg(-\frac{256205552272074402170491}{1422853212726121176000} + \frac{1440968}{34749}\zeta_3 - \frac{481761665447}{36522936450}\zeta_2 \Bigg). \tag{F.50}$$

Appendix G
3-loop Moments for Transversity

We obtain the following fixed moments of the fermionic contributions to the 3-loop transversity anomalous dimension $\gamma_{qq}^{(2),TR}(N)$

$$\hat{\gamma}_{qq}^{(2),TR}(1) = C_F T_F \left[-\frac{8}{3} T_F (1 + 2n_f) - \frac{2008}{27} C_A + \frac{196}{9} C_F + 32 (C_F - C_A) \zeta_3 \right], \tag{G.1}$$

$$\hat{\gamma}_{qq}^{(2),TR}(2) = C_F T_F \left[-\frac{184}{27} T_F (1 + 2n_f) - \frac{2084}{27} C_A - 60 C_F + 96 (C_F - C_A) \zeta_3 \right], \tag{G.2}$$

$$\hat{\gamma}_{qq}^{(2),TR}(3) = C_F T_F \left[-\frac{2408}{243} T_F (1 + 2n_f) - \frac{19450}{243} C_A - \frac{25276}{243} C_F + \frac{416}{3} (C_F - C_A) \zeta_3 \right], \tag{G.3}$$

$$\hat{\gamma}_{qq}^{(2),TR}(4) = C_F T_F \left[-\frac{14722}{1215} T_F (1 + 2n_f) - \frac{199723}{2430} C_A - \frac{66443}{486} C_F + \frac{512}{3} (C_F - C_A) \zeta_3 \right], \tag{G.4}$$

$$\hat{\gamma}_{qq}^{(2),TR}(5) = C_F T_F \left[-\frac{418594}{30375} T_F (1 + 2n_f) - \frac{5113951}{60750} C_A - \frac{49495163}{303750} C_F + \frac{2944}{15} (C_F - C_A) \zeta_3 \right], \tag{G.5}$$

S. Klein, *Charm Production in Deep Inelastic Scattering*, Springer Theses,
DOI: 10.1007/978-3-642-23286-2, © Springer-Verlag Berlin Heidelberg 2012

$$\hat{\gamma}_{qq}^{(2),TR}(6) = C_F T_F \Bigg[-\frac{3209758}{212625} T_F (1+2n_f) - \frac{3682664}{42525} C_A - \frac{18622301}{101250} C_F + \frac{1088}{5} (C_F - C_A)\zeta_3 \Bigg], \quad \text{(G.6)}$$

$$\hat{\gamma}_{qq}^{(2),TR}(7) = C_F T_F \Bigg[-\frac{168501142}{10418625} T_F (1+2n_f) - \frac{1844723441}{20837250} C_A - \frac{49282560541}{243101250} C_F + \frac{8256}{35} (C_F - C_A)\zeta_3 \Bigg] \quad \text{(G.7)}$$

$$\hat{\gamma}_{qq}^{(2),TR}(8) = C_F T_F \Bigg[-\frac{711801943}{41674500} T_F (1+2n_f) - \frac{6056338297}{66679200} C_A - \frac{849420853541}{3889620000} C_F + \frac{8816}{35} (C_F - C_A)\zeta_3 \Bigg] \quad \text{(G.8)}$$

These moments ($N = 1..8$) agree with the corresponding terms obtained in [52]. The newly calculated moments read

$$\hat{\gamma}_{qq}^{(2),TR}(9) = C_F T_F \Bigg[-\frac{20096458061}{1125211500} T_F (1+2n_f) - \frac{119131812533}{1285956000} C_A - \frac{24479706761047}{105019740000} C_F + \frac{83824}{315} (C_F - C_A)\zeta_3 \Bigg] \quad \text{(G.9)}$$

$$\hat{\gamma}_{qq}^{(2),TR}(10) = C_F T_F \Bigg[-\frac{229508848783}{12377326500} T_F (1+2n_f) - \frac{4264058299021}{45008460000} C_A - \frac{25800817445759}{105019740000} C_F + \frac{87856}{315} (C_F - C_A)\zeta_3 \Bigg] \quad \text{(G.10)}$$

$$\hat{\gamma}_{qq}^{(2),TR}(11) = C_F T_F \Bigg[-\frac{28677274464343}{1497656506500} T_F (1+2n_f) - \frac{75010870835743}{778003380000} C_A - \frac{396383896707569599}{1537594013340000} C_F + \frac{1006736}{3465} (C_F - C_A)\zeta_3 \Bigg] \quad \text{(G.11)}$$

$$\hat{\gamma}_{qq}^{(2),TR}(12) = C_F T_F \Bigg[-\frac{383379490933459}{19469534584500} T_F (1+2n_f) - \frac{38283693844132279}{389390691690000} C_A - \frac{1237841854306528417}{4612782040020000} C_F + \frac{1043696}{3465} (C_F - C_A)\zeta_3 \Bigg] \quad \text{(G.12)}$$

$$\hat{\gamma}_{qq}^{(2),TR}(13) = C_F T_F \Bigg[-\frac{66409807459266571}{3290351344780500} T_F(1+2n_f) - \frac{65714936443750201 21}{65807026895610000} C_A - \frac{367133190154071415700 17}{1317456678450112200 00} C_F + \frac{14011568}{45045}(C_F - C_A)\zeta_3 \Bigg]. \tag{G.13}$$

The fixed moments of the constant terms $a_{qq,Q}^{(3),\mathrm{TR}}(N)$ of the unrenormalized OME, see Eq. 9.15, are given by

$$\begin{aligned} a_{qq,Q}^{(3),\mathrm{TR}}(1) = {} & T_F C_F C_A \left(-\frac{26441}{1458} + \frac{8}{3} B_4 - 24\zeta_4 + \frac{481}{27}\zeta_3 - \frac{61}{27}\zeta_2 \right) \\ & + T_F C_F^2 \left(\frac{15715}{162} - \frac{16}{3} B_4 + 24\zeta_4 - \frac{278}{9}\zeta_3 + \frac{49}{3}\zeta_2 \right) \\ & + T_F^2 C_F \left(-\frac{6548}{729} - \frac{256}{27}\zeta_3 - \frac{104}{27}\zeta_2 \right) \\ & + n_f T_F^2 C_F \left(-\frac{15850}{729} + \frac{112}{27}\zeta_3 - \frac{52}{27}\zeta_2 \right), \end{aligned} \tag{G.14}$$

$$\begin{aligned} a_{qq,Q}^{(3),\mathrm{TR}}(2) = {} & T_F C_F C_A \left(\frac{1043}{162} + 8B_4 - 72\zeta_4 + \frac{577}{9}\zeta_3 + \frac{\zeta_2}{3} \right) \\ & + T_F C_F^2 \left(\frac{10255}{54} - 16B_4 + 72\zeta_4 - \frac{310}{3}\zeta_3 + 33\zeta_2 \right) \\ & + T_F^2 C_F \left(-\frac{1388}{81} - \frac{256}{9}\zeta_3 - 8\zeta_2 \right) \\ & + n_f T_F^2 C_F \left(-\frac{4390}{81} + \frac{112}{9}\zeta_3 - 4\zeta_2 \right), \end{aligned} \tag{G.15}$$

$$\begin{aligned} a_{qq,Q}^{(3),\mathrm{TR}}(3) = {} & T_F C_F C_A \left(\frac{327967}{21870} + \frac{104}{9} B_4 - 104\zeta_4 + \frac{40001}{405}\zeta_3 + \frac{121}{81}\zeta_2 \right) \\ & + T_F C_F^2 \left(\frac{1170943}{4374} - \frac{208}{9} B_4 + 104\zeta_4 - \frac{1354}{9}\zeta_3 + \frac{3821}{81}\zeta_2 \right) \\ & + T_F^2 C_F \left(-\frac{52096}{2187} - \frac{3328}{81}\zeta_3 - \frac{904}{81}\zeta_2 \right) \\ & + n_f T_F^2 C_F \left(-\frac{168704}{2187} + \frac{1456}{81}\zeta_3 - \frac{452}{81}\zeta_2 \right), \end{aligned} \tag{G.16}$$

$$a_{qq,Q}^{(3),\mathrm{TR}}(4) = T_F C_F C_A \left(\frac{4400353}{218700} + \frac{128}{9} B_4 - 128\zeta_4 + \frac{52112}{405}\zeta_3 + \frac{250}{81}\zeta_2 \right)$$
$$+ T_F C_F^2 \left(\frac{56375659}{174960} - \frac{256}{9} B_4 + 128\zeta_4 - \frac{556}{3}\zeta_3 + \frac{4616}{81}\zeta_2 \right)$$
$$+ T_F^2 C_F \left(-\frac{3195707}{109350} - \frac{4096}{81}\zeta_3 - \frac{1108}{81}\zeta_2 \right)$$
$$+ n_f T_F^2 C_F \left(-\frac{20731907}{218700} + \frac{1792}{81}\zeta_3 - \frac{554}{81}\zeta_2 \right), \tag{G.17}$$

$$a_{qq,Q}^{(3),\mathrm{TR}}(5) = T_F C_F C_A \left(\frac{1436867309}{76545000} + \frac{736}{45} B_4 - \frac{736}{5}\zeta_4 + \frac{442628}{2835}\zeta_3 + \frac{8488}{2025}\zeta_2 \right)$$
$$+ T_F C_F^2 \left(\frac{40410914719}{109350000} - \frac{1472}{45} B_4 + \frac{736}{5}\zeta_4 - \frac{47932}{225}\zeta_3 + \frac{662674}{10125}\zeta_2 \right)$$
$$+ T_F^2 C_F \left(-\frac{92220539}{2733750} - \frac{23552}{405}\zeta_3 - \frac{31924}{2025}\zeta_2 \right)$$
$$+ n_f T_F^2 C_F \left(-\frac{596707139}{5467500} + \frac{10304}{405}\zeta_3 - \frac{15962}{2025}\zeta_2 \right), \tag{G.18}$$

$$a_{qq,Q}^{(3),\mathrm{TR}}(6) = T_F C_F C_A \left(\frac{807041747}{53581500} + \frac{272}{15} B_4 - \frac{816}{5}\zeta_4 + \frac{172138}{945}\zeta_3 + \frac{10837}{2025}\zeta_2 \right)$$
$$+ T_F C_F^2 \left(\frac{14845987993}{36450000} - \frac{544}{15} B_4 + \frac{816}{5}\zeta_4 - \frac{159296}{675}\zeta_3 + \frac{81181}{1125}\zeta_2 \right)$$
$$+ T_F^2 C_F \left(-\frac{5036315611}{133953750} - \frac{8704}{135}\zeta_3 - \frac{35524}{2025}\zeta_2 \right)$$
$$+ n_f T_F^2 C_F \left(-\frac{32472719011}{267907500} + \frac{3808}{135}\zeta_3 - \frac{17762}{2025}\zeta_2 \right), \tag{G.19}$$

$$a_{qq,Q}^{(3),\mathrm{TR}}(7) = T_F C_F C_A \left(\frac{413587780793}{52509870000} + \frac{688}{35} B_4 - \frac{6192}{35}\zeta_4 + \frac{27982}{135}\zeta_3 + \frac{620686}{99225}\zeta_2 \right)$$
$$+ T_F C_F^2 \left(\frac{12873570421651}{29172150000} - \frac{1376}{35} B_4 + \frac{6192}{35}\zeta_4 - \frac{8454104}{33075}\zeta_3 \right.$$
$$\left. + \frac{90495089}{1157625}\zeta_2 \right)$$

$$+T_F^2 C_F\left(-\frac{268946573689}{6563733750}-\frac{22016}{315}\zeta_3-\frac{1894276}{99225}\zeta_2\right)$$
$$+n_f T_F^2 C_F\left(-\frac{1727972700289}{13127467500}+\frac{1376}{45}\zeta_3-\frac{947138}{99225}\zeta_2\right), \quad (G.20)$$

$$a_{qq,Q}^{(3),\mathrm{TR}}(8) = T_F C_F C_A\left(-\frac{91321974347}{112021056000}+\frac{2204}{105}B_4-\frac{6612}{35}\zeta_4+\frac{87613}{378}\zeta_3\right.$$
$$\left.+\frac{11372923}{1587600}\zeta_2\right)$$
$$+T_F C_F^2\left(\frac{1316283829306051}{2800526400000}-\frac{4408}{105}B_4+\frac{6612}{35}\zeta_4-\frac{9020054}{33075}\zeta_3\right.$$
$$\left.+\frac{171321401}{2058000}\zeta_2\right)$$
$$+T_F^2 C_F\left(-\frac{4618094363399}{105019740000}-\frac{70528}{945}\zeta_3-\frac{2030251}{99225}\zeta_2\right)$$
$$+n_f T_F^2 C_F\left(-\frac{29573247248999}{210039480000}+\frac{4408}{135}\zeta_3-\frac{2030251}{198450}\zeta_2\right), \quad (G.21)$$

$$a_{qq,Q}^{(3),\mathrm{TR}}(9) = T_F C_F C_A\left(-\frac{17524721583739067}{1497161413440000}+\frac{20956}{945}B_4-\frac{20956}{105}\zeta_4\right.$$
$$\left.+\frac{9574759}{37422}\zeta_3+\frac{16154189}{2041200}\zeta_2\right)$$
$$+T_F C_F^2\left(\frac{1013649109952401819}{2041583745600000}-\frac{41912}{945}B_4\right.$$
$$\left.+\frac{20956}{105}\zeta_4-\frac{85698286}{297675}\zeta_3+\frac{131876277049}{1500282000}\zeta_2\right)$$
$$+T_F^2 C_F\left(-\frac{397003835114519}{8506598940000}-\frac{670592}{8505}\zeta_3-\frac{19369859}{893025}\zeta_2\right)$$
$$+n_f T_F^2 C_F\left(-\frac{2534665670688119}{17013197880000}+\frac{41912}{1215}\zeta_3-\frac{19369859}{1786050}\zeta_2\right),$$
$$(G.22)$$

$$
\begin{aligned}
a_{qq,Q}^{(3),\mathrm{TR}}(10) = T_F C_F C_A \Biggl(& -\frac{176834434840947469}{7485807067200000} + \frac{21964}{945} B_4 - \frac{21964}{105}\zeta_4 \\
& + \frac{261607183}{935550}\zeta_3 + \frac{618627019}{71442000}\zeta_2 \Biggr) \\
+ T_F C_F^2 \Biggl(& \frac{116694997971413741 21}{22457421201600000} - \frac{43928}{945} B_4 + \frac{21964}{105}\zeta_4 \\
& - \frac{3590290}{11907}\zeta_3 + \frac{137983320397}{1500282000}\zeta_2 \Biggr) \\
+ T_F^2 C_F \Biggl(& -\frac{50558522757917663}{1029298471740000} - \frac{702848}{8505}\zeta_3 - \frac{4072951}{178605}\zeta_2 \Biggr) \\
+ n_f T_F^2 C_F \Biggl(& -\frac{321908083399769663}{2058596943480000} + \frac{43928}{1215}\zeta_3 - \frac{4072951}{357210}\zeta_2 \Biggr),
\end{aligned}
\tag{G.23}
$$

$$
\begin{aligned}
a_{qq,Q}^{(3),\mathrm{TR}}(11) = T_F C_F C_A \Biggl(& -\frac{43650800048962705083 7}{1177517451670560000 0} + \frac{251684}{10395} B_4 - \frac{251684}{1155}\zeta_4 \\
& + \frac{3687221539}{12162150}\zeta_3 + \frac{149112401}{16038000}\zeta_2 \Biggr) \\
+ T_F C_F^2 \Biggl(& \frac{1779793111791108189401}{328799103812625600000} - \frac{503368}{10395} B_4 + \frac{251684}{1155}\zeta_4 \\
& - \frac{452259130}{1440747}\zeta_3 + \frac{191230589104127}{1996875342000}\zeta_2 \Biggr) \\
+ T_F^2 C_F \Biggl(& -\frac{6396997235105384423}{124545115080540000} - \frac{8053888}{93555}\zeta_3 - \frac{514841791}{21611205}\zeta_2 \Biggr) \\
+ n_f T_F^2 C_F \Biggl(& -\frac{40628987857774916423}{249090230161080000} + \frac{503368}{13365}\zeta_3 - \frac{514841791}{43222410}\zeta_2 \Biggr),
\end{aligned}
\tag{G.24}
$$

$$
\begin{aligned}
a_{qq,Q}^{(3),\mathrm{TR}}(12) = T_F C_F C_A \Biggl(& -\frac{245210883820358086333}{4783664647411650000} + \frac{260924}{10395} B_4 - \frac{260924}{1155}\zeta_4 \\
& + \frac{3971470819}{12162150}\zeta_3 + \frac{85827712409}{8644482000}\zeta_2 \Biggr)
\end{aligned}
$$

$$
\begin{aligned}
&+T_F C_F^2\left(\frac{23963837217146225516 10173}{427438834956413 2800000}-\frac{521848}{10395}B_4+\frac{260924}{1155}\zeta_4\right.\\
&\qquad\left.-\frac{468587596}{1440747}\zeta_3+\frac{198011292882437}{1996875342000}\zeta_2\right)\\
&+T_F^2 C_F\left(-\frac{1124652164258976877487}{210481244486112 60000}-\frac{8349568}{93555}\zeta_3-\frac{535118971}{21611205}\zeta_2\right)\\
&+n_f T_F^2 C_F\left(-\frac{7126865031281296825487}{42096248897222520000}+\frac{521848}{13365}\zeta_3-\frac{535118971}{43222410}\zeta_2\right),
\end{aligned}
\tag{G.25}
$$

$$
\begin{aligned}
a_{qq,Q}^{(3),\mathrm{TR}}(13) = {} & T_F C_F C_A\left(-\frac{430633219615523278883051}{646751460330055 0800000}+\frac{3502892}{135135}B_4\right.\\
&\qquad\left.-\frac{3502892}{15015}\zeta_4+\frac{327241423}{935550}\zeta_3+\frac{15314434459241}{1460917458000}\zeta_2\right)\\
&+T_F C_F^2\left(\frac{70680445585608577308861582893}{122080805651901196900800000}-\frac{7005784}{135135}B_4\right.\\
&\qquad\left.+\frac{3502892}{15015}\zeta_4-\frac{81735983092}{243486243}\zeta_3+\frac{449066258795623169}{4387135126374000}\zeta_2\right)\\
&+T_F^2 C_F\left(-\frac{19689788786597173029 5303}{3557133031815302940000}-\frac{112092544}{1216215}\zeta_3\right.\\
&\qquad\left.-\frac{93611152819}{3652293645}\zeta_2\right)\\
&+n_f T_F^2 C_F\left(-\frac{1245167831299024242467303}{7114266063630605880000}+\frac{7005784}{173745}\zeta_3\right.\\
&\qquad\left.-\frac{93611152819}{7304587290}\zeta_2\right).
\end{aligned}
\tag{G.26}
$$

References

1. T. Muta, in *Foundations of Quantum Chromodynamics*, 2nd edn. World Scientific Lecture Notes in Physics, vol. 57 (World Scientific, Singapore, 1998)
2. F.J. Yndurain, *The Theory of Quark and Gluon Interactions*, 4th edn. (Springer, Berlin, 2006), p. 474

3. M.J.G. Veltman, *Diagrammatica: The Path to Feynman Rules*, (Cambridge University Press, Cambridge, 1994), p. 284
4. G. 't Hooft, M.J.G. Veltman, Diagrammar, CERN Yellow Rep. 73–9 (1973)
5. E.G. Floratos, D.A. Ross, C.T. Sachrajda, Nucl. Phys. B **129**, 66 (1977) [Erratum-ibid.] B139, 545 (1978)
6. R. Mertig, van W.L. Neerven, Z. Phys. C **70**, 637 (1996) hep-ph/9506451
7. E.G. Floratos, D.A. Ross, C.T. Sachrajda, Nucl. Phys. B **152**, 493 (1979)
8. R. Hamberg, van W.L. Neerven, Nucl. Phys. B **379**, 143 (1992)
9. M. Buza, Y. Matiounine, J. Smith, R. Migneron, van W.L. Neerven, Nucl. Phys. B **472**, 611 (1996) hep-ph/9601302
10. M. Buza, Y. Matiounine, J. Smith, van W.L. Neerven, Nucl. Phys. B **485**, 420 (1997) hep-ph/9608342
11. Y. Matiounine, J. Smith, van W.L. Neerven, Phys. Rev. D **57**, 6701 (1998) hep-ph/9801224
12. Y. Matiounine, J. Smith, van W.L. Neerven, Phys. Rev. D **58**, 076002 (1998) hep-ph/9803439
13. M. Abramowitz, I.A. Stegun, *Handbook of Mathematical Functions*, (Dover Publications Inc., New York, 1972), p. 1046
14. N. Nielsen, *Handbuch der Theorie der Gammafunktion*, (Chelsea Publishing Company, New York, 1965), p. 328 ; first published: (Teubner, Leipzig, 1906), 326 p.
15. L. Slater, *Generalized Hypergeometric Functions*, (Cambridge University Press, Cambridge, 1966), p. 273
16. W. Bailey, *Generalized Hypergeometric Series*, (Cambridge University Press, Cambridge, 1935), p. 108
17. G. Andrews, R. Askey, R. Roy, *Special Functions, Encyclopedia of Mathematics and its Applications* vol 71, (Cambridge University Press, Cambridge, 2001), p. 663
18. E. Barnes, Proc. Lond. Math. Soc. 6(2) 141 (1908)
19. E. Barnes, Quart. J. Math. **41**, 136 (1910)
20. H. Mellin, Math. Ann. **68**, 305 (1910)
21. R. Paris, D.D. Kaminski, *Asymptotics and Mellin-Barnes Integrals*, (Cambridge University Press, Cambridge, 2001), p. 438
22. V.A. Smirnov, Evaluating Feynman integrals. Springer Tracts Mod. Phys. **211**, 1 (2004)
23. J. Blümlein, S. Kurth, Phys. Rev. D **60**, 014018 (1999) hep-ph/9810241
24. J.A.M. Vermaseren, Int. J. Mod. Phys. A **14**, 2037 (1999) hep-ph/9806280
25. J. Blümlein, Comput. Phys. Commun. **159**, 19 (2004) hep-ph/0311046
26. M.E. Hoffman, J. Algebra, **194**, 477 (1997)
27. M.E. Hoffman, Nucl. Phys. Proc. Suppl. **135**, 215 (2004) math/0406589
28. J.M. Borwein, D.M. Bradley, D.J. Broadhurst, P. Lisonek, Trans. Am. Math. Soc. **353**, 907 (2001) math/9910045
29. L. Euler, Novi Comm. Acad. Sci Petropolitanae **1**, 140 (1775)
30. D. Zagier, *Proceedings First European Congress of Mathematics*, vol. II (Paris 1994), p. 497
31. A. Devoto, D.W. Duke, Riv. Nuovo Cim. **7N6**, 1 (1984)
32. J. Blümlein, H. Kawamura, Nucl. Phys. B **708**, 467 (2005) hep-ph/0409289
33. J. Blümlein, Collection of Polylog-Integrals, unpublished.
34. K.S. Kölbig, SIAM J. Math. Anal. **17**, 1232 (1986)
35. E. Remiddi, J.A.M. Vermaseren, Int. J. Mod. Phys. A **15**, 725 (2000) hepph/ 9905237
36. J. Blümlein, J. Comput. Phys. Commun. **133**, 76 (2000) hep-ph/0003100
37. J. Blümlein, S.O. Moch, . Phys. Lett. B **614**, 53 (2005) hep-ph/0503188
38. E. Carlson, Sur une classe de séries de Taylor, PhD Thesis, Uppsala, 1914
39. E. Titchmarsh, *Theory of Functions*, (Oxford University Press, Oxford, 1939), Chapt. 9.5.
40. K. Knopp, *Theorie und Anwendung der unendlichen Reihen*, (Springer, Berlin, 1947), p. 583
41. E. Landau, S.-Ber. Königl, Bayerische Akad. Wiss. München math.-naturw. Kl. **36**, 151 (1906)
42. A. Gonzalez-Arroyo, C. Lopez, F.J. Yndurain, Nucl. Phys. B **153**, 161 (1979)

43. J. Blümlein, J. Comput. Phys. Commun. **180**, 2218 (2009) hep-ph/0901.3106
44. I. Bierenbaum, J. Blümlein, S. Klein, Nucl. Phys. B **780**, 40 (2007) hepph/ 0703285
45. I. Bierenbaum, J. Blümlein, S. Klein, C. Schneider, Nucl. Phys. B **803**, 1 (2008) hep-ph/ 0803.0273
46. S.A. Larin, P. Nogueira, van T. Ritbergen, J.A.M. Vermaseren, Nucl. Phys. B **492**, 338 (1997) hep-ph/9605317
47. A. Retey, J.A.M. Vermaseren, . Nucl. Phys. B **604**, 281 (2001) hep-ph/0007294
48. S. Moch, J.A.M. Vermaseren, A. Vogt, Nucl. Phys. B **688**, 101 (2004) hepph/ 0403192
49. A. Vogt, S. Moch, J.A.M. Vermaseren, Nucl. Phys. B **691**, 129 (2004) hepph/ 0404111
50. J.A. Gracey, Phys. Lett. B **322**, 141 (1994) hep-ph/9401214
51. S. Moch, J.A.M. Vermaseren, A. Vogt, Nucl. Phys. B **646**, 181 (2002) hepph/ 0209100
52. J.A. Gracey, Nucl. Phys. B**662**, 247 (2003), hep-ph/0304113; Nucl. Phys. B**667**, 242 (2003), hep-ph/0306163; JHEP **10**, 040 (2006), hep-ph/0609231; Phys. Lett. B**643**, 374 (2006), hep-ph/0611071

MIX
Papier aus verantwortungsvollen Quellen
Paper from responsible sources
FSC® C105338

If you have any concerns about our products,
you can contact us on
ProductSafety@springernature.com

In case Publisher is established outside the EU,
the EU authorized representative is:
Springer Nature Customer Service Center GmbH
Europaplatz 3, 69115 Heidelberg, Germany

Printed by Libri Plureos GmbH
in Hamburg, Germany